Antisemitism and Islam

macmillan

Editors
James Renton
Department of English and History
Edge Hill University
Ormskirk, United Kingdom

Ben Gidley
School of Social Sciences,
History and Philosophy
Birkbeck, University of London
United Kingdom

ISBN 978-1-137-41299-7 ISBN 978-1-137-41302-4 (eBook)
DOI 10.1057/978-1-137-41302-4

Library of Congress Control Number: 2016962733

Printed on acid-free paper

This Palgrave Macmillan imprint is published by Springer Nature
The registered company is Macmillan Publishers Ltd.
The registered company address is: The Campus, 4 Crinan Street, London, N1 9XW, United Kingdom

For Clemente,
and in memory of John Klier

Acknowledgements

This book has taken quite some time to reach publication, and we have accumulated many people to thank along the way. It started life in 2008 with a conference—the first of its kind—on antisemitism and Islamophobia. The conference was generously funded by Edge Hill University, the Centre for Urban and Community Research at Goldsmiths, the British Academy, the Kessler Foundation and the Wingate Foundation, and was kindly hosted by the Department of Hebrew and Jewish Studies at University College London.

Many people, including most of the contributors to this book, made vital intellectual contributions to the conference—presenting or responding to papers, chairing sessions, asking difficult questions. We especially thank Michael Berkowitz, Matti Bunzl, Cathie Carmichael, Bryan Cheyette, Alex Drace-Francis, Robert Fine, François Guesnet, David Hirsh, the late Sam Johnson, Ivan Davidson Kalmar, Benjamin Kaplan, Vladimir Levin, Ruth Mandel, David Motadel, Max Silverman and the late David Cesarani. Others also did the literal heavy lifting, without which the conference could not have happened: Ofra Koffman, Madli Maruste, Dafna Steinberg, Gil Toffell and Mira Vogel.

The conference was dedicated to the memory of Professor John Klier, who passed away in 2007 but who had been crucial in the early stages of the conference organisation, and whose path-breaking and rigorous scholarship continues to provide a model for us. We take this opportunity to remember him again.

We are grateful to the team at Palgrave Macmillan—Angharad Bishop, Claire Mence, Rowan Milligan, Emily Russell—for their patience and diligence in bringing this project to fruition. For her important contribution, we also thank Monica Gonzalez-Correa.

Bryan Cheyette and Ivan Davidson Kalmar have inspired much of the framing and the thinking about the issues that are at the heart of this book, and have also been warm and generous supporters during the process. We are especially grateful to them.

Finally, but most importantly, our wonderful families—Monica and Vanessa, Clemente, Nathan and Seth—have been with us throughout the journey, and we are thankful for that.

It gives us no pleasure to note that, in the years we have been working on this project, it has become no less relevant. Violence, terror, suspicion, surveillance, exclusion and hatred have become normalised features of the European public landscape; the fault-lines between Muslims and Jews become more tense, while the shared task of combating European racisms becomes more urgent. We hope this book makes a small contribution to the understanding required to find a way out.

Contents

Notes on Contributors

Fiaz Ahmed is an Honorary Research Fellow in the Department of Anthropology at the University of Durham, UK.

Gil Anidjar is Professor in the Department of Religion, the Department of Middle Eastern, South Asian, and African Studies (MESAAS), and the Institute for Comparative Literature and Society (ICLS) at Columbia University, New York, USA.

Robert D. Crews is an Associate Professor of History at Stanford University, CA, USA.

Yulia Egorova is a Reader in the Department of Anthropology and Director of the Centre for the Study of Jewish Culture, Society and Politics, at Durham University, UK.

Ben Gidley is a Senior Lecturer in the Department of Psychosocial Studies at Birkbeck, University of London, UK.

Sander L. Gilman is Distinguished Professor of the Liberal Arts and Sciences and Professor of Psychiatry at Emory University, Atlanta, GA, USA.

Daniel A. Gordon is Senior Lecturer in European History at Edge Hill University, Ormskirk, UK.

Marko Attila Hoare is an Associate Professor in the Faculty of Arts and Sciences at Kingston University, London, UK.

Andrew Jotischky is Professor of Medieval History at Royal Holloway, University of London, UK.

James Renton is Reader in History at Edge Hill University, Ormskirk, UK.

François Soyer is Associate Professor of Late Medieval and Early Modern History at the University of Southampton, UK.

David J. Wertheim is Director of the Menasseh ben Israel Institute for Jewish Social and Cultural Studies, Amsterdam, The Netherlands.

List of Figures

CHAPTER 1

Introduction: The Shared Story of Europe's Ideas of the Muslim and the Jew—A Diachronic Framework

James Renton and Ben Gidley

In our activity alone do we find the sustaining illusion of an independent existence as against the whole scheme of things of which we form a helpless part.

—*Joseph Conrad, Nostromo: A Tale of the Seabord (1904)*

[B]y defending myself against the Jew, I am fighting for the work of the Lord.

—*Adolf Hitler, Mein Kampf, Vol. I (1925)*

Twenty days after the beginning of the events that led to the murders in the offices of *Charlie Hebdo* magazine and the kosher supermarket in Paris, French President François Hollande delivered a speech. It is Holocaust Memorial Day, 27 January 2015, and he is standing at a lectern at the Mémorial de la Shoah accompanied by ministers from his government. His goal: to argue

J. Renton (✉)
Edge Hill University, Ormskirk, UK

B. Gidley
Birkbeck, University of London, UK

J. Renton, B. Gidley (eds.), *Antisemitism and Islamophobia in Europe*, DOI 10.1057/978-1-137-41302-4_1

that the recent killings of Jews belonged to an ancient, persistent and singular narrative of anti-Jewish prejudice that, in recent French history, began with the Holocaust, and had not ended—indeed, it was escalating. 'Three weeks ago', he pronounced to the gathered throng, 'four men died in a kosher shop for the same motive that families were rounded up at the Vel d'Hiv in 1942, that the faithful of rue Copernic were attacked in 1980, that pedestrians on rue des Rosiers were assassinated in 1982, that the young Ilan Halimi was attacked in 2006, that the children of the Ozar HaTorah school were massacred in Toulouse in 2012.' He went on to contend:

> The responsibility of the authorities of the Republic—they are here, all of them assembled—is therefore to do everything, in order for Jews to be completely at home, in France, so that they never feel threatened or isolated. To fight an enemy, one has first to identify it and name it: antisemitism. It has changed its face but it has not lost its millennial roots.[1]

Hollande claimed that this fight was not the exclusive concern of the Jews; it was a challenge that belonged to all of humanity: 'The democratic nations chose to inscribe 27 January on the memory of humanity, what did they mean by doing so? That 27 January is a universal event, which not only concerns the Jews, but the whole world.' To emphasise the point, he tried to distance the fight against antisemitism from the states of the West by enlisting the words of the totem of anticolonial radicalism, Frantz Fanon: 'When you hear bad things being said about Jews, prick up your ears, they are being said about you.'

For all of his focus on antisemitism, Hollande attempted to present himself and his government as opponents of racial prejudice in general. In addition to anti-Jewish persecution, he referred to 'the anti-Muslim acts that also have multiplied in recent weeks'. And in his solution to this crisis of racism in France, Hollande declared that the government would put forward 'a global plan to fight against racism and antisemitism'. His words were soon put into action. On 17 April, Prime Minister Manuel Valls, beneath the slogan 'The Republic mobilised against racism and antisemitism', announced the government's plan for what the President called 'une Grande cause nationale' for 2015, with a fund of 100 million euros over three years and its own Twitter hashtag, '#PlanAntiRacisme'.[2]

The claim to universalism of the French government's plan was misleading, however. The term 'racism' was arguably deployed in order to assert the state's concern for Muslims alongside Jews, but without recognising, or promoting the significance of, Islamophobia. This stance of silent erasure was not new: much of the French political elite have largely resisted using the term Islamophobia since it came into circulation in international civil society in the late 1990s.[3] The acrobatics performed to claim concern for Muslims while prioritising the struggle against antisemitism was clear from the content of Valls' speech. France, he argued,

> has become aware that we must fight with determination against everything that divides us and separates us. This is what I wanted to sum up ... [at] the National Assembly on 13 January, by simply saying that French Jews must no longer be scared to be Jewish. And that French Muslims must no longer be ashamed to be Muslim.

The location of the speech was meant to illustrate the same point of a shared story: Créteil in Le Val-de-Marne, which, the official government summary explained, possessed 'the most important Jewish community of the region and, at the same time, one of the most important mosques in France'.[4] But—and this is the most revealing point—the town was also the location where, on 1 December 2014, three men broke into the home of a couple who, it was widely reported, were believed to be wealthy because they were Jewish. While one of the robbers stole money from a cashpoint, another sexually assaulted the woman.[5] Valls described this as 'One more trauma, one trauma too many'.[6] The selection of the site was, therefore, a nod to thinking about Muslims alongside Jews, and diversity in general. Yet it was only a nod. The present violence to be noted is against Jews. Muslims merely risked shame—the shame of being themselves.

However, there is an enemy that the French government, with its partners, deems to be more perilous and significant than antisemitism: the Muslim enemy. In his speech on 13 January for the victims of the *Charlie Hebdo* attacks, Valls said: 'France is at war with terrorism, jihadism and radical Islamism.' Certainly, he added that 'France is not at war with a religion, France is not at war with Islam and Muslims'.[7] Yet the fact remains that Muslims are the enemy, even if not Muslims are the target. Antisemitism pales in significance in the face of this foe: on 21 January, Valls announced 425 million euros of additional spending for 'la lutte contre le terrorisme' over three years, more than four times the budget for 'la lutte contre le racisme et l'antisémitisme', and a total expenditure of 940 million. The government began or expanded a raft of measures and initiatives, including an additional 60 million euros for 'la prévention de la radicalisation' programme.[8] And on 24 July, the government adopted a new intelligence law that allows the use of surveillance techniques in 'le domaine de la prévention' that in the eyes of the law were previously reserved for judicial investigations, including computer data capture and surveillance on private premises; as well as access to telecommunications operators for preventative individual monitoring; and data gathering from travel and transport companies. The government's press dossier argued that the law was needed in response to 'a protean threat undoubtedly without precedent'. The safety of France, the dossier asserted, required the state to 'anticipate, detect, analyse, understand and thwart the threats' that confronted the country.[9] Following the attacks in Paris in November 2015, the government's desire to extend the law to facilitate the war on Islam(ism) surpassed the limits of the possible, and could only be satisfied by suspending the rule of law itself: a state of emergency. After the murder of over eighty people in Nice on 14 July 2016, Valls reiterated that France was at war, a war that had been forced on the country by terrorism; and Hollande announced that the government would extend the state of emergency.[10]

These developments in France belong to a global process that began in earnest after the attacks on the World Trade Center on 11 September 2001. Antisemitism and so-called Islamism—not Islamophobia—are twin and, in the Western official mind, connected enemies of the West. The latter opposition is illustrated quite clearly in a 2015 French government world map that depicts the political geography of the 'international coalition' against Islamic State[11]: a politico-military incarnation of the Islamic enemy that seeks to establish a borderless Muslim state in western Asia, and thereby smash the Middle East state system sponsored by the West since 1918.

Genealogies of Coupling and Divergence

How can it make sense to talk about a relationship between antisemitism and Islamophobia in this context, in which the figures of the persecuted Jew and the political Muslim are on opposite sides of a war waged by the West? Anthropologist Matti Bunzl gave us one possibility with his 2007 pamphlet *Anti-Semitism and Islamophobia: Hatreds Old and New in Europe*.[12] We can, he suggested, compare and contrast the present European fear of Islam with the past European fear of Jews—in short, a comparative study of two temporally separate racisms, which has also been put forward by others such as Ilan Halevi, Nasar Meer and Tehseen Noorani.[13] Yet why bother with this kind of exercise? Of course we can see similarities: the association with political and cultural subversion; and the ideas of a protean, invisible, morally corruptive, degenerative and fanatical enemy. Nevertheless, as Bunzl has argued, it would be misleading to ignore the huge differences.

For example, there have been earlier wars on terror, which have often focused on racialised suspect minorities (the nineteenth-century panic about Fenians; the globally co-ordinated war on anarchism of the 1890s–1900s, which gave birth to many modern policing techniques; the first red scare in the wake of the Bolshevik revolution; the second red scare during the Cold War[14]), but the global infrastructure of surveillance, incarceration and killing that is today focused on Muslims has no precise precedent. The closest we come to it in history, in essence though not in scale, is the structure of the anti-Jewish surveillance and control apparatus of the Nazi state and then empire, particularly the work in the late 1930s of the Sicherheitsdienst (SD) internal intelligence office, AMT II 112, which tracked Jewish conspiracies around the globe.[15] However, the aim of the Nazis was not, ultimately, to control or defeat one single category of Jew—a political Jew, analogous to the West's political Muslim of today. For the Nazis, all Jews were the political enemy. And from the summer of 1941, the Nazi elite aimed to murder every single Jew—nothing less.[16]

Even so, if we return for a moment to the similarities between Western antisemitism and Islamophobia, we are left with the fact of unique traits held in common. Something about the Nazi Jewish enemy and the contemporary Western Muslim enemy demands complete surveillance—the power to see beneath the veil permanently and everywhere—an imperative that is not apparent with any other racialised enemy in history. To put it another way, few in the West speak or have spoken of the fanatical Gypsy, the protean menace of the Hindu, the world conspiracy of the Irish Catholic. We can, however, attach the Jew or Muslim interchangeably to these terms or goals and find ourselves with recognisable notions in Western thought. This interchangeability requires explanation, but it cannot be done without acknowledging and understanding the contemporary opposition, the sharp separation, of antisemitism and Islamophobia as political categories in Western states, international institutions of power and global civil society. If we look at the problem of the antisemitism–Islamophobia relationship in this sense of simultaneous epistemological connectedness and political divergence, we must, for all of the problematic Enlightenment echoes, search for the origins of both this singular coupling and its separation. To do this requires the recovery and telling of one story, a relational story—not an exercise in comparative racism studies, or the telling of separate albeit parallel stories. We have to excavate and concentrate on a shared story of evolution; in short, we need a diachronic framework, in which we can identify moments of beginning, change, separation. This is the aim of our book.

Writing before the post-2008 crises that have engulfed Europe, Bunzl proposed that 'the Jewish Question' that structured modern antisemitism belonged to the age of the nation-state, since Jews were its constitutive outsider. Yet in the age of a supra-national Europe, he argued, the nation's 'Jewish Question' has been superseded by a pan-European 'Muslim Question'. Such shifting configurations of statehood are, as several contributions to this book show, key to the changing nature of racism: we propose that the passage from empires to nation-states and the instability of nation-states in a period of intensified globalization are particularly important moments in history. However, as we will argue, it is inadequate to pair one racism with one configuration of the state form and another racism with a different political structure; the point is that both racisms change over time as the state form changes.

Adopting a comparative approach similar to Bunzl's, Nasar Meer published recently an important collection of essays that explores the relationship between antisemitism and Islamophobia as racisms.[17] It is an inquiry into the connections between the two phenomena that is often comparative in practice, but aims to establish the shared essence of the notions of the Jew and the Muslim as racialised religious subjects. The relationship between race-thinking and figures of religion is a critical aspect of our subject, and Meer's volume has provided a significant contribution to developing our conception of this issue. Yet the limitation of this kind of approach to studying the conceptual ties between antisemitism and Islamophobia *across* time is that we lose the critical importance of change *over* time—of an unfolding chronology. The relationship between Christian notions of Jews and Muslims is not the same today as it was one hundred, three hundred, or one thousand years ago—just as the Islamophobia of the war on terror is not the only historical form of anti-Muslim hatred, and genocidal Nazi antisemitism is not the only historical form of antisemitism.

Terms

As Hollande explained in his speech at the Mémorial de la Shoah, an enemy needs to be named. But it also requires definition. A referent, after all, possesses no intrinsic meaning. Hollande noted himself that the 'face' of antisemitism has changed, although he went on to insist that its 'millennial roots' remain. The problem, however, is that the word 'antisemitism' does not belong to pre-modernity. It was, instead, the product of a very distinct context of political, cultural and economic strife in central Europe at the end of the nineteenth century. In an intellectual culture shaped by racial nationalist thought, and driven by a desire for racial purity as political panacea, self-declared Jew haters deployed the word as the name for their political movement: the Antisemiten-Liga of Berlin, founded in 1879. This is the deeply problematic origin of Hollande's name for the anti-Jewish enemy.

Emerging from an entirely different time, place and motivation, the label Islamophobia first came to prominence in the United Kingdom at the end of the 1990s. The term was popularised by a report, *Islamophobia: A Challenge for Us All*, published by the race equality think-tank the Runnymede Trust in 1997.[18] The report followed a similar one by the Trust three years earlier on antisemitism, which, significantly, included what appears to be the first published recognition of 'Islamophobia' by non-Muslims.[19] The aim of the 1997 report was to examine the state of attitudes towards Muslims in a contemporary Britain marked by multi-culturalism and its discontents, globalisation, European integration and the consciously post-secular strife that followed the sentencing to death of the novelist Salman Rushdie in 1989 by Iran's Ayatollah Khomeini. Whereas much of late nineteenth-century Western Europe celebrated racial national individuality and demarcation, if not separation, the end of the twentieth century was full of ambivalence regarding the nation-state, was unsure about secularism and celebrated a post-racial diversity. The yawning chasm between the two moments is apparent from the employment of race in the word 'antisemitism', with its invocation of the notion of the Semite, and the preoccupation with religion in the named phobia of an abstract Islam, without any reference to Muslims. And whereas the original antisemites were proud to mobilise publicly and name themselves, Islamophobes, who more often than not will reject this label, were named by their opponents as a badge of shame.

Comparative analyses of anti-Jewish and anti-Muslim racisms can easily become locked into endless debates about the differences between these two terms. Yet this kind of discussion confuses the etymology of labels with the content of fields of knowledge. We use antisemitism and Islamophobia in this book merely as recognisable monikers for prejudice against Jews and Judaism, Muslims and Islam; we attribute no fixed meaning to them. The contributors to this volume make it clear that ideas in Europe concerning Jews and Muslims, Judaism and Islam, have evolved in complex ways across time and space. We can go further and say that while we use this terminology, which denotes prejudice and/or animosity, our interests in this book are far wider. Antisemitism and Islamophobia are useful as recognisable labels, but they cannot encapsulate the breadth of what is required to understand even the subjects to which they pertain. To comprehend disdain in Europe for Jews, Judaism, Muslims and Islam, we must investigate the entire intellectual universe to which it belongs: ideas about Jews and Muslims in all their fullness and complexity, and, as Hannah Arendt showed us more than sixty years ago, their wider contexts.[20]

Europe

The idea of Europe is a part of that intellectual universe, the space of Europe a key element of those wider contexts. The history of the continent could not be told without including both a Jewish and a Muslim presence, which have touched—often in profound ways—every single country in the continent. Although Εὐρώπη was used as a geographical term as early as the sixth century BCE, the concept of Europe as a cultural or geopolitical entity is, as Norman Davies has noted, 'a relatively modern idea'.[21] What defined Europe has historically often been what lay outside it—which increasingly meant the Muslim world, especially from the Ottoman empire's seizure of Constantinople in 1453 and the Muslim defeat in Iberia in 1492[22]—and minorities within—including the Jews and Roma. The paradoxical centrality to, and exclusion from, Europe of Muslims and Jews has been central to how the European imagination has figured them. Considerable scholarship, in particular that of Edward Said, has shown how this othering has often proceeded through the contrast between an occidental Europe and its Orientalism; but, as we will argue, that is only part of the story.

Davies continues that the idea of Europe only 'gradually replaced the earlier concept of "Christendom" in a complex intellectual process lasting from the fourteenth to the eighteenth centuries'.[23] However, the idea of Europe has never fully broken free from Christendom. In the nineteenth century, the age of modern nationalism, the Jews in particular became paradigmatic of what could not be assimilated; the 'Jewish Question' referred to the problematic of how to deal with the internal other.[24] Zygmunt Bauman suggested that European modernity developed two solutions to this question: using Levi-Strauss' terminology, one solution was anthropophagic—consuming difference, assimilating others by remaking them in the image of Christian Europeans—while the other was anthropoemic—expulsion or elimination.[25] However, as Marko Attila Hoare shows in his account of the Balkans in this volume, these solutions were not just used in relation to Jewish others in the age of emerging nations, but also Muslim minorities. And the population transfers that created a 'Christian' Greece and a 'Muslim' Turkey as late as the 1920s demonstrate how religious identity has been central to nationalism within Europe. This is no less true today with, for example, the vociferous objections to Turkish membership of the European Union in various member states. As Sander Gilman argues in his contribution to this book, the avowed secularism of contemporary Europe remains normatively Christian; the concept of Europe—and how it frames its others—remains bound up with theology.

Theology

As is often noted, Edward Said spoke of the 'Islamic branch' of Orientalism as 'a strange, secret sharer of Western anti-Semitism'.[26] Further, he argued that the Arab superseded and became the Jew (or a previous incarnation of the Jew): '[t]he transference of a popular anti-Semitic animus from a Jewish to an Arab target was made smoothly, since the figure was essentially the same'.[27] The timing of this separation, which was not entirely clear in Said's account (he writes of the bifurcation of 'the Jew of pre-Nazi Europe' into the Jewish hero and its 'Arab Oriental' shadow, and discusses 1973 as a watershed),[28] has attracted the attention of some scholars,[29] including in this volume; indeed, the process of uncoupling is a critical part of our archaeology of the present moment. Central to Said's analysis was not, however, the moment of divergence, but the precursor to supersession: the time of intimate sharing—the idea of the Semites.[30] This notion posited that Jews and Arabs were defined by a joint linguistic and racial heritage. The category of the Semites emerged with the Enlightenment and belonged to its task of hunting for human origins through language and history. Yet it would be a mistake to see the Semites as an Enlightenment invention, an idea summoned ex nihilo by European scholars. As James Renton argues in his chapter in this book on the end of the Semites, this category was, for all of the racial language that attended it, at base a theological construct. Ernest Renan, the 'Orientalist' most associated with the Semite concept, was ultimately concerned with writing the history of Jesus and Christianity; Jews and Arabs were, in his assessment, bound by their origins in the desert landscape of prophecy—the home of Abrahamic monotheism.[31]

Theodor Adorno and Max Horkheimer pointed to the theological, or to be more precise, Christian basis of antisemitism that is shrouded by the conceit of secularism—Valls' precious 'la laïcité, la laïcité, la laïcité!',[32] which sounds less convincing with each utterance. However, they also saw modernity—the 'alliance between enlightenment and domination'—as a corruption or mutation of religious thinking.[33] The extent and nature of the religious architecture of Western thought are only recently being uncovered, against fierce criticism. For all of his discussion of religious figures in modernity, Talal Asad's anthropology of secularism insists on a fundamental discontinuity between medieval and modern.[34] Gil Anidjar, Ivan Kalmar, Susannah Heschel, Giorgio Agamben and David Nirenberg have led the charge against the insistence on this separation.[35] Anidjar put forward what we can call a diachronic telling of the Christian basis of modernity with his history of blood and its centrality in Western political, economic and social thought. In similar and connected fashion, his previous two volumes revealed the philosophical relationship between the Jew and the Arab as Christian enemies, and the career of the Semites.[36] Agamben's genealogies of contemporary Western political philosophy traverse the lines of entrenched temporal borders—stretching from Roman law and Greek thought to twentieth-century central Europe. And he too has pointed to the determining significance of medieval Christian theological traditions in modernity.[37] In her work on Nazi Germany, Heschel has shown the critical role played by Christian thought even in the European state that re-invented itself, and attempted to re-engineer the continent, on the basis of twentieth-century race science.[38] Returning to our specific point of interest, Kalmar's seminal work on Orientalism demonstrated that both Jews and Muslims were joint subjects from the discourse's medieval beginnings.[39] In addition, he has shown the importance in Orientalism of the foundational Christian notion of connected Jewish and Muslim understandings of God as unforgiving and despotic.[40] And David Nirenberg has argued for the Christian origins of what he calls 'anti-Judaism'—not in the sense of a timeless, unchanging prejudice, but as a pivotal thought framework in which Jews and Judaism are the projection and vehicle through which Christian political and social crises are negotiated.[41]

Nirenberg does not consider that Christian views of Muslims can be analysed together with 'anti-Judaism'. Nonetheless, he makes clear that a religious frame of thought has continued across the boundary of the periods commonly labelled medieval, early and late modern. As Anidjar demonstrates, the Jewish or Muslim Questions are products of a Christian Question: the racialised, blood identity of Christendom—of Europe—from the time of the Crusades onwards. As our book also shows from the opening chapter by Andrew Jotischky, the Christian beginnings of a shared Judeo-Muslim epistemology are evident from this moment; in other words, the Christian, Jewish and Muslim Questions are of a piece, and of a time. This is not to say that they were the same, as is also demonstrated in the chapter by François Soyer on early modern Iberia and the specifics of the Jewish doctor as a conspiratorial agent. Yet the shared epistemology was there, and we maintain that it accounts for a great deal through to the Nazis' Jewish enemy, and the West's Islamic world conspiracy today. The era of European empires, including the Enlightenment, shifted the contexts, significance and coding of Christendom's Muslim and Jewish Questions, but the overarching architectural foundations and schema remained. As the book also explains, the First World War and the beginnings of the Israel–Palestine conflict finally exploded the explicit category of the Semite and a Muslim–Jewish framework. From 1918, European notions of Muslims and Jews entered a new period that ended the explicit shared framework that began with the Crusades. This is the moment that has preoccupied much of the commentary since 11 September 2001. It is, however, very different to what went before, and constitutes a late and rather short—though hugely significant—chapter in our story.

This diachronic framework is informed by the loose periodisation of Orientalism put forward by Kalmar in his monograph *Early Orientalism* and his collaboration with Derek Penslar. They identify four moments: the Saracen, from the beginnings of Islam, in which Muslim civilisation was not thought to inhabit a geographical space separate from Christendom; the Turkish, starting in the late fourteenth century, which tied Islam to the Orient and the Ottoman empire—drawing on ancient Greek understandings of the Persian East; the Arab, from the early nineteenth century, which witnessed a shift away from the figure of the Turk, prompted by the weakness of the Ottomans; and the post-colonial—a period in which Islam becomes a largely political subject and European Jews cease to be Oriental, a process that begins with the alliance between Zionism and the British empire from 1917.[42]

While our book is indebted to this historical framing, our lens is somewhat different—inevitably. For all of the significance of Orientalism in European ideas of the Jew and the Muslim, it is only a part of the story. Before the imagined geographical division between a Christian European civilisation and that of Islamic lands came the theological separation—of Christians from Jews, first of all, and then from Muslims. As the foundation of European thinking about Jews and Muslims, the Christian theological frame, broadly conceived, is not only the starting point of the story that we explore in this volume, but it continues in significance, despite the manifold changes, however momentous, in European thought since the birth of Islam. In addition, the history of European empire—again in the widest sense of the term—and its consequences determined much of how European conceptions of Muslims and Jews evolved over time and place. These overlapping historical frameworks, of the intellectual and the political, constitute the principal windows through which we have approached our subject, rather than being driven by internal changes in the figures of the Jew and the Muslim themselves.

Structure

The chapters in this book cover roughly a millennium. We have structured our investigation around the key moments and locations in the history of the relationship between European Christian understandings of Jews and Muslims. We begin with a section on 'Christendom'. This opens with Jotischky's chapter on the Crusades, which he contends was a watershed in terms of ideas and the beginnings of *Christian* violence. Jotischky places Christendom's Islamic other alongside its Jewish other, to explore what is and is not specific about each exclusion. The Crusades present an ideal opportunity for this exploration, as not only, as far as the evidence is able to show, was there no systematic persecution of Jews within Christendom prior to the First Crusade, but the campaigns were also accompanied by violence both against Jews *within* Europe (for example as the First Crusade moved eastwards through German lands) and against Muslim civilians *beyond* Europe's edges, in the Levant.

We move on to Iberia, as the space of the great expulsions, where Soyer considers what differentiated the Jewish threat in the Christian imagination. Jews and Muslims were both violently expelled from the peninsula in the fifteenth and sixteenth centuries. Yet rather than creating a pure, homogenous Latin Christian space, this expulsion opened up new spaces of ambiguity. 'The number of converts and their descendants, *judeoconversos* and *moriscos*, was so great', Soyer writes, 'that they did not assimilate and disappear into the wider Christian population.' These groups became suspect populations, and as such subject to conspiracy theories, including those centring on malevolent medical doctors. However, one group—the *judeoconversos*—became the main target of such accusations, while the other—the *moriscos*—were apparently mostly spared. Soyer explains this by showing the specificity of the Jewish Question in relation to the Moorish Question: Judeophobic conspiracy theories were already inscribed in Christian culture centuries before the expulsions.

The section on 'Christendom' is followed by two chapters on 'Empire', with an essay on the Russian empire by Robert Crews, and James Renton's discussion of the British and French empires in the Middle East. For Crews, 'Muslims and Jews became stock figures in an ominous pantheon of ethnic and religious groups that seemed to pose a threat to the [Russian] imperial political order', positioned as analogous to each other in some ways but unique in others. Both were ethnographically racialised and both were marked as other to Christian order. However, Jews were connected to diffuse and metaphysical fears to do with urbanisation and globalisation, while Muslims were positioned as a colonial and geopolitical problem of imperial rule in localised internal or border regions. Both chapters in this section demonstrate the importance of contingency and specific political contexts in driving change in notions of Jews and Muslims. The First World War, in particular, and the beginnings of the collapse of the colonial world order led to the splintering of Christendom's Jew and Muslim.

The penultimate section is thus called 'Divergence', with chapters on distinguishing features of our present moment from Sander Gilman, Hoare and Gil Anidjar. Gilman's chapter explores the contemporary concept of the Muslim as an echo of the post-Enlightenment Jew. Gilman demonstrates again that the place of Jews and Muslims in Europe is fundamentally structured by the implicit normative Christianity of European secularism. And, as Jewishness has been the paradigmatic figure for Christianity's alterity-within, it provides, Gilman argues, a 'template' for Islam; both have an intimate familial relationship with their

'Abrahamic' cousin Christianity, so that Christendom, including in its supposedly secular form, has worked hard at the task of boundary-making in relation to them. The 'minor differences' (as Freud would put it) between Christendom and its Abrahamic others have been the locus of this boundary-making work: specifically, ritual practices such as head-covering and circumcision.

Analysing the Balkans, Hoare presents a history of anti-Jewish and anti-Muslim prejudice shaped by Orthodox Christianity and, critically, political circumstances that encompass independence struggles, the Holocaust, Communism and the conflicts of the 1990s. This history culminates in a clear division by the end of the twentieth century between antisemitism and Islamophobia; by this point Serb and Croat nationalists were 'more likely to identify with Israel on an anti-Muslim basis than they were to indulge in antisemitism'. Returning to a broader trans-European, or rather international, context, Anidjar considers the origins of the West's contemporary war on antisemitism. He conceives of anti-antisemitism as a war that spans the world, and operates through a vast apparatus including national and international law, global institutions, schools, universities, non-government organisations, media and entertainment, and more. Yet as much as the war on antisemitism is a central part of global politics since 9/11, as we have already discussed, it has no obvious history, or self-understanding, Anidjar argues. The answer to its origins and essence, Anidjar suggests, lies in the unacknowledged 'negative portrait' of the figure of the anti-antisemite—the 'survivor'; that antonym is the Muslim.

We have named the final part of the book 'Response'; it is the beginning of a consideration of how we can understand the shared impact of antisemitism and Islamophobia among Jews, Muslims and their advocates. Opening this section, Daniel Gordon examines the history of the fight against anti-Jewish and anti-Muslim racisms in France. French anti-racism has often been portrayed as splitting into two camps. However, Gordon's excavation of this history demonstrates that the ideological divergence of antisemitism and Islamophobia in the second half of the twentieth century is not reflected in this activist public space. Instead, he shows a subterranean connection between the two traditions of antiracism, opened up by what we might call a *cosmopolitan* understanding of racism, which highlights how colonial histories have shaped antisemitism within Europe, while the racialisation of Jews in the continental metropoles has shaped the management of Muslim difference in empires abroad. Shifting our attention to how Jews and Muslims themselves are affected in related ways by antisemitism and Islamophobia, David Wertheim has written a comparative study of the troubled biographies of two prominent figures: one from the twenty-first-century Netherlands—Ayaan Hirsi Ali, a Muslim-born former politician who left Islam and then abandoned Europe—and one from nineteenth-century German lands, Heinrich Heine, a Jewish writer who converted to Christianity and also left his home, migrating to Paris. Wertheim demonstrates that antisemitism and Islamophobia both place similar complex and contradictory demands on their subjects that are impossible to fulfil. He focuses on the demand for assimilation, and how it is undermined by the simultaneous insistence on authenticity.

The book concludes with an analysis of the impact of antisemitism and Islamophobia on how Muslims and Jews see each other. Written by Yulia Egorova and Fiaz Ahmed, the concluding chapter is based on forty in-depth interviews with British Muslims and Jews, and participant observation of the meetings of two Jewish–Muslim dialogue initiatives between 2013 and 2015. Egorova and Ahmed argue that this ethnographic research demonstrates the influence on their subjects of discourses in wider British society concerning Muslims, Jews and other minority groups. And, significantly, they contend that an individual's attitudes towards the other community are affected by their own lived experience of discrimination, and their sense of in/security.

The concluding section of the book thus underscores our principal argument: that there has indeed been a relationship between antisemitism and Islamophobia in Europe since the Crusades, but that this shared story has evolved in complex ways over time and space. To begin to understand this complexity, we need to approach the subject from a multiplicity of perspectives, including those of Jews and Muslims themselves, and those who claim to defend them in the face of prejudice.

Notes

1. François Hollande, 'Discours au Memorial de la Shoah', 27 January 2015, site de la Présidence de la République, http://www.elysee.fr/declarations/article/discours-au-memorial-de-la-shoah-2/ (Hollande 2015).
2. '#PlanAntiRacisme: "Éveiller les consciences. Agir. Ne plus rien laisser passer"', Dossier de Presse, 17 April 2015, http://www.gouvernement.fr/planantiracisme-eveiller-les-consciences-agir-ne-plus-rien-laisser-passer (Gouvernement.fr 2015b).

3. Abdellali Hajjat and Marwan Mohammed, *Islamophobie: Comment les élites françaises fabriquent le "problème musulman"* (Paris: La Découverte, 2013), pp. 226–31 (Hajjat and Mohammed 2013).
4. '#PlanAntiRacisme'.
5. Ondine Millot, 'Il m'a dit: "Dis-moi où est le fric sinon je te bute"', *Libération*, 10 December 2014, http://www.liberation.fr/societe/2014/12/10/il-m-a-dit-dis-moi-ou-est-le-fric-sinon-je-te-bute_1161113 (Millot 2014).
6. '#PlanAntiRacisme'.
7. 'Hommage aux victimes des attentats', Assemblée nationale, XIVe législature, Session ordinaire de 2014–2015, Compte rendu intégral, Première séance du mardi 13 janvier 2015, http://www.assemblee-nationale.fr/14/cri/2014-2015/20150106.asp (Assemblée nationale 2015).
8. '#Antiterrorisme: Manuel Valls annonce des mesures exceptionnelles', 21 January 2015, http://www.gouvernement.fr/antiterrorisme-manuel-valls-annonce-des-mesures-exceptionnelles (Gouvernement.fr 2015a).
9. 'Projet de loi renseignement. "Protéger les Français dans le respect des libertés"', dossier de presse, Conseil des ministres du 19 mars 2015, http://www.gouvernement.fr/sites/default/files/document/document/2015/03/dp-loi-renseignement_v3-bat.pdf; 'LOI n° 2015-912 du 24 juillet 2015 relative au renseignement', https://www.legifrance.gouv.fr/affichTexte.do?cidTexte=JORFTEXT000030931899&categorieLien=id (Gouvernement.fr 2015c; Legifrance.gouv.fr 2015).
10. 'Discours de Manuel VALLS, Premier Ministre, Attentat de Nice', 15 July 2016, http://www.gouvernement.fr/sites/default/files/document/document/2016/07/20160715_discours_de_manuel_valls_premier_ministre_-_attentat_de_nice.pdf; François Hollande, 'Déclaration à la suite des événements de Nice', 15 July 2016, http://www.elysee.fr/declarations/article/declaration-al- a-suite-des-evenements-de-nice-2/ (Gouvernement.fr 2016; Hollande 2016).
11. 'L'intervention de la France contre #Daech', infographic, 2 October 2014, http://www.gouvernement.fr/partage/1920-l-intervention-de-la-france-contre-daech (Gouvernement.fr 2014).
12. Matti Bunzl, *Anti-Semitism and Islamophobia: Hatreds Olds and New in Europe* (Chicago: University of Chicago Press, 2007) (Bunzl 2007).

13. Ilan Halevi, *Islamophobie et Judéophobie: L'Effet Miroir* (Paris: Syllepse, 2015); Nasar Meer and Tehseen Noorani, 'A Sociological Comparison of Anti-Semitism and Anti-Muslim Sentiment in Britain', *The Sociological Review*, 56, no. 2 (2008), 195–219 (Halevi 2015; Meer and Noorani 2008).
14. See, for example, David Glover, *Literature, Immigration, and Diaspora in Fin-de-Siècle England: A Cultural History of the 1905 Aliens Act* (Cambridge: Cambridge University Press, 2012); Paddy Hillyard, *Suspect Community: People's Experience of the Prevention of Terrorism Acts in Britain* (London: Pluto Press, 1993); Mary Hickman et al., *'Suspect Communities?' Counter-Terrorism Policy, the Press, and the Impact on Irish and Muslim Communities in Britain* (London: London Metropolitan University, 2011) (Glover 2012; Hillyard 1993; Hickman et al. 2011).
15. Saul Friedländer, *Nazi Germany and the Jews, Vol. I: The Years of Persecution, 1933–1939* (London: Weidenfeld & Nicolson, 1997), pp. 199–201 (Friedländer 1997).
16. Christopher R. Browning with contributions by Jürgen Matthäus, *The Origins of the Final Solution: The Evolution of Nazi Jewish Policy, September 1939–March 1942* (Lincoln, NE: University of Nebraska Press, 2004) (Browning 2004).
17. Nasar Meer (ed.), *Racialization and Religion: Race, Culture and Difference in the Study of Antisemitism and Islamophobia* (London: Routledge, 2014) (Meer 2014).
18. On the Runnymede report, see AbdoolKarim Vakil, 'Is the Islam in Islamophobia the Same as the Islam in Anti-Islam; Or, When is it Islamophobia Time?' and Chris Allen, 'Islamophobia: From K.I.S.S. to R.I.P.', in *Thinking Through Islamophobia: Global Perspectives*, eds. S. Sayyid and AbdoolKarim Vakil (London: Hurst, 2010), chs. 4 and 6 (Vakil 2010; Allen 2010).
19. Runnymede Commission on Antisemitism, *A Very Light Sleeper: The Persistence and Dangers of Antisemitism* (London: Runnymede Trust, 1994); Hajjat and Mohammed, *Islamophobie*, p. 82 (Runnymede Commission on Antisemitism 1994).
20. Hannah Arendt, *The Origins of Totalitarianism* (Oregon: Harcourt, c.1976 [1st edn., 1951]), part II (Arendt c. 1976).
21. Norman Davies, *Europe: A History* (London: Pimlico, 1997), p. 7 (Davies 1997).
22. Ivan Kalmar, *Early Orientalism: Imagined Islam and the Notion of Sublime Power* (London: Routledge, 2012), chs. 3–4 (Kalmar 2012).

23. Ibid.
24. Aamir R. Mufti, *Enlightenment in the Colony: The Jewish Question and the Crisis of Postcolonial Culture* (Princeton, NJ: Princeton University Press, 2007) (Mufti 2007).
25. Zygmunt Bauman, *Postmodernity and its Discontents* (Cambridge: Polity, 1997) (Bauman 1997).
26. Edward W. Said, *Orientalism: Western Conceptions of the Orient-Reprinted with a New Afterword* (London: Penguin, 1995), pp. 27–8 (Said 1995).
27. Ibid., p. 286.
28. Ibid., pp. 285–6.
29. See, in particular, Gil Anidjar, *Semites: Race, Religion, Literature* (Stanford: Stanford University Press, 2008), pp. 20–1, ch. 1 (Anidjar 2008).
30. Said, *Orientalism*, esp. his initial discussion of Renan, pp. 130–50.
31. Ernest Renan, *Histoire des origines du Christianisme*, 7 vols (Paris: Michel Lévy frères/ Calmann- Lévy, 1863–1882), the first volume of which was *Vie de Jésus*; and *Histoire du peuple d'Israël*, 5 vols (Paris: Calmann-Lévy, 1887–1893) (Renan 1863–1883, 1887–1893).
32. 'Hommage aux victimes des attentats'.
33. Theodor W. Adorno and Max Horkheimer, *Dialectic of Enlightenment*, trans. John Cumming (London: Verso, 1997), p. 176 (Adorno and Horkheimer 1997).
34. Talal Asad, *Formations of the Secular: Christianity, Islam, Modernity* (Stanford: Stanford University Press, 2003) (Asad 2003).
35. See, most recently, Gil Anidjar, 'The Violence of Violence: Response to Talal Asad's "Reflections on Violence, Law, and Humanitarianism"', *Critical Inquiry*, 41, no. 2 (Winter 2015), 435–42 (Anidjar 2015).
36. Gil Anidjar, *Blood: A Critique of Christianity* (New York: Columbia University Press, 2014); *The Jew, the Arab: A History of the Enemy* (Stanford: Stanford University Press, 2003); *Semites* (Anidjar 2003, 2014).
37. Giorgio Agamben, *The Kingdom and the Glory: For a Theological Genealogy of Economy and Government*, trans. Lorenzo Chiesa with Matteo Mandarini (Stanford: Stanford University Press, 2011), originally published in Italian in 2007 (Agamben 2011).
38. Susannah Heschel, *The Aryan Jesus: Christian Theologians and the Bible in Nazi Germany* (Princeton, NJ: Princeton University Press, 2008) (Heschel 2008).

39. Ivan Davidson Kalmar and Derek J. Penslar (eds.), *Orientalism and the Jews* (Hanover, NH: University Press of New England, 2005), Introduction and ch. 1; Ivan Kalmar, *Early Orientalism*; 'Benjamin Disraeli: Romantic Orientalist', *Comparative Studies in Society and History*, 47, no. 2 (2005), 348–71; 'Anti-Semitism and Islamophobia: The Formation of a Secret', *Human Architecture: Journal of the Sociology of Self-Knowledge*, 7, no. 2 (Spring 2009), 135–44; 'Arabizing the Bible: Racial Supersessionism in Nineteenth Century Christian Art and Biblical Scholarship', in *Orientalism Revisited: Art, Land and Voyage*, ed. Ian Netton (London: Routledge, 2013), pp. 176–86 (Kalmar and Penslar 2005; Kalmar 2005, 2009, 2013).
40. Kalmar, *Early Orientalism*, ch. 1.
41. David Nirenberg, *Anti-Judaism: The Western Tradition* (New York: W.W. Norton, 2013) (Nirenberg 2013).
42. Kalmar and Penslar, *Orientalism and the Jews*, pp. xxiii–xxxv; Kalmar, *Early Orientalism*, chs. 3–4.

References

Adorno, Theodor W., and Max Horkheimer. 1997. *Dialectic of Enlightenment*, trans. John Cumming. London: Verso.

Agamben, Giorgio. 2011 (originally published in Italian in 2007). *The Kingdom and the Glory: For a Theological Genealogy of Economy and Government*, trans. Lorenzo Chiesa with Matteo Mandarini. Stanford: Stanford University Press.

Allen, Chris. 2010. Islamophobia: From K.I.S.S. to R.I.P. In *Thinking Through Islamophobia: Global Perspectives*, ed. S. Sayyid, and AbdoolKarim Vakil. London: Hurst.

Arendt, Hannah. c. 1976. *The Origins of Totalitarianism*. Orlando: Harcourt [1st edn., 1951].

Anidjar, Gil. 2003. *The Jew, the Arab: A History of the Enemy*. Stanford: Stanford University Press.

———. 2008. *Semites: Race, Religion, Literature*. Stanford: Stanford University Press.

———. 2014. *Blood: A Critique of Christianity*. New York: Columbia University Press.

———. 2015. The Violence of Violence: Response to Talal Asad's "Reflections on Violence, Law, and Humanitarianism". *Critical Inquiry*, 41, no. 2 (Winter), 435–442.

Asad, Talal. 2003. *Formations of the Secular: Christianity, Islam, Modernity*. Stanford: Stanford University Press.

Assemblée nationale. 2015. Hommage aux victimes des attentats, XIVe législature, Session ordinaire de 2014–2015, Compte rendu intégral, Première séance du mardi 13 janvier. http://www.assemblee-nationale.fr/14/cri/2014-2015/20150106.asp

Bauman, Zygmunt. 1997. *Postmodernity and its Discontents*. Cambridge: Polity.

Browning, Christopher R. with contributions by Jürgen Matthäus. 2004. *The Origins of the Final Solution: The Evolution of Nazi Jewish Policy, September 1939–March 1942*. Lincoln, NE: University of Nebraska Press.

Bunzl, Matti. 2007. *Anti-Semitism and Islamophobia: Hatreds Olds and New in Europe*. Chicago: University of Chicago Press.

Davies, Norman. 1997. *Europe: A History*. London: Pimlico.

Friedländer, Saul. 1997. *Nazi Germany and the Jews, Vol. I: The Years of Persecution, 1933–1939*. London: Weidenfeld & Nicolson.

Glover, David. 2012. *Literature, Immigration, and Diaspora in Fin-de-Siècle England: A Cultural History of the 1905 Aliens Act*. Cambridge: Cambridge University Press.

Gouvernement.fr. 2014. L'intervention de la France contre #Daech, Infographic, 2 October. http://www.gouvernement.fr/partage/1920-l-intervention-de-la-france-contre-daech

———. 2015a. #Antiterrorisme: Manuel Valls annonce des mesures exceptionnelles, 21 January 2015. http://www.gouvernement.fr/antiterrorisme-manuel-valls-annonce-des-mesures-exceptionnelles

———. 2015b. "Éveiller les consciences. Agir. Ne plus rien laisser passer". Dossier de presse, 17 April. http://www.gouvernement.fr/planantiracisme-eveiller-les-consciences-agir-ne-plus-rien-laisser-passer

———. 2015c. 'Projet de loi renseignement. "Protéger les Français dans le respect des libertés"'. Dossier de presse, Conseil des ministres du 19 mars. http://www.gouvernement.fr/sites/default/files/document/document/2015/03/dp-loi-renseignement_v3-bat.pdf

———. 2016. Discours de Manuel VALLS, Premier Ministre, Attentat de Nice, 15 July. http://www.gouvernement.fr/sites/default/files/document/document/2016/07/20160715_discours_de_manuel_valls_premier_ministre_-_attentat_de_nice.pdf

Hajjat, Abdellali, and Marwan Mohammed. 2013. *Islamophobie: Comment les élites françaises fabriquent le "problème musulman"*. Paris: La Découverte.

Halevi, Ilan. 2015. *Islamophobie et Judéophobie: L'Effet Miroir*. Paris: Syllepse.

Heschel, Susannah. 2008. *The Aryan Jesus: Christian Theologians and the Bible in Nazi Germany*. Princeton, NJ: Princeton University Press.

Hickman, Mary, et al. 2011. *Suspect Communities?* In *Counter-Terrorism Policy, the Press, and the Impact on Irish and Muslim Communities in Britain*. London: London Metropolitan University.

Hillyard, Paddy. 1993. *Suspect Community: People's Experience of the Prevention of Terrorism Acts in Britain*. London: Pluto Press.

Hollande, François. 2015. Discours au Memorial de la Shoah, 27 January. Site de la Présidence de la République. http://www.elysee.fr/declarations/article/discours-au-memorial-de-la-shoah-2/

———. 2016. Déclaration à la suite des événements de Nice, 15 July. http://www.elysee.fr/declarations/article/declaration-a-la-suite-des-evenements-de-nice-2/

Kalmar, Ivan. 2005. Benjamin Disraeli: Romantic Orientalist. *Comparative Studies in Society and History* 47(2): 348–371.

———. 2009. Anti-Semitism and Islamophobia: The Formation of a Secret. *Human Architecture: Journal of the Sociology of Self-Knowledge* 7(2) (Spring): 135–144.

———. 2012. *Early Orientalism: Imagined Islam and the Notion of Sublime Power*. London: Routledge.

———. 2013. Arabizing the Bible: Racial Supersessionism in Nineteenth Century Christian Art and Biblical Scholarship. In *Orientalism Revisited: Art, Land and Voyage*, ed. Ian Netton, 176–186. London: Routledge.

Kalmar, Ivan, and Derek J. Penslar (ed). 2005. *Orientalism and the Jews*. Hanover, NH: University Press of New England.

Legifrance.gouv.fr.2015. 'LOI n° 2015-912 du 24 juillet 2015 relative au renseignement', https://www.legifrance.gouv.fr/affichTexte.do?cidTexte=JORFTEXT000030931899&categorieLien=id

Meer, Nasar (ed). 2014. *Racialization and Religion: Race, Culture and Difference in the Study of Antisemitism and Islamophobia*. London: Routledge.

Meer, Nasar, and Tehseen Noorani. 2008. A Sociological Comparison of Anti-Semitism and Anti-Muslim Sentiment in Britain. *The Sociological Review* 56(2): 195–219.

Millot, Ondine. 2014. Il m'a dit: "Dis-moi où est le fric sinon je te bute". *Libération*, 10 December. http://www.liberation.fr/societe/2014/12/10/il-m-a-dit-dis-moi-ou-est-le-fric-sinon-je-te-bute_1161113

Mufti, Aamir R. 2007. *Enlightenment in the Colony: The Jewish Question and the Crisis of Postcolonial Culture*. Princeton, NJ: Princeton University Press.

Nirenberg, David. 2013. *Anti-Judaism: The Western Tradition*. New York: W.W. Norton.

Renan, Ernest. 1863–1883. *Histoire des origines du Christianisme*, 7 vols. Paris: Michel Lévy frères/Calmann- Lévy.

———. 1887–1893. *Histoire du peuple d'Israël*. Paris: Calmann-Lévy.

Runnymede Commission on Antisemitism. 1994. *A Very Light Sleeper: The Persistence and Dangers of Antisemitism*. London: Runnymede Trust.

Said, Edward W. 1995. *Orientalism: Western Conceptions of the Orient- Reprinted with a New Afterword*. London: Penguin.

Vakil, AbdoolKarim. 2010. Is the Islam in Islamophobia the Same as the Islam in Anti-Islam; Or, When is it Islamophobia Time? In *Thinking Through Islamophobia: Global Perspectives*, ed. S. Sayyid, and AbdoolKarim Vakil. London: Hurst.

PART I

Christendom

CHAPTER 2

Ethnic and Religious Categories in the Treatment of Jews and Muslims in the Crusader States

Andrew Jotischky

The tone for the treatment of Jews and Muslims by crusaders was set even before the First Crusade reached its destination. The departure of some groups of crusaders to the East in 1096 was preceded in some towns in Normandy and western parts of Germany—Rouen, Worms, Mainz and Cologne especially—by mob violence against settled Jewish communities.[1] The capture of some cities in Syria and Palestine by the crusaders between 1098 and ca. 1110, moreover, was followed by massacres of Muslim (mostly Egyptian) defenders and civilian inhabitants. The most notorious case occurred in Jerusalem in July 1099, when a contemporary Frankish account, citing Apocalypse 14.20, spoke of blood rising as high as the knees of the horsemen and the bridles of their horses.[2] None of these massacres appears to have fulfilled the intentions of the papacy in

A. Jotischky (✉)
Royal Holloway, University of London, Egham, UK

J. Renton, B. Gidley (eds.), *Antisemitism and Islamophobia in Europe*, DOI 10.1057/978-1-137-41302-4_2

the preaching of the Crusade, and collectively they have been explained as examples of how crusading took on a momentum of its own in the hands of the participants.[3] The First Crusade can thus be said to have marked a turning point in the treatment of Jews and Muslims. Prior to 1096, although outbreaks of violence against Jewish communities had occurred, there had been no systematic persecution of Jews within the Christian West, and there is little evidence of any interest in legal or theological discourses in singling them out for special treatment.[4]

For most Europeans, Muslims were a distant reality before 1099. Only in the Iberian peninsula had there been any sustained contact with Islamic communities, and Islam itself, though it attracted attention from a few Western scholars, was barely understood in the West.[5]

Outbreaks of violence continued to be a feature of Crusade preaching and mobilisation in the West: in 1147 the preaching of the Second Crusade was accompanied by further mob attacks on Jews in Germany, and the same happened in London and York in 1189–90.[6] In the Crusader States, however—in other words, the territories in Syria and Palestine that came under the control of Franks from 1097 onwards—there are no recorded incidents of such massacres after 1110. The Crusader States established in the wake of the First Crusade in 1095 lasted some two hundred years on the east Mediterranean littoral, occupying more or less the area from south-east Turkey to the Red Sea and including at their greatest extent the present-day states of Israel/Palestine, Lebanon and parts of western and northern Syria and eastern Turkey. The European settlers constituted a minority, albeit the dominant military force, among the indigenous peoples. More to the point, perhaps, they found the indigenous peoples far more diverse and difficult to categorise than they appear to have expected. In preaching the Crusade, Pope Urban II probably referred in general terms to the indigenous Christians of the Holy Land,[7] but it was left to the first generation of Western settlers after the Crusade to negotiate the variety of religious and ethnic groups in newly conquered territory and to fit them into a scheme of law and government.[8]

One difficulty was that the indigenous peoples resisted easy categorisation. The majority of the indigenous population were Arabic speaking, but may have been fairly evenly divided between Christians and Muslims. The indigenous Muslims, however, were themselves a subaltern people who had been living under Seljuq Turkish rule since the mid-eleventh century. The Muslim communities of the region were neither monocultural nor static in identity throughout the period, and there was a continuous influx of new ethnic groups, such as the Kurdish Ayyubids, in the twelfth and thirteenth centuries. Though the Seljuqs were the dominant political force among Muslims in the region as a whole, they were, like the crusaders, an ethnically distinct conquering military aristocracy. The Jewish communities, most of whom probably spoke Arabic, were also divided between different traditions and enjoyed no overall leadership.[9] In addition, there were indigenous peoples who spoke Syriac or Armenian. This complexity perhaps explains why the crusaders initially treated all non-Franks as enemies or potential enemies—most obviously in the massacre of the defenders of Jerusalem and other towns that resisted surrender, but subsequently within the legal framework of the new polities established after the Crusade. The consensus among historians of the Crusader States has tended to be that in terms of legal status, distinctions were made not according to religion or language group, but according to the political fact of conquest: one was either a Frank or a non-Frank.[10] However, the earliest laws promulgated in the kingdom of Jerusalem, the Concordat of Nablus (1120), overlaid this state of affairs with a degree of religious segregation by prohibiting sexual unions between Franks and Muslims. Muslim women who consented to sex with a Frank were to have their noses cut off; Frankish men who raped Muslim women were to be castrated; while Frankish women who had consensual sex with Muslim men were to be treated as adulterers, and therefore liable to death. These laws, the penalties for which are taken from Byzantine precedents rather than contemporary Western laws, demonstrate the desire to keep Franks and Muslims in separate social spheres. In the concern they betrayed to ensure that there could be no mixed biological issue from such unions, moreover, they reinforced legal separation on grounds of ethnicity.[11]

The law alone, however, is an inadequate guide to attitudes. For one thing, the evidence it provides is patchy at best. With the exception of the Nablus decrees, the law codes used in the Crusader States in the twelfth century were lost in 1187 and had to be reconstructed from oral memory in the thirteenth century. As the most recent historiography shows, this has resulted in a distorted understanding of the legal situation of indigenous peoples under Frankish rule.[12] Thus, for example, although the authoritative law code of the thirteenth century, the Assises de Jerusalem, prohibits those who were not Latin Christians from testifying in court, we are not in a position to say whether this had been the case before 1187.[13] There is also another consideration. Given that most laws tend to be reactive rather than creative, in the sense that they respond to given situations as they arise, and given that the preamble to the code of Nablus hints at such precedents, we can surmise that, twenty years after its foundation, the kingdom of Jerusalem was already facing the question of the consequences of sexual contact between Western Christian settlers and indigenous Muslims—and, thus, of mixed-race children. Such a supposition, however, cannot be tested fully against surviving evidence. Although there are indications that mixed-race marriages took place between Latins and indigenous Christians, we do not know for certain of any consensual relations between Latins and Muslims. Nevertheless, the question of children born from sexual unions between Christians and Muslims, whether consensual or not, had clearly preoccupied Latins even before the Council of Nablus, for chroniclers writing about the events of the First Crusade described episodes where Western women had been taken as captives by Muslims and either raped or forced into sexual relationships by their captors.[14] Similarly, secular literature written for the purposes of entertainment in twelfth- and thirteenth-century Western Europe alludes to sexual unions between Muslims and Western Christians in tones that suggest that they were frequent enough to be part of the expected landscape of the exotic.[15] The likelihood that European settlers in the East, both men and women, had indulged in sexual relations with Muslims was regarded by Western observers as one reason for the weakness of the Crusader States at the end of the twelfth century.[16]

The decrees of Nablus were ecclesiastical canons passed at a Church council presided over by the highest religious authority in the Crusader States, the patriarch of Jerusalem. Such ecclesiastical legislation allows us to tap into a richer vein of thinking and writing about the issue of religious and ethnic otherness being mined by theologians as well as lawyers. In this context, one question that has so far gone unanswered about the legislation of the Council of Nablus concerns language and definition. What exactly was meant by the term 'Saracenus' (Muslim) and what did the framers and users of law in the Crusader States understand by it? Was it intended to define a religious or an ethnic/racial category? Indeed, did twelfth- and thirteenth-century Europeans in the Crusader States distinguish between these categories and, if not, what are the implications for our understanding of their attitudes towards their subject peoples?[17] The evidence that might enable us to determine the question is limited. One important clue is provided by the presence of a Muslim who had converted to Christianity, Walter Mahomet, as one of the prominent landowners in the kingdom of Jerusalem around the time of the Council.

Similarly, we know of other ethnic non-Westerners, such as Armenians and Byzantines, who were significant landowners in the kingdom of Jerusalem in the twelfth century.[18] These examples suggest that ethnicity by itself was not a bar to promotion or social status.

The Council of Nablus says nothing about Jews, whom the framers of the law do not seem to have been regarded as a concern, probably because their numbers were small and they were, at least in the first quarter of the twelfth century, limited to certain areas of the kingdom of Jerusalem. By the third quarter of the twelfth century, however, Jews and Muslims were routinely considered in the same legal category in ecclesiastical law in Christendom as a whole. The first such instance of specific coupling occurred in 1179, in a decree of the Third Lateran Council, and by 1190 it featured as a stock item in canon law collections.[19] This did not necessarily result in legal changes in the kingdom of Jerusalem; in fact, there is no evidence that decrees of the Council regarding Jews, such as limitations on bearing witness in court, were ever enacted there. In this respect, however, canon law was far behind theologians, who had since the late eighth century associated Judaism with Islam as a 'mistaken religion'—in other words, a deviant form of true religion or, if one prefers, a heresy.[20]

A genre of writing observable for the first time in the late twelfth century, in which Jews and Muslims occupy an uncertain place as both religious deviants and ethnically categorised 'peoples', was of particular importance to the Crusader States: the pilgrimage account of the Holy Land. And it is with two examples of this literature that the remainder of this chapter will be concerned. Within this genre, one group stands out for its interest in its contemporary inhabitants, topography and flora and fauna. The earliest example of such accounts, dating from 1168/87 and subsequently copied in various redactions, is the anonymous *Tractatus de locis et statu sancte terre*.[21] This treatise, once thought to have been commissioned or written by the patriarch of Jerusalem, Haymarus Monachus, as a report on the state of the kingdom of Jerusalem for Pope Innocent III, is now accepted as the work of a Western visitor to the kingdom of Jerusalem with a particular interest in geography and ethnography.[22] Material from this treatise was re-used by two early thirteenth-century Western visitors. Jacques de Vitry, Bishop of Acre, expanded considerably on the information in the *Tractatus* in his *Historia Orientalis*, a work encompassing both the history and geography of the Holy Land written in the early 1220s, while Thietmar's pilgrimage account of 1217 also used the sections on animals, plants and ethnic groups.[23]

Let us take the *Tractatus* first. After summary descriptions of the Christian peoples living in the Holy Land, the main holy places, the ecclesiastical hierarchy and monasteries, the author moves on to discuss the regional divisions of landholding, geographical and topographical features, animals and plants, the etymology of place names, then finally the non-Christian inhabitants: Jews, Samaritans and two groupings of Muslims—Assassins and Bedouin. It is surely a significant indication of the author's own sense of hierarchy that non-Christians are placed not only after Christians but at the end of the account, after places, features and ecclesiastical organisation. The intent is surely to treat them as lesser or even non-participants in the state of the Holy Land. The first point to note is that the strict legal divide between Frank and non-Frank is less important to the author than the identification of peoples by a process of categorisation that is broadly conceived along confessional lines. Here, however, one looks in vain for a systematic methodology of determining the nature of a given category. Why, for example, are the Bedouin treated as a separate religious/ethnic category, even though the term describes neither an ethnic nor a religious group in the same way as does the category 'Assassins' (Nizari Ismailis)? In the case of the Bedouin, the categorisation appears to have been made according to urban–rural comparisons of social organisation rather than anything else. Bedouin are referred to frequently in Western pilgrimage literature: sometimes, as for example in the report of Wilbrand of Oldenbourg (1211–12), as an Arab people to be distinguished from Turks; sometimes with particular reference to inherent abilities such as horsemanship; sometimes simply as dangerous brigands.[24]

A further indication of the lack of system in the text is the description of the Jews. Under this category the author mixes religious beliefs and practices in a single sentence—'They keep the Old Testament to the letter, and use Hebrew script'—and follows this with a further subdivision according to belief, distinguishing Jews from Sadducees, who do not believe in the Resurrection. However, the main observation about Jews appears to be derived from observation of a supposed ethnic peculiarity: 'Jewish men are stubborn, more unwarlike than women, who are slaves wherever they live, and who suffer a flux of blood every month.'[25] The complexities in this rather confusing single sentence are considerable. In fact, it is not as determined as it may appear by ethnic categorisation. The reference to servile status, though on one level certainly an allusion to the legal status of Jews in much of Western Europe as 'royal serfs', has a theological origin, and derives from the prevailing notion that Jews had become 'enslaved to Christ' through the Resurrection, both in its effective action and through their failure to acknowledge it.[26] The contention that the Jews were unwarlike, a charge that was incidentally also levelled at Arabic-speaking indigenous Christians, follows the logic of this theological assertion.[27] While it can be read as simply a reference to the status of the Jews as subject peoples without the military power to resist dominant powers and thus to form self-governing polities, it also has deeper implications for our understanding of medieval perceptions of Judaism. The ideology of medieval kingship was based on the premise that the office of king had been established by God for the Jewish people, and the anointing of David provided a model for Western coronation rituals; moreover, even the most fleeting familiarity with the Old Testament was sufficient to remind Christians that kingship had throughout Jewish history been sustained by force of arms. What divided the military heroics of the Jewish past from their feebleness of today was, of course, their denial of Christ, which rendered them unworthy of God's favour and gave rise to the destruction of their political autonomy. Medieval Christian discourses on Judaism thus distinguished between two different Jewish communities, each specific to their own historical moment in time: that before the Incarnation and contemporary Judaism, which was destined to remain weak and servile.[28] Ethnic characterisation is thus explained by the prevailing theological system.

The idea that Jewish males were subject to periodic bleeding similar to women's menstrual cycles, bizarre though it might appear, was to become an issue of theological debate in the University of Paris in the early fourteenth century, when it was argued that there were 'natural' differences between Jews and Christians analogous to those between the sexes.[29] In the 1220s, however, Jacques de Vitry, developing the theme linking the Jews' unwarlike nature to femininity, and in particular to the monthly effusion of blood, attributed the phenomenon to divine punishment for Cain's murder of Abel. He seems here to be conflating an idea borrowed from the *Sermo in laude dominici sepulchri* by Peter the Venerable (Abbot of Cluny 1122–56), according to which the sacrifice of Abel, which pleased the Lord, was held to symbolise Christian worship, in contrast to the sacrifice of Cain, which prefigured Jewish and Canaanite burnt offerings. Peter, however, saw the punishment assigned to the Jews not as the monthly bleeding suffered by individual Jewish males, but also as the loss of autonomy and expulsion from their homeland.[30]

Jacques de Vitry takes Peter's analysis of Jewish helplessness a stage further by eliding the theological with the physical. The monthly bleeding equates Jews with women, and consequently explains their lack of military capacity. Since the bleeding was imposed for the sinfulness of Cain, their political subjection to others, both Christians and Muslims, is theologically justifiable. The means by which this justified penalty is imposed, however, is by making Jewish men similar in physical (in)capacity to women. In this formulation, it is important for Jacques's argument that Jews who lived under Muslim rule were just as subservient to and just as despised by them as those who lived under Christian dominion were hated by Christians.[31] Here, too, we see an echo in his treatment of the Jews of his characterisation of another people despised for their lack of warlike tendencies, the Arabic-speaking Christians. Jacques warned his readers against indigenous Christians of the Holy Land who lived under Islamic rule: they could not be trusted, because although they claimed to be Christian, they spoke Arabic, which was the language of the Saracens. This made them, in some sense, a marginal people, who belonged neither to one side nor the other.[32] The Jews—or at least some groups among them—likewise resisted strict categorisation by using both Hebrew and Arabic, and sometimes Aramaic. The militarily weak and politically subservient slip from one language to another; in contrast, the national or ethnic groupings of whom Jacques approves, the Armenians and Georgians, are not only militarily powerful enough to maintain a measure of political independence, they are also easily identifiable by their use of a language unique to them.[33]

A slightly different method of hooking an ethnographic observation to a theological point can be seen in Jacques's discussion of the Essenes, a subgroup in his treatment of the Jews of the Holy Land. His critique of the Essenes begins with their matrimonial practices. Most Essenes remain unmarried, according to Jacques, because the men do not trust the women to keep their marriage vows. Those who do marry, however, are permitted to share their wives with other men if the women are unable to conceive with their husbands. This is because the purpose of marriage is not the satisfaction of sexual desire, but the procreation of children. He then leaves the subject of marriage and goes on to remark that some of the Essenes do not believe in either the punishment or the salvation of the soul after death.[34] There is more to comment on here than space will allow in a short chapter, but it is worth remarking on the association of these ideas in Jacques's critique. Criticism of marriage customs is typical of

Latin writing about peoples who lived on the margins of Christian society, even when those peoples were themselves Christians. Thus, the tribal marriage customs of the Welsh, Irish and lowland Scots, which also apparently involved greater freedom within marriage than was the norm in Latin Christendom, are singled out by Gerald of Wales and Aelred of Rievaulx in the later twelfth century as characteristics of a 'barbaric' society that was only Christian on the surface.[35] More particularly, the accusation that men share their wives allows critics to compare them to 'beasts', who lack the rational mind necessary to govern their senses, and who are thus incapable of monogamy. Thence it follows, by an association of ideas, that those whose social organisation is at the level of animals are incapable of higher intellect, and are thus unable to conceive of the afterlife of the soul.

Although Jacques calls the Essenes a people 'detestable to God and contemptible to men', it is far from clear how we are to map his description onto any racial or religious group active in his own day. The Essenes were known in the medieval West from the accounts of their way of life and history by Josephus, and featured in Latin patristic literature as the precursors of, and even exemplars for, the first monks.[36] Did Jacques imagine that Essene communities still existed in the Judaean desert? Or, more likely, was he applying a defunct label to contemporary rural communities of Jews, in much the same way as some medieval chroniclers demonised the Turks and other perceived enemies of Christendom by giving them biblical labels such as 'Philistines'?[37] A clue as to the origins of this uncertainty is provided by the *Tractatus*, which includes as a separate category, after the Jews, the Assassins, or *Esseie*. The author of the *Tractatus* defines these as people descended from Jews, but who no longer practise Jewish rites.[38] The Assassins were, of course, an Ismaili Shia Islamic group with strongholds in Syria, properly called Nizari, who were particularly noted, as the *Tractatus* explains, for acts of political terrorism that took the form of assassinations of prominent individuals.[39] Although they had no connection with Jews or Judaism, the similarity of nomenclature (*Esseie–Assassini*) led the author of the *Tractatus* to assume a historical connection, and appears also to have led Jacques de Vitry to the false assumption that the Essenes were still, under the name Assassins, an active subgroup within the Jewish community.[40] This assumption was not without its problems for Jacques, for he was fully aware that the Assassins were Muslims and, indeed, in a separate long chapter had already dealt at some length with their location, organisational structures and recent history as it pertained to the kingdom of Jerusalem.[41] In his chapter on the

Jews, however, he simply asserts that the Assassins took their origins from the Essenes.[42] Playing around with etymology enabled Jacques to construct a specific category in which could be grouped a series of assertions, both ethnographic and theological, that applied to no such actual group of people. For good measure, he includes Sadducees—a biblical category determined by a set of distinctive beliefs—and Samaritans, a few communities of which were probably still in existence in the early thirteenth century, under the same general heading. Having done so, he takes the opportunity to embark on a critique of contemporary Judaism, with the implication that, although they may differ in specific theological beliefs, Essenes, Nizari Ismailites, Sadducees and Samaritans are all subgroups of the genus Jews.[43]

The connection between Nizari Ismaelites and Jews appears bizarre at first, but it becomes fully explicable when we examine Jacques's treatment of Islam. Despite his method of categorising the native subjects of the Crusader States according to religious practice and belief, Jacques has no separate discussion of the Muslim subjects of the Crusader States as a separate people comparable to his entries for the Jews, Armenians, Syrians or Georgians. Instead, he devotes a separate series of chapters at the beginning of his *Historia Orientalis* to the life of Muhammad, the origins of Islam and the nature of Islamic beliefs and practices. In common with Christian polemic against Islam since the eighth century, Jacques saw Islam as, in origin, a Christian heresy rather than a separate theological system.[44] Moreover, following standard treatments of heresy, which demanded the demonisation of the heresiarch's character and morals, Jacques demolishes Muhammad's claims to a prophetic ministry. Muhammad himself is portrayed as sexually licentious and violent. His teaching, Jacques explains, derived from the company he kept after his expulsion from Mecca, when the only people who would receive him were the inhabitants of a semi-deserted city populated only by uneducated and uncivilised men, some of whom were Jews, but who had lost contact with their own teachings. These people constituted Muhammad's first public audience. His doctrines were further developed by contact with a renegade Christian monk who had been expelled from his community for heresy. The religious practices that Muhammad advocated were thus drawn from such ritual practices as the leaderless and untaught Jews among whom he sought refuge had remembered and continued among themselves, including the prohibition on certain foods and the practice of circumcision. For Jacques, then, there was a logic behind the coupling of Jews and Muslims in his construct of Essenes/Assassins. Very little of his critique of Muhammad or Islam is original. The attack on the Prophet's morality was a standard feature of Christian anti-Islamic polemic at least since the appearance of the *Risala*, a pre-eleventh-century Arabic work from Spain attributed to 'Abd al-Masih ibn Ishaq al Kindi. The *Risala* was known in the West from the 1140s, when it was translated into Latin, along with the Qur'an, as part of the project undertaken by Peter the Venerable of bringing greater knowledge of Islam into Western monastic discourse.[45] Jacques's version of the origins of Islam appears to share much in common with two other sources, the anonymous *Scriptum Gregorio nono missum*, which was itself copied and pasted into his *Chronica Maiora* by Matthew Paris in ca. 1240, and Godfrey of Viterbo's *Pantheon sive memoria sanctorum*.[46] It differs sufficiently from these, however, to indicate that Jacques did not know the *Risala* directly.

A further element in Jacques's critique that is common to his treatment of both Jews and Muslims is the theme of the 'unconnected community'. The origins of Islam are sought in the semi-deserted city of scarcely civilised people who, having no religious leadership, have become idolaters. These people, who included Jews and ignorant and uncultivated pagans, formed Muhammad's first followers after his flight from Mecca: 'Seeing that these poor and uncivilized people were easy to lead, Muhammad built a temple in the city, in which he preached his fantasies and empty beliefs to the ignorant.'[47] In this matrix of religious error, the loss of doctrinal authority stems from loss of habitat: both lead to lack of spiritual rootedness. Similarly, in his discussion of the Jews, Jacques emphasises the scattering of the Jewish tribes across the Near East, with the result that they lost whatever religious guidance they had, and in consequence developed beliefs and traditions incompatible with their own scriptures. The Jews were eternal exiles whose destiny was to live alongside other peoples but separate from them, and distinguished from them by the mark of Cain.[48]

It is also striking that both the *Tractatus* and, under its influence, Jacques de Vitry treat the rootless Bedouin as a separate ethnic category within the Crusader States. In both sources, the distinguishing feature of Bedouin life, nomadism, is treated as a cause for suspicion. According to the *Tractatus*, Bedouin are wild by nature and live outdoors, having neither home nor homeland. They can be found wandering in many different provinces, guided by Fortune—by implication, rather than by Reason. Thus they cannot be trusted: sometimes they are friendly to us, sometimes to the Saracens, depending on whom they fear more at the time.[49] Thietmar describes them as 'ugly and poorly dressed'; they are good horsemen but own no land but many flocks, which they raise for food and for sale; they are 'the best of robbers', but they also sell 'Christians to Saracens and Saracens to Christians'.[50] This last reference implies slavery, but is more likely to allude to the selling of information, for example about military manoeuvres. The author of the *Tractatus* remarks that they are friendly towards the Franks while the Franks appear to be in the ascendancy over the Muslims, but when the situation is reversed, they help the Muslims and sell out the Christians.[51] They are mendacious and inconstant, Jacques adds. Their religious practices, likewise, are mixed up: although most of them follow Islamic teaching, they were in origin Christians, and some elements of Christian practice, such as in their forms of prayer, are still in evidence among them. By implication, a society so demonstrably lacking in organic historical integrity will inevitably become spiritually deviant.

The Jews, in principle, shared one further characteristic with another of the categories with which they are placed in close juxtaposition by the author of the *Tractatus* and Jacques de Vitry. Within Western Christian kingdoms, Jews had the legal status of 'royal serfs'. They were the personal property of the king and, although they thereby came under his protection, they were also susceptible to arbitrary taxation or other treatment. This does not appear to have been the case in the kingdom of Jerusalem at any point in its two-hundred-year existence. The legislation governing Jewish conduct and dress promulgated by the papacy in 1179 and 1215 was never, so far as the evidence indicates, applied in the Crusader States.[52] Yet the conceptual and legal category of 'royal serfs' did exist in the kingdom of Jerusalem—applied to the Bedouin rather than to the Jews. A charter issued by King Baldwin IV in 1178 enumerated and described a Bedouin tribe, the BeniKarkas, comprising 103 families, that was evidently subject to both gift and sale. Since the Bedouin within the borders of the kingdom were recognised as belonging to no particular glebe or manorial estate, they could be protected only by the crown or, as seems to have happened by 1180, by the Templars or Hospitallers.[53] Despite the apparent autonomy attributed to them by both the *Tractatus* and other Western observers—an autonomy that was often simply a form of lawlessness—they were in fact a people with whom many of the conceptual attributes associated with Jews could also be linked.

What can we conclude from the evidence offered here about what was important to Western observers about their Jewish and Muslim subjects?

Both the *Tractatus* and Jacques de Vitry's *Historia Orientalis* were written for Western readers who, though ignorant of the Holy Land, had access to plentiful writings on Jews and Judaism, and Muslims and Islam, from theological perspectives. The context in which they should be seen is a genre that includes pilgrimage and travel accounts, and reports of natural marvels; Jacques, indeed, proceeds to treat the flora, fauna, minerals and topography of the Holy Land at length after having dealt with its human inhabitants.[54] These texts are an attempt to place Jews and Muslims within an ethnographic framework according to which ethnic or national traits, and even history, explain and qualify religious beliefs and practices. The attitudes they show towards non-Christians, however, are part of a broader discourse. It has been observed that the other side of the coin of the 'renaissance' of learning in the twelfth century and the 'discovery of Europe' was the exclusion of those who did not belong to this new world. A newly self-confident and articulate society came to define itself increasingly through a discourse of belonging in which the Church functioned as a holistic anthropological unit.[55] Jews and Muslims, in denying the logic of the Church as the universal earthly expression of God's charity, were thus perceived as defying an emerging set of ideas about what constituted humanity. The solution reached in Jacques de Vitry's re-working of the earlier sketchy material in the *Tractatus* was to place them within a penumbra of civilisation, occupying a marginal territory between the fully realised humanity of the Church and animals, which lived by instinct rather than by reason.[56] Such a framework is conceptually flawed and, of course, laden with internal inconsistencies; nevertheless, it constitutes a way of thinking about questions of race and ethnicity within the context of contemporary experience that marks an important stage in a European journey towards more systematic antisemitism and Islamophobia.

Acknowledgement I am grateful to Professor Conrad Hirschler for reading and commenting on an early draft of this essay.

Notes

1. There is a large literature on this incident, among which see R. Chazan, *European Jewry and the First Crusade* (Berkeley: University of California Press, 1987); idem, 'The Facticity of Medieval Narrative: A Case Study of the Hebrew First Crusade Narratives', *AJS Review*, 16 (1991), 31–56; J. Cohen, 'The

Persecutions of 1096—From Martyrdom to Martyrology: The Sociocultural Context of the Hebrew Crusade Chronicles' (in Hebrew), *Zion*, 59 (1994), 169–208; B.Z. Kedar, 'Crusade Historians and the Massacres of 1096', *Jewish History*, 12 (1998), 11–31 (Chazan 1987, 1991; Cohen 1994; Kedar 1998a).

2. Raymond of Aguilers, *Historia Francorum qui ceperunt Iherusalem, Receuil des Historiens des Croisades [RHC]*, Historiens Occidentals III (Paris, 1858), p. 300. For a recent "traditional" narrative account of the Jerusalem massacre in English, see T. Asbridge, *The First Crusade. A New History* (London, 2004), pp. 316–19. For two recent critical studies of the 'massacres', see D. Hay, 'Gender Bias and Religious Intolerance in Accounts of the "Massacres" of the First Crusade', in *Tolerance and Intolerance. Social Conflict in the Age of the Crusades*, eds. M. Gervers and J.M. Powell (Syracuse, 1991), pp. 3–10, and B.Z. Kedar, 'The Jerusalem Massacre of July 1099 in the Western Historiography of the Crusades', *Crusades*, 3 (2004), 15–76 (Raymond of Aguilers 1858; Asbridge 2004; Hay 1991; Kedar 2004).
3. For example, K. Elm, 'Die Eroberung Jerusalems im Jahre 1099. Ihre Darstellung, Beurteilung und Deutung in den Quellenzur Geschichte des Ersten Kreuzzugs', in *Jerusalem im Hoch- und Spätmittelalter: Konflikte und Konfliktbewältigung—Vorstellung und Vergegenwärtigungen*, eds. D. Bauer, K. Herbers and N. Jaspert (Frankfurt, 2001), pp. 42–54 (Elm 2001).
4. For a brief survey of the development of the Church's attitudes to Jews from c.1000 onwards, see R. Chazan, *The Jews of Medieval Western Christendom 1000–1500* (Cambridge: Cambridge University Press, 2006), pp. 44–76 (Chazan 2006).
5. For discussion of a pre-Crusade Western *vita Machometi*, see Sini Kangas, '*Inimicus Dei et sanctae Christianitatis*? Saracens and their Prophet in Twelfth-Century Crusade Propaganda and Western Travesties of Muhammad's Life', in *The Crusades and the Near East: Cultural Histories*, ed. C. Kostick (London, 2011), pp. 131–60 (Kangas 2011).
6. R. Chazan, 'Ephraim of Bonn's Sefer Zechirah', *Revue des etudes juives*, 132 (1973), 119–26; R.B. Dobson, *The Jews of Medieval York and the Massacre of March 1190*, Borthwick Papers 45 (York, 1974); idem, 'The Medieval York Jewry Reconsidered', *Jewish Culture and History*, 3 (2000), 7–20. The anti-Jewish violence in

London was reported at the time by Roger of Hoveden, *Chronica*, III, ed. W. Stubbs, Rolls Series (London, 1870), p. 12 (Chazan 1973; Dobson 1974, 2000; Roger of Hoveden 1870).

7. On the preaching of the Crusade and the difficulty of reconstructing the initial papal appeal, see H.E.J. Cowdrey, 'Pope Urban II's Preaching of the First Crusade', *History*, 55 (1970), 177–88 (Cowdrey 1970).
8. On the status of Muslims, see especially B.Z. Kedar, 'The Subjected Muslims of the Frankish Levant', in *Muslims Under Latin Rule 1100–1300*, ed. J.M. Powell (Princeton, 1990), pp. 135–74; on Christians, A. Jotischky, 'Ethnographic Attitudes in the Crusader States. The Franks and the Indigenous Orthodox People', in *East and West in the Crusader States. Context, Contacts, Confrontations II*, eds. K. Ciggaar and H. Teule (Leuven, 2003), pp. 1–19; C. MacEvitt, *The Crusades and the Christian World of the East. Rough Tolerance* (Philadelphia, 2008); in general, J. Prawer, 'Social Classes in the Crusader States: The "Minorities"', in *A History of the Crusades*, gen ed. K. Setton, 6 vols (Philadelphia and Madison, 1958–1989), *Vol V. The Impact of the Crusades on the Near East*, eds. N.P. Zacour and H.W. Hazard (Madison, 1985), pp. 59–116 (Kedar 1990; Jotischky 2003; MacEvitt 2008; Prawer 1985).
9. On Jews in the Holy Land before 1099, see J. Prawer, *The History of the Jews in the Latin Kingdom of Jerusalem* (Oxford, 1988), pp. 1–18 (Prawer 1988).
10. Thus J. Prawer, *Crusader Institutions* (Oxford, 1980), pp. 201–2 (Prawer 1980a).
11. B.Z. Kedar, 'On the Origins of the Earliest Laws of Frankish Jerusalem: The Canons of the Council of Nablus (1120)', *Speculum*, 74 (1999), 310–35 (Kedar 1999).
12. MacEvitt, *The Crusades and the Christian World*, pp. 139–42, P. Edbury, *John of Ibelin and the Kingdom of Jerusalem* (Woodbridge, 1997), esp. pp. 105–26 (Edbury 1997).
13. *Le Livre de Jean d'Ibelin*, in *RHCLois*, I, 114.
14. The best-known piece of contemporary evidence for mixed marriages is Fulcher of Chartres, *Historia Hierosolymitana*, III, 37, ed. H. Hagenmeyer (Heidelberg, 1913), p. 748; for recent discussion of the phenomenon, see MacEvitt, *The Crusades and the Christian World*, pp. 77–8. Examples of Christian women being taken prisoner on the First Crusade are Albert of Aachen, *Historia Hierosolymitana*,

I, 12, 21, IV, 45, VII, 40, VIII, 18–19, ed. and trans. Susan Edgington (Oxford, 2007), pp. 24, 42, 318, 610–12. Yvonne Friedman, *Encounter Between Enemies. Captivity and Ransom in the Latin Kingdom of Jerusalem* (Leiden, 2002), p. 169, remarks that 'the sexual abuse of female captives was more or less taken for granted', and this is repeated in the observation of Natasha Hodgson, *Women, Crusading and the Holy Land in Historical Narrative* (Woodbridge, 2007), p. 43, that women taken captive by Muslims were usually assumed to have been raped or seduced. Friedman, *Encounter Between Enemies*, pp. 182–3, describes episodes in which Christian women preferred to remain with their Muslim captors because of the suspicion that would attach to them on their return to their own people (Fulcher of Chartres 1913; Albert of Aachen 2007; Friedman 2002; Hodgson 2007).

15. For example, *La Chanson d'Aspremont, chanson de geste du XIIe siècle*, ed. L. Brandin, 2 vols (Paris, 1924), lines 7833–46, 7916–30, 7993–8015; *La fille du comte de Pontieu*, ed. C. Brunel (Paris, 1926), I, lines, pp. 278–80, and the discussion in N. Daniel, *Heroes and Saracens* (Edinburgh, 1984), pp. 72–7 (Brandin 1924; Brunel 1926; Daniel 1984).
16. Hodgson, *Women, Crusading and the Holy Land*, pp. 138–9.
17. On 'Saracenus', see N. Daniel, *Islam and the West. The Making of an Image* (Oxford, 1993 edn), p. 100; J.V. Tolan, *Saracens. Islam in the Medieval European Imagination* (New York, 2002), pp. 127–8.The first usage of the term in a Latin source appears to be Jerome, *Commentarii in Ezechielem*, ed. F. Glorie, Corpus Christianorum Series Latina 75 (Turnhout, 1964). For the decrees of Nablus, see G.D. Mansi, *Sacrorum conciliorum nova et amplissima collectio*, XXI (Venice, 1776), cols 222–6, and now Kedar, 'Origins of the Earliest Laws', pp. 331–5, esp. 333–4. For recent work on questions of identity, ethnicity and race in medieval sources, see R. Bartlett, 'Medieval and Modern Concepts of Race and Ethnicity', *Journal of Medieval and Early Modern Studies*, 31 (2001), 39–56; W. Pohl, 'Telling the Difference: Signs of Ethnic Identity', in *Strategies of Distinction: The Construction of Ethnic Communities 300–800*, eds. W. Pohl and H. Reimitz, The Transformation of the Roman World 2 (Leiden, 1998), pp. 17–70 (Daniel 1993; Tolan 2002; Jerome 1964; Mansi 1776; Bartlett 2001; Pohl 1998).

18. Walter Mahomet appears as a witness to a charter of 1115, R. Röhricht (ed.), *Regesta regni Hierosolymitani* (Innsbruck, 1893), no. 76b; H.E. Mayer, *Die Kanzlei der lateinischen Könige von Jerusalem*, MGH Schriften 40, 2 vols (Hanover, 1996), II, p. 888; MacEvitt, *The Crusades and the Christian World*, pp. 150–1 (Röhricht 1893; Mayer 1996).
19. William of Tyre, *Chronicon*, XXI, 25, ed. R.B.C. Huygens, Corpus Christianorum Continuatio Medievalis, 63 (Turnhout, 1986), p. 998; S. Kuttner, 'Concerning the Canons of the Third Lateran Council', *Traditio*, 13 (1957), 505–6 (William of Tyre 1986; Kuttner 1957).
20. *Alcuini Epistolae*, CLXXII, ed. E. Dummler, Monumenta Germaniae Historica, Epistolae IV, pp. 284–5, a letter dated to 799, in which Alcuin refers to a staged disputation between a Jew and a Christian in Pavia, but says that he is unaware of an extant text of such a disputation between a Christian and a Muslim.
21. B.Z. Kedar, 'The *Tractatus de locis et statu sancte terre Ierosolimitane*', in *The Crusades and their Sources. Essays Presented to Bernard Hamilton*, eds. J. France and W. Zajac (Aldershot, 1998), pp. 111–33, with edition of the text from BL Royal MS 14.C X at pp. 122–31. R. Röhricht, *Bibliotheca geographica Palestinae* (Berlin, 1890), pp. 44–5, listed twenty-eight manuscripts containing the text in Latin and seven in old French, but Kedar, 'The *Tractatus*', p. 114, n.12, has identified a further two (Kedar 1998b; Röhricht 1890).
22. Kedar, 'The *Tractatus*', p. 120.
23. Jacques de Vitry, *Historia Orientalis*, LXXV–XCIII, ed. F. Moschus (Douai, 1597), pp. 137–223; idem, *Historia Orientalis*, ed. and trans. J. Donnadieu (Turnhout, 2008), pp. 294–418. On the dating of the *Historia Orientalis*, see C. Canuyer, 'La date de redaction de l'*Historia Orientalis* de Jacques de Vitry, 1160–1240, évêque d'Acre', *Revue d'histoire ecclésiastique*, 78 (1982), 65–72, arguing for a date of 1219–21, and *contra*, J. Donnadieu, 'L'*Historia Orientalis* de Jacques de Vitry, tradition manuscriteet histoire du texte', *Sacris Eruditi*, 45 (2006), 443–80. Thietmar, *Liber peregrinationis*, VIII, XXIX, ed. J.C.M. Laurent (Hamburg, 1857), pp. 22, 51–4. Thietmar's passages on the Jews and Muslims are based on the *Tractatus*, though with some points of difference in detail. Burchard of Mt Sion's

account (1274–85), *Peregrinatores medii aevi quatuor*, XI, ed. J.C.M. Laurent (Leipzig, 1864), pp. 19–100, owes something to the *Tractatus* and its derivatives, though he omits the Jews altogether and includes some points of detail on Bedouin dress and customs that are not found in earlier accounts. See also Jotischky, 'Ethnographic Attitudes', pp. 1–19; idem, 'Penance and Reconciliation in the Crusader States: Matthew Paris, Jacques de Vitry and the Eastern Christians', in *Retribution, Repentance and Reconciliation*, eds. K. Cooper and J. Gregory, Studies in Church History 40 (Woodbridge, 2004), pp. 74–83 (de Vitry 1597, 2008; Canuyer 1982; Donnadieu 2006; Thietmar 1857; Burchard of Mt Sion's account (1274–85) 1864; Jotischky 2004).

24. Denys Pringle (ed.), 'Wilbrand of Oldenburg's Journey to Syria, Lesser Armenia, Cyprus and the Holy Land (1211–12): A New Edition', *Crusades*, 11 (2012), 122; Thietmar, *Liber peregrinationis* VIII, 22, XVIII, 41; *The Ways and Pilgrimages of the Holy Land*, trans. Denys Pringle, *Pilgrimage to Jerusalem and the Holy Land, 1187–1291* (Farnham, 2012), p. 215, Burchard of Mt Sion, *Descriptio Terrae Sanctae*, XIII, 89 (Pringle 2012a, b).
25. Kedar, 'The *Tractatus*', pp. 124–5.
26. Dominique Iogna-Prat, *Order and Exclusion. Cluny and Christendom Face Heresy, Judaism and Islam (1100–1150)*, trans. Graham Robert Edwards (Ithaca, 2002), pp. 283–4 (Iogna-Prat 2002).
27. Kedar, 'The *Tractatus*', pp. 124–5; cf. William of Tyre, *Chronicon*, XVII, 17, p. 785, XXII, 12, p. 843.
28. Iogna-Prat, *Order and Exclusion*, 318, discussing Peter the Venerable, *Adversus Iudeorum in veteratam duritiem*, ed. Y. Friedman, CCCM, 58 (Turnhout, 1985), in which the 'Jewish present' is identified as the Talmud, in contrast to the Old Testament. See also A.S. Abulafia, *Christians and Jews in the Twelfth Century Renaissance* (London, 1995), pp. 127–9 (Abulafia 1995).
29. G. Dahan, *Les intellectuels chrétiens et les juifs* (Paris, 1990), pp. 528–9; C. Fabre-Vassas, *La bête singuliaire: les juifs, les chrétiens et le cochon* (Paris, 1993), pp. 140, 228 (Dahan 1990; Fabre-Vassas 1993).
30. *Sermo in laude dominici sepulchri*, in G. Constable (ed.) 'Sermones tres', *Revue Bénédictine*, 64 (1954), p. 252.

31. Jacques Vitry, *Historia Orientalis*, LXXXII, 160, ed. Donnadieu, p. 324: 'They are held in hatred and contempt as much by the Muslims among whom they live as by the Christians'.
32. Jacques Vitry, *Historia Orientalis*, LXXV, 139, ed. Donnadieu, pp. 294–6; Jotischky, 'Ethnographic Attitudes', pp. 7–8.
33. Jacques Vitry, *Historia Orientalis*, LXXIX, 153–4, ed. Donnadieu, pp. 318–24.
34. Jacques Vitry, *Historia Orientalis*, LXXXII, 157–8, ed. Donnadieu, pp. 324–6.
35. Giraldus Cambriensis, *Descriptio Kambriae*, II, vi, ed. J.F. Dimock, in *Giraldi Cambriensis Opera*, eds. J.S. Brewer, J.F. Dimock and G.F. Warner, 8 vols, VI (London, 1868), pp. 213–14; see also Robert Bartlett, *Gerald of Wales, 1146–1223* (Oxford, 1982), pp. 161–97; Walter Daniel, *Vita Aelredi*, XXXVIII, ed. and trans. F.M. Powicke (Oxford, 1950), p. 45; cf. Bernard of Clairvaux, *Vita S. Malachiae Episcopi*, in *Sancti Bernardi Opera*, eds. J. Leclercq, H. Rochais and C.H. Talbot, 9 vols (Rome, 1957–77), III, p. 325, on the Irish, and John of Salisbury, *Letters*, eds. W.J. Millor, H.E. Butler, and C.N.L. Brooke, 2 vols (London, 1979–86), I, p. 87, on the Welsh (Cambriensis 1868; Bartlett 1982; Daniel 1950; Bernard of Clairvaux 1957–1977; John of Salisbury 1979–1986).
36. Jospehus, *Jewish War*, II, 120–58, ed. and trans. H. St J. Thackeray, Loeb Classical Library (Cambridge, MA and London, 1955), pp. 368–83; cf. idem, *Antiquities*, XVIII, 18–22, ed. and trans. L.H. Feldman, Loeb Classical Library (Cambridge, MA and London, 1965), pp. 14–21; Jerome, *Epistolae*, XXII, *Patrologia Latina* 30, cols 211–13. For examples of the reception of Josephus' work in the medieval West, see M. Balfour, 'Moses and the Princess: Josephus' *Antiquitates Judicae* and the *chansons de geste*', *Medium Aevum*, 64 (1995), 1–16; Fausto Parente, 'Sulla doppia trasmissione, filologica ed ecclesiastica, del testo di Flavio Giuseppe: un contributo alla storia della ricenzione della sua opera nel mondo cristiano', *Rivista di storie e letteratura religiosa*, 36 (2000), 3–51 (Jospehus 1955, 1965; Balfour 1995; Parente 2000).
37. S. Schein, *Gateway to the Heavenly City. Crusader Jerusalem and the Catholic West 1099–1187* (Aldershot, 2005), pp. 26–7; J. Prawer, 'Jerusalem in Christian and Jewish Perspectives of the Early Middle Ages', *Settimane di studio del centro Italiano di studi*

sull'alto medioevo, XXVI. *Gli Ebrei nell'Alto Medioevo* (Spoleto, 1980), pp. 742–50 (Schein 2005; Prawer 1980b).

38. 'The *Tractatus*', p. 130: 'Others are the Essenes, who are commonly known as Assassins. These are descended from the Jews, but they do not observe Jewish rites in everything.'
39. B. Lewis, *The Assassins. A Radical Sect in Islam* (London, 2003 edn); for European perceptions of the Assassins, see F. Daftary, *The Assassin Legends: Myths of the Isma'ilis* (London, 2001) (Lewis 2003; Daftary 2001).
40. Jacques Vitry, *Historia Orientalis*, LXXXII, 158, ed. Donnadieu, p. 326: 'From these [Essenes], the Assassins are said to have had their origins.' However, what Jacques says of the Essenes actually bears little resemblance to the gist of Josephus' observations, and there is no direct evidence to suggest that he was familiar with Josephus. Heinz Schreckenberg, *Die Flavius-Josephus Tradition in antike und Mittelalter*, Arbeiten zur Literatur und Geschichte des Hellenistischen Judentums, 5 (Leiden, 1972), does not list Jacques among medieval Latin authors to whom Josephus was known. I am grateful to Conrad Hirschler for pointing out that the Jewish origin of the Fatimids was one of the standard accusations levelled against them by Arab Sunni authors. Is this a coincidence, or does Jacques's use of the trope indicate some form of transcultural transfer of knowledge (Schreckenberg 1972)?
41. Jacques Vitry, *Historia Orientalis*, XIV, 40–4, ed. Donnadieu, pp. 152–8.
42. Jacques Vitry, *Historia Orientalis*, LXXXII, 158, ed. Donnadieu, p. 326.
43. Jacques Vitry, *Historia Orientalis*, LXXXII, 158–9, ed. Donnadieu, pp. 324–32. The chapter heading is 'About the Essenes, Sadducees, Samaritans and Other Jews Dispersed through Various Parts of the World'; on Samaritans, see B.Z. Kedar, 'The Samaritans in the Frankish Period', in *The Samaritans*, ed. A.D. Crown (Tubingem, 1989), pp. 82–94 (Kedar 1989).
44. The first systematic treatment of Islam by a Christian writer, the *De haeresibus* by John of Damascus (d. 749), Patrologia Graeca 94, cols 578–789, treated Islam as an aberrant form of contemporary Christian doctrine. John disseminated the legend that Muhammad had been instructed by a renegade Christian monk, Sergius, and this formed the basis of subsequent Western critiques of the

prophet and of Islam, as repeated by Jacques de Vitry, *Historia Orientalis*, VI, p. 19. On the origins and dissemination of the Sergius legend, see Barbara Roggema, 'The Legend of Sergius-Bahira. Some Remarks on its Origin in the East and its Traces in the West', in *East and West in the Crusader States. Context, Contacts, Confrontations* I, eds. K. Ciggaar and H. Teule (Louvain, 1999), pp. 107–23 (Roggema 1999).

45. The Risala was translated into Latin by Samson of Cordova as *Liber Apologeticus*, ed. Enrique Florez, *Espana Sagrada*, XI, 327–8. On Peter's work, see J. Kritzeck, *Peter the Venerable and Islam* (Princeton, 1964), and now Iogna-Prat, *Order and Exclusion*, pp. 265–366 (Kritzeck 1964).
46. Matthew Paris, *Chronica Maiora* III, ed. H. Luard, Rolls Series (London, 1876), pp. 344–61. The previously unedited passage from Godfrey of Viterbo's *Pantheon, sive memoria sanctorum* is in E. Cerulli, *Il 'Libro della scala' e la questione delle fonte arabo-espagnole della Divina Commedia* (Vatican City, 1949), pp. 417–19 (Matthew Paris 1876).
47. Jacques Vitry, *Historia Orientalis*, VI, 19, ed. Donnadieu, p. 122.
48. Jacques Vitry, *Historia Orientalis*, LXXXII, 160, ed. Donnadieu, p. 328.
49. Kedar, 'The *Tractatus*', p. 131.
50. Thietmar, *Peregrinatio*, XXVIII, p. 52.
51. Kedar, 'The *Tractatus*', p. 131; repeated in M. Paris, *Itinerary*, 8c, in D. Pringle, *Pilgrimage to Jerusalem and the Holy Land 1187–1291* (Farnham, 2012), p. 201.
52. Prawer, *History of the Jews*, p. 104.
53. J. Delaville le Roulx (ed.), *Cartulaire générale de l'ordre des Hospitaliers de Saint-Jean de Jérusalem, 1100–1310*, 4 vols (Paris, 1894–1906), I, pp. 132–4; E. Strehlke (ed.), *Tabulae ordinis Theutonici* (Berlin, 1869), no. 3, 3–4 (Delaville le Roulx 1894–1906; Strehlke 1869).
54. J. de Vitry, *Historia Orientalis*, LXXXIV-XCIII, 165–223, ed. Donnadieu, pp. 336–418.
55. Iogna-Prat, *Order and Exclusion*, pp. 359–60; R.I. Moore, *The Formation of a Persecuting Society. Power and Deviance in Western Europe, 950–1250* (Oxford, 1987), pp. 152–3 (Moore 1987).
56. Iogna-Prat, *Order and Exclusion*, p. 361.

References

Abulafia, A.S. 1995. *Christians and Jews in the Twelfth Century Renaissance*. London. Routledge.

Albert of Aachen. 2007. *Historia Hierosolymitana*, ed. and trans. Susan Edgington. Oxford: Oxford University Press.

Asbridge, T. 2004. *The First Crusade. A New History*. London: Free Press.

Balfour, M. 1995. Moses and the Princess: Josephus' *Antiquitates Judicae* and the *chansons de geste*. *Medium Aevum* 64: 1–16.

Bartlett, Robert. 1982. *Gerald of Wales, 1146–1223*. Oxford: Clarendon Press.

Bartlett, R. 2001. Medieval and Modern Concepts of Race and Ethnicity. *Journal of Medieval and Early Modern Studies* 31: 39–56.

Bernard of Clairvaux. 1957–1977. *Vita S. Malachiae Episcopi*. In *Sancti Bernardi Opera*, ed. J. Leclercq, H. Rochais, and C.H. Talbot, 9 vols, III. Rome.

Brandin, L. 1924. *La Chanson d'Aspremont, chanson de geste du XIIe siècle*, 2 vols. Paris.

Brunel, C. (ed). 1926. *La fille du comte de Pontieu*. Paris.

Burchard of Mt Sion's account (1274–85). 1864. *Peregrinatores medii aevi quatuor*, XI, ed. J.C.M. Laurent. Leipzig.

Cambriensis, Giraldus. 1868. *Descriptio Kambriae*, II, vi, ed. J.F. Dimock, in *Giraldi Cambriensis Opera*, ed. J.S. Brewer, J.F. Dimock, and G.F. Warner, 8 vols, VI. London.

Canuyer, C. 1982. La date de redaction de l'*Historia Orientalis* de Jacques de Vitry, 1160–1240, évêque d'Acre. *Revue d'histoire ecclésiastique* 78: 65–72.

Chazan, R. 1973. Ephraim of Bonn's Sefer Zechirah. *Revue des études juives* 132: 119–126.

———. 1987. *European Jewry and the First Crusade*. Berkeley: University of California Press.

———. 1991. The Facticity of Medieval Narrative: A Case Study of the Hebrew First Crusade Narratives. *AJS Review* 16: 31–56.

———. 2006. *The Jews of Medieval Western Christendom 1000–1500*. Cambridge: Cambridge University Press.

Cohen, J. 1994. The Persecutions of 1096—From Martyrdom to Martyrology: The Sociocultural Context of the Hebrew Crusade Chronicles (in Hebrew). *Zion* 59: 169–208.

Cowdrey, H.E.J. 1970. Pope Urban II's Preaching of the First Crusade. *History* 55: 177–188.

Daftary, F. 2001. *The Assassin Legends: Myths of the Isma'ilis*. London: I.B. Tauris.

Dahan, G. 1990. *Les intellectuels chrétiens et les juifs*. Paris: Editions du Cerf.

Daniel, Walter. 1950. *Vita Aelredi*, ed. and trans. F.M. Powicke. Oxford.

Daniel, N. 1984. *Heroes and Saracens*. Edinburgh: Edinburgh University Press.

———. 1993. *Islam and the West. The Making of an Image*, 1993 edn. Oxford: Oxford University Press.

Delaville le Roulx, J.(ed). 1894–1906. *Cartulaire générale de l'ordre des Hospitaliers de Saint-Jean de Jérusalem, 1100–1310*, 4 vols, I. Paris.

Dobson, R.B. 1974. *The Jews of Medieval York and the Massacre of March 1190.* Borthwick Papers 45, York.

———. 2000. The Medieval York Jewry Reconsidered. *Jewish Culture and History* 3: 7–20.

Donnadieu, J. 2006. L'Historia Orientalis de Jacques de Vitry, tradition manuscrite et histoire du texte. *Sacris Eruditi* 45: 443–480.

Edbury, P. 1997. *John of Ibelin and the Kingdom of Jerusalem.* Woodbridge: Boydell.

Elm, K. 2001. Die Eroberung Jerusalems im Jahre 1099. Ihre Darstellung, Beurteilung und Deutung in den Quellenzur Geschichte des Ersten Kreuzzugs. In *Jerusalem im Hoch- und Spätmittelalter: Konflikte und Konfliktbewältigung—Vorstellung und Vergegenwärtigungen*, ed. D. Bauer, K. Herbers, and N. Jaspert. Frankfurt: Campus.

Fabre-Vassas, C. 1993. *La bête singulière: les juifs, les chrétiens et le cochon.* Paris: Gallimard.

Friedman, Yvonne. 2002. *Encounter Between Enemies. Captivity and Ransom in the Latin Kingdom of Jerusalem.* Leiden.

Fulcher of Chartres. 1913. *Historia Hierosolymitana*, III, 37, ed. H. Hagenmeyer. Heidelberg.

Hay, D. 1991. Gender Bias and Religious Intolerance in Accounts of the "Massacres" of the First Crusade. In *Tolerance and Intolerance. Social Conflict in the Age of the Crusades*, ed. M. Gervers and J.M. Powell. Syracuse: Syracuse University Press.

Hodgson, Natasha. 2007. *Women, Crusading and the Holy Land in Historical Narrative.* Woodbridge: The Boydell Press.

Iogna-Prat, Dominique. 2002. *Order and Exclusion. Cluny and Christendom Face Heresy, Judaism and Islam (1100–1150)*, trans. Graham Robert Edwards. Ithaca: Cornell University Press.

Jacques de Vitry. 1957. *Historia Orientalis*, ed. F. Moschus. Douai.

———. 2008. *Historia Orientalis*, ed. and trans. J. Donnadieu. Turnhout.

Jerome. 1964. *Commentarii in Ezechielem*, ed. F. Glorie, Corpus Christianorum Series Latina 75. Turnhout.

John of Salisbury. 1979–1986. *Letters*, ed. W.J. Millor, H.E. Butler, and C.N.L. Brooke, 2 vols, I. London.

Jospehus. 1955. *Jewish War*, II, 120–158, ed. and trans. H. St J. Thackeray, Loeb Classical Library. Cambridge, MA and London.

———. 1965. *Antiquities*, XVIII, 18–22, ed. and trans. L.H. Feldman, Loeb Classical Library. Cambridge, MA and London.

Jotischky, A. 2003. Ethnographic Attitudes in the Crusader States. The Franks and the Indigenous Orthodox People. In *East and West in the Crusader States. Context, Contacts, Confrontations II*, ed. K. Ciggaar and H. Teule. Leuven: Peeters.

———. 2004. Penance and Reconciliation in the Crusader States: Matthew Paris, Jacques de Vitry and the Eastern Christians. In *Retribution, Repentance and Reconciliation*, ed. K. Cooper and J. Gregory, Studies in Church History 40. Woodbridge.

Kangas, Sini. 2011. *Inimicus Dei et sanctae Christianitatis?* Saracens and their Prophet in Twelfth-Century Crusade Propaganda and Western Travesties of Muhammad's Life. In *The Crusades and the Near East: Cultural Histories*, ed. C. Kostick. London: Routledge.

Kedar, B.Z. 1989. The Samaritans in the Frankish Period. In *The Samaritans*, ed. A.D. Crown. Tubingen: J.C.B. Mohr.

———. 1990. The Subjected Muslims of the Frankish Levant. In *Muslims Under Latin Rule 1100–1300*, ed. J.M. Powell. Princeton.

———. 1998a. Crusade Historians and the Massacres of 1096. *Jewish History* 12: 11–31.

———. 1998b. The *Tractatus de locis et statu sancte terre Ierosolimitane*. In *The Crusades and their Sources. Essays Presented to Bernard Hamilton*, ed. J. France and W. Zajac. Aldershot: Ashgate.

———. 1999. On the Origins of the Earliest Laws of Frankish Jerusalem: The Canons of the Council of Nablus (1120). *Speculum* 74: 310–335.

———. 2004. The Jerusalem Massacre of July 1099 in the Western Historiography of the Crusades. *Crusades* 3: 15–76.

Kritzeck, J. 1964. *Peter the Venerable and Islam.* Princeton: Princeton University Press.

Kuttner, S. 1957. Concerning the Canons of the Third Lateran Council. *Traditio* 13: 505–506.

Lewis, B. 2003 edn. *The Assassins. A Radical Sect in Islam*, 2003 edn. London: Phoenix.

MacEvitt, C. 2008. *The Crusades and the Christian World of the East. Rough Tolerance.* Philadelphia: University of Pennsylvania Press.

Mansi, G.D. 1776. *Sacrorum conciliorum nova et amplissima collectio*, XXI. Venice.

Mayer, H.E. 1996. *Die Kanzlei der lateinischen Könige von Jerusalem*, MGH Schriften 40, 2 vols, II. Hanover.

Moore, R.I. 1987. *The Formation of a Persecuting Society. Power and Deviance in Western Europe, 950–1250.* Oxford: Blackwell.

Parente, Fausto. 2000. Sulla doppia transmissione, filologicae ed ecclesiastica, del testo di Flavio Giuseppe: un contributo alla storia della ricenzione della sua opera nel mondo cristiano. *Rivista di storie e letteratura religiosa* 36: 3–51.

Paris, Matthew. 1876. *Chronica Maiora* III, ed. H. Luard, Rolls Series. London.

Pohl, W. 1998. Telling the Difference: Signs of Ethnic Identity. In *Strategies of Distinction: The Construction of Ethnic Communities 300–800*, ed. W. Pohl and H. Reimitz, The Transformation of the Roman World 2. Leiden.
Prawer, J. 1980a. *Crusader Institutions.* Oxford: Oxford University Press.
———. 1980b. Jerusalem in Christian and Jewish Perspectives of the Early Middle Ages. *Settimane di studio del centro Italiano di studi sull'alto medioevo*, XXVI. *Gli Ebrei nell'Alto Medioevo.* Spoleto.
———. 1985. Social Classes in the Crusader States: The "Minorities". In *A History of the Crusades*, gen. ed. K. Setton, 6 vols. Philadelphia and Madison, *Vol V. The Impact of the Crusades on the Near East*, ed. N.P. Zacour and H.W. Hazard. Madison: University of Wisconsin Press.
———. 1988. *The History of the Jews in the Latin Kingdom of Jerusalem.* Oxford.
Pringle, Denys (ed.). 2012a. Wilbrand of Oldenburg's Journey to Syria, Lesser Armenia, Cyprus and the Holy Land (1211–12): A New Edition. *Crusades* 11: 122.
Pringle, Denys, trans. 2012b. The Ways and Pilgrimages of the Holy Land. In *Pilgrimage to Jerusalem and the Holy Land, 1187–1291.* Farnham: Ashgate.
Raymond of Aguilers. 1858. *Historia Francorum qui ceperunt Iherusalem, Receuil des Historiens des Croisades [RHC]*, Historiens Occidentals III. Paris.
Roger of Hoveden. 1870. *Chronica*, III, ed. W. Stubbs, Rolls Series. London.
Roggema, Barbara. 1999. The Legend of Sergius-Bahira. Some Remarks on its Origin in the East and its Traces in the West. In *East and West in the Crusader States. Context, Contacts, Confrontations* I, ed. K. Ciggaar and H. Teule. Louvain: Peeters.
Röhricht, R. 1890. *Bibliotheca geographica Palestinae.* Berlin.
———. (ed.). 1893. *Regesta regni Hierosolymitani.* Innsbruck.
Schein, S. 2005. *Gateway to the Heavenly City. Crusader Jerusalem and the Catholic West 1099–1187.* Aldershot: Ashgate.
Schreckenberg, Heinz. 1972. *Die Flavius-Josephus Tradition in antike und Mittelalter.* Arbeiten zur Literatur und Geschichte des Hellenistischen Judentums, 5. Leiden: Brill.
Strehlke, E. (ed.). 1869. *Tabulae ordinis Theutonici.* Berlin.
Thietmar. 1857. *Liber peregrinationis*, VIII, XXIX, ed. J.C.M. Laurent. Hamburg.
Tolan, J.V. 2002. *Saracens. Islam in the Medieval European Imagination.* New York: Columbia University Press.
William of Tyre. 1986. *Chronicon*, ed. R.B.C. Huygens, Corpus Christianorum Continuatio Medievalis, 63. Turnhout: Brepols.

CHAPTER 3

Antisemitism, Islamophobia and the Conspiracy Theory of Medical Murder in Early Modern Spain and Portugal

François Soyer

> A multitude of doctors, surgeons and apothecaries have been arrested in Lisbon and other parts of the kingdom, not counting those who have fled (leaving their wives imprisoned), and they have all confessed to wilfully perpetrating many murders of [Old] Christian noblemen and men of the Church. In some cases the exact numbers are known because they killed one out of every twelve patients. One of them, who was burnt at the stake in Évora, confessed that he had killed one hundred and fifty Old Christians, including eighteen noblemen and he was also found to possess a book attacking our holy faith.[1]

In his antisemitic diatribe entitled *Breve discurso contra a heretica perfidia do judaismo*, first printed in Lisbon in 1622, the Portuguese author Vicente da Costa Mattos railed against many secret conspiracies. He believed that

F. Soyer (✉)
University of Southampton, UK

J. Renton, B. Gidley (eds.), *Antisemitism and Islamophobia in Europe*, DOI 10.1057/978-1-137-41302-4_3

these conspiracies were being orchestrated by the descendants of Jews—generically known as 'New Christians'—to achieve the complete destruction of both the Catholic Church and the Iberian monarchies of Spain and Portugal. One of the most striking claims made in the *Breve discurso* is that the allegedly false converts were systematically infiltrating the medical professions to be able to murder genuine Catholics not descended from Jews—so-called Old Christians—and particularly to assassinate high-status individuals in both the ecclesiastical hierarchy and the secular aristocracy. Vicente da Costa Mattos' strident warning about the evil deeds of secret Jewish serial-killer doctors was no innovation on his part. The fear of Jewish doctors was widely shared in medieval and early modern Europe and was a recurrent theme in anti-Jewish and antisemitic polemics and propaganda. In the early modern Iberian context, however, the fear of medical murder by doctors from marginalised religious communities acquired a new and sinister dimension.

The sizeable Jewish and Muslim populations residing under Christian rule in Spain and Portugal were either expelled or forced to convert by various royal edicts promulgated in the late fifteenth and early sixteenth centuries. The end of the toleration of religious pluralism in the Iberian Peninsula did not, however, bring about the religious uniformity and harmony wished by the Catholic rulers of Spain and Portugal. The number of converts and their descendants, *judeoconversos* and *moriscos*, was so great that they did not assimilate and disappear into the wider Christian population. Moreover, the reluctant converts and their descendants were commonly suspected of remaining faithful to Judaism and Islam while publicly pretending to be Christian. The resulting social tensions between the 'New' and 'Old' Christians led the rulers of Spain and Portugal to establish tribunals of the Holy Office of the Inquisition in 1480 and 1536, respectively. With the support of their respective monarchs, the Spanish and Portuguese inquisitors waged a judicial campaign to suppress religious dissension, starting with judaising *judeoconversos*, but later also targeting the *moriscos* and other groups of alleged 'heretics'.

This chapter seeks to explore the nature of the conspiracy theory prevalent in early modern Spain and Portugal relating to the medical murder of Christians by secret Jews and Muslims. It compares how this myth affected the *judeoconversos* and the *moriscos* differently and asks why one group—the *judeoconversos*—became the principal target of such accusations while the other—the *moriscos*—was apparently mostly spared. To achieve this, it examines the strength, especially in the Iberian Peninsula, of the medieval tradition that accused Jewish doctors of poisoning their Christian patients. It then moves on to the early modern period to examine how such a belief developed a new and far more menacing character in the wake of the expulsion of the Jews from the Iberian Peninsula between 1492 and 1498 and the fears surrounding the judaising of *judeoconversos*. Finally, it examines the impact of the conspiracy theory of medical murder on the *morisco* communities from their forced conversion between 1502 and 1526 until their ultimate expulsion between 1609 and 1614.

The Jewish Medical Conspiracy in Spain and Portugal (I): The Medieval Legacy

The xenophobic fear that specific ethnic or religious groups might infiltrate the medical professions to murder patients from other, rival groups is certainly as old as the emergence of the medical professions themselves. According to the first-century CE Roman philosopher and naturalist Pliny the Elder, the Roman statesman and moralist Cato the Elder (234–149 BCE), a fierce adversary of the Hellenisation of Roman culture, accused Greek physicians of having 'conspired among themselves to murder all barbarians [i.e. non-Greeks] with their medicine'.[2] Likewise, in Islamic Spain, the twelfth-century jurist Muhammad ibn Ahmad ibn 'Abdun expressed the view that Muslims should not consent to be treated by either Jews or Christians since these subject *dhimmis* ('protected peoples' living under Islamic rule) must be assumed to be hostile to Muslims, and any Muslim treated by Jewish or Christian doctors consequently risked his or her life.[3]

Through their mastery of Greek (especially Galenic) medical culture, Jews occupied a prominent position in the medical professions in medieval Christendom, where medical practitioners (of any faith) were in short supply. Although they did not have a monopoly on medical practice in the Christian kingdoms of medieval Spain and Portugal, Jews played a notable role in the medical professions, and the number of Jewish physicians was certainly disproportional to the demographic size of Jewish communities.[4]

The professional success of Jewish medical practitioners inevitably caused anxiety in a tense religious and social context where the ecclesiastical hierarchy was increasingly concerned by interaction between Christians and Jews or Muslims. Moreover, as Joshua Trachtenberg has demonstrated, the Jews became the target of a campaign of demonisation from the twelfth century onwards, which saw them accused of practising magic, sorcery and the ritual murder of children (the infamous Blood Libel).[5]

In the twelfth century, the use of Jewish doctors by Christian patients was firmly condemned in the compendium of Canon Law assembled by Gratian—the *Decretum Gratiani*—and Christians who did seek remedies from Jews were theoretically threatened with excommunication.[6] Similar injunctions, often re-stating the interdiction made in the *Decretum Gratiani*, were issued by various Church councils held in Spain during the fourteenth century, but their very existence only seems to confirm that they remained largely ignored in practice.[7] Secular laws were more explicit in the manner in which they linked restrictions on Jewish medical practice and the fear of medical murder. In the thirteenth-century *Siete Partidas*, a legal compendium collated and composed at the behest of King Alfonso X of Castile, Christians were forbidden from receiving medicinal potions or emetics prepared by Jews and such remedies could only be prepared 'by the hand of a Christian', although their elaboration could take place under the supervision of a Jewish doctor.[8] Another law, promulgated in 1397 by Queen Maria of Aragón, forbade Christians from seeking treatment from Jewish doctors and openly expressed the fear of religiously motivated medical murder:

> Since the perfidious Jews are thirsty for Christian blood, as enemies would be, and it is dangerous for Christians to obtain any medical help from Jewish doctors when they are sick ... we ordain and establish that no Jew, in any case of a Christian's infirmity, should dare to exercise his office unless a Christian doctor will take part in the cure.[9]

The concept of medical murder was rapidly incorporated into the list of accusations made against Jews prior to 1492. In one of his firebrand sermons, the formidable Dominican preacher Saint Vincent Ferrer (1350–1419), who exerted himself to incite strong anti-Jewish sentiments among his Christian audience, warned the latter to avoid Jewish doctors, surgeons and apothecaries precisely because of the danger that they represented to their patients:

> They wish a greater evil upon us Christians than we wish upon the Devil .... A certain Jewish doctor who was about to die said [the following] to those who were present and weeping: 'Do not cry, for it does not pain me to die. This is because I have caused the death of over five hundred Christians thanks to my medicine.'[10]

Just like Vicente Ferrer, the authors of major Spanish anti-Jewish polemics in the fourteenth and fifteenth centuries played on the fear of murderous Jewish doctors, including Alfonso de Valladolid (c. 1270–1346), Alfonso Chirino (c. 1365–1429) and the notorious Alonso de Espina (?–1469).[11] Alfonso de Valladolid, himself a Jewish convert to Christianity, expounded at length on the reasons why Jewish medical practitioners were to be avoided. In addition to claiming that the dietary restrictions followed by Jews rendered them ignorant of the medicinal properties of numerous non-kosher foods, Alfonso de Valladolid asserted that the Jews did not possess any concept of sin, thus rendering Jewish doctors particularly dangerous:

> It is clear that [the Jewish doctor] is not [afraid of sinning] because the Jewish wisemen said that the best physicians will go to Hell. And the reason is that the physicians' intentions are generally to acquire wealth and renown and not to serve God. And this is why they sometimes prolong an illness or make it worse or precipitate the patient's death. And if such is the nature of healers in general, it is even more so in the case of the Jews, for, by and large, they are not afraid of sinning, as has been said ....[12]

Beyond the legal texts and religious polemics, one legend contributed more than any other factor to popularising and perpetuating the myth of Jewish medical murder: the alleged murder of King Enrique III by his personal Jewish physician Meir ben Solomon Alguadex on 25 December 1406. The legend held that Meir Alguadex was accused of poisoning Enrique III for unspecified reasons, although the narrative implies that the murder was religiously motivated and that Meir was correspondingly executed as a regicide by being drawn and quartered. In many accounts of the legend, the charge of medical murder was combined with that of host desecration. Meir Alguadex, with various Jewish accomplices, was also accused of having plotted to desecrate a consecrated host in the town of Segovia and to assassinate with poison the bishop of Segovia, Juan de Tordesillas.[13] It is difficult to trace the precise origins of the legend. Interestingly, the alleged poisoning of Enrique III by Meir Alguadex is not mentioned in the *Crónica de Juan II*, an official history of the reign of Enrique's son and successor composed by various authors. It may well have been the product of oral lore in fifteenth-century Spain before being mentioned in the *Fortalitium fidei* by Alonso de Espina, an author whose explicit anti-Jewish objective renders his account highly suspect.[14]

Whatever its origins were, the legend quickly came to be accepted as a historical fact in Spain and Portugal and was repeated in a stunning variety of printed works throughout the early modern period. Unsurprisingly, it is mentioned in all the major vernacular antisemitic polemics: the *Breve discurso* of Vicente da Costa Mattos (1622), the *Centinela contra Judíos* of Francisco de Torrejoncillo (1674) and the *Mayor fiscal contra Judíos* of Antonio de Contreras (1736).[15] Beyond these texts, however, the legend of Meir Alguadex was recounted in a multitude of printed works with no explicit polemical aims. References to the alleged murder of Enrique III by his Jewish physician mostly appear in historical works such as Diego de Colmenares' history of the town of Segovia (1637) and the major ecclesiastical history authored by the Hieronymite monk Pablo de San Nicolas (1744), but they can also be found in such diverse works as the playwright Tirso de Molina's comedy *Prospera fortuna de Ruy Lopez Avalos y adversa de Ruy López de Avalos: Primera parte* (c. 1635) and the 1721 'moralised encyclopaedia' produced by the Capuchin Fray Martín de Torrecilla.[16]

The Jewish Medical Conspiracy in Spain and Portugal (II): The *Judeoconverso* Doctor

The well-established medieval concept that Jewish medical practitioners engaged in the religiously motivated murder of Christian patients and the widespread propagation in the Iberian Peninsula of the legend of Meir Alguadex as the archetypal murderous Jewish doctor created secure foundations for the survival of the notion into the early modern period. The edicts of expulsion of the Jews from Spain in 1492, Portugal in 1497 and finally the small Pyrenean kingdom of Navarre in 1498 did not end the Jewish presence in the Iberian Peninsula. On the contrary, thousands of Jews converted to Christianity (in Portugal the conversion was actually forced on almost all the Jews by King Manuel I and his officials), thus creating a substantial population of *judeoconversos*. The religious beliefs of the *judeoconversos* were unquestionably complex, and increasingly so as the decades and centuries passed after 1492. They doubtless varied enormously across individuals and family groups and across the centuries, defying straightforward categorisation but spanning a wide spectrum extending between genuinely devout Catholics and zealous Jews, to include *judeoconversos* who syncretised various elements of Catholicism and Judaism and followed only vestiges of Jewish tradition.

For the majority of the population that was not descended from Jews (the 'Old Christians'), the *judeoconversos* were the object of considerable fears. The dramatic circumstances of their conversion to Christianity, the continued practice of Judaism (even adulterated or vestigial) by some and the constant flow of confessions obtained from prisoners by the Inquisition's tribunals (often the fruit of torture or psychological duress) led many people to believe that the secret practice of Judaism was a generalised and persistent phenomenon among *judeoconversos*. As late as 1727, the economist Francisco Máximo de Moya Torres bemoaned the 'old evil' of 'Judaism' afflicting Spain:

> A great deal of Judaism is to be found in Spain in spite of the indefatigable zeal with which the Holy Tribunal [of the Inquisition] stands guard, as it is proven by its trials and numerous autos de fé .... These vile people are concealed, infecting so much .... Their usual life is one of profit and usury: their professions are those of doctors, rentiers, merchants, confectioners, and all their trades are those of idlers. They are clever and shrewd and take vengeance on Christian blood.[17]

The subtleties and complexity of the nature of *judeoconverso* religious beliefs, which have exercised modern historians and been the cause of much debate, were largely overlooked by many 'Old Christians', for whom all 'judaising' *judeoconversos* (or 'judaising apostates') were as Jewish as their pre-conversion Jewish ancestors.

Since many Spaniards and Portuguese feared *judeoconversos* to be nothing but Jews pretending to be Christians, it was to be expected (and entirely logical) that the fear of Jewish doctors would be replaced by that of *judeoconverso* doctors. The survival of the myth of medical murder during the centuries after 1492–98 is not, however, simply the result of the endurance of the medieval tradition. In the sixteenth century, two new factors played a major role in exacerbating such fears as well as guaranteeing their continuance in the form of the myth of the murderous *judeoconverso* doctor:

1. The fear that judaising *judeoconversos* were secretly taking advantage of their new identity as Christian to murder unsuspecting Catholic patients.
2. The creation of an elaborate conspiracy theory based on a set of two forged letters between Jews in Spain and Jews in the Ottoman Empire.

The first of these developments was a direct and inevitable consequence of the forced conversions and the mistrust of the 'Old Christian' population. Prior to 1492–98, Jews had been a distinguishable section of the Iberian population. In both Spain and Portugal, they were legally forced to reside in segregated areas, compelled to display prominent symbols on their clothing and subject to sumptuary legislation (banning them from riding horses or bearing swords). Although some wealthy Jews could purchase exemptions from the Crown, these restrictions marked out the Jews. Apart from the shortage of trained doctors in many areas, there was no reason why a Christian seeking to avoid treatment by a Jewish medical practitioner could not do so. After their baptism in 1492–98 (or even before in the late fourteenth and early fifteenth centuries), however, the *judeoconversos* became part of Catholic society and were no longer subject to segregationist legislation. Many patients, therefore, could no longer be certain of whether or not the medical practitioner tending to their needs was an 'Old' or a 'New' Christian. Given the widely held perception of the majority of the *judeoconversos* as secretly judaising heretics, this new state of affairs created a situation in which the wildest antisemitic fantasies could be unleashed.

The perceived link between the 'secret identity' of the *judeoconversos* and their role as medical practitioners soon became an integral part of an elaborate antisemitic conspiracy theory that appeared in the second half of the sixteenth century and was based on two forged letters. One of these letters was supposedly written and sent by the *Jews* of Toledo to the Jews of Constantinople in 1492, complaining of their fate and persecution by Old Christians, and the other was the reply to the latter, advising them to make use of their new public identity as Christians and infiltrate Christian institutions in order to gain their revenge. Among the professions that the *judeoconversos* were advised to infiltrate by the Jews in Constantinople were the medical professions.

The origins of these two letters, which were analysed and convincingly demonstrated to be forgeries by the French historian Isidore Loeb in the nineteenth century, is obscure.[18] Manuscript copies of them were certainly circulating from the 1550s onwards. In the 1580s, these two letters and the claim that *judeoconversos* were seeking to use their medical knowledge to kill Christian patients were mentioned by Andrés de Noronha, a Portuguese bishop who, after occupying the bishopric of Portalegre in Portugal, was appointed by Philip II to the See of Plasencia in Spain. Seemingly convinced of their genuineness and claiming that the letters had been brought to his attention in the 1560s by an inquisitor from the inquisitorial tribunal of Llerena in Spain, Bishop Andrés de Noronha urged the Habsburg sovereign to pay attention to the letters and take immediate action to prevent *judeoconversos* from becoming medical practitioners.[19] Given their importance as 'evidence' of a 'Jewish plot' threatening Spain and Portugal and linking the *judeoconversos* of the Iberian Peninsula with Jewish communities outside of it, it is hardly surprising that the letters came to feature prominently in the myth of Jewish medical murder. They rapidly found their way into antisemitic propaganda and polemics, of which they became a recurring and integral part.

Just as was the case with the legend of Rabbi Meir Alguadex's supposed poisoning of Enrique III, the widespread diffusion of the forged letters can be observed in the fact that they were copied and reprinted verbatim even in works with no declared antisemitic aim. They appeared, for instance, in the collection of intriguing stories entitled *La Silva curiosa* and compiled by Julián de Medrano, which printed in the town of Zaragoza in 1580.

The fear of judaising *judeoconversos* led to the introduction of statutes of racial purity (*limpieza de sangre*) in numerous (though not all) cathedral chapters and religious orders, preventing the converted descendants of Jews from seeking to enter into these. Given the level of fear generated by the possibility of a crypto-Jewish medical plot, it is not surprising that similar statutes seeking to exclude candidates 'tainted' by 'Jewish blood' from studying medicine, surgery or as apothecaries were gradually introduced by both various university colleges where medicine was taught and colleges of surgeons and apothecaries. Even before the appearance of the infamous forged letters, the inquisitorial *instrucciones* compiled at the command of Inquisitor General Tomas de Torquemada in 1484 explicitly stipulated that the sons and grandsons of individuals convicted of heresy by the Inquisition, which at that time can only have meant *judeoconversos* convicted of judaising, should not be allowed to become doctors, surgeons, apothecaries or 'bleeders' (*sangradores*).[20] As early as 1506, the college of surgeons in Barcelona adopted statutes of *limpieza* and the fifteenth-century statutes of the college of apothecaries in Valencia not only banned those 'of Jewish descent' from training as apothecaries, but in 1529 extended the ban to affect any Old Christian married to a woman of Jewish ancestry. The statutes of the Valencian college stipulated that any individual of Jewish ancestry who fraudulently sought to be examined would be fined five hundred ducats and condemned to perpetual exile from the city. Similar racial statutes were adopted or confirmed by the

Spanish Crown at the colleges of apothecaries in the cities of Barcelona, Zaragoza and Seville during the sixteenth century.[21]

Even though the statutes of racial purity were not universally implemented, they gave the libel of Jewish medical murder official recognition and indubitably played a major role in ensuring its wide diffusion and bolstering its credibility. At the end of the sixteenth century, the Catalan friar Juan Benito de Guardiola praised the faculty of medicine of the University of Barcelona for its strict adherence to its statute of *limpieza* and rejection of *judeoconversos*, and as an example to be emulated by other faculties of medicine in the Iberian Peninsula.[22] Another boost to their social standing came from the fact that the Spanish and Portuguese Crowns confirmed such statutes (as Philip II did with those of the college of apothecaries in Valencia in 1564). In Portugal, royal action was not limited to the passive confirmation of racial statutes targeting *judeoconverso* medical practitioners, but actively sought to assist and implement racial discrimination in its own employment of medical staff. A special bursary was established by King Sebastian of Portugal in 1568 (subsequently confirmed by King Philip II of Portugal in 1604 and 1606) to grant bursaries to thirty students of 'Old Christian' stock to study medicine and surgery at the faculty of Medicine of the University of Coimbra.[23] An edict issued by the Crown in December 1585 instructed all municipalities, charitable institutions (*misericórdias*) and hospitals to immediately oust any New Christian doctor when an Old Christian medical practitioner was available and willing to accept employment with them. This racial preference was expanded in 1599 to include doctors employed by the supreme royal law court (*Casa da Suplicação*) and the appellate law court (*Casa do Cível*).[24] Moreover, on 30 March 1581, Pope Gregory XIII (1572–85) issued the Bull *Multos adhuc ex Christianis*, reiterating once more the prohibition of the *Decretum Gratiani* on Christians seeking medical treatment from Jews.[25] Although not aimed specifically at the Iberian *judeoconversos* but at Jews more generally, the papal decree was used as ammunition by antisemitic propagandists in the Iberian Peninsula.

To render the tale even more terrifying, antisemitic authors developed the story to include rumours and details intended to reinforce its verisimilitude. In his polemical defence of the statutes of racial purity preventing New Christians from acceding to positions in the ecclesiastical hierarchy of the Cathedral of Toledo, printed in 1573, Bishop Diego de Simancas (who wrote under the alias of Diego Velázquez) alluded to the (forged) letters, warning that Jews had counselled their secret co-religionists in the

Iberian Peninsula that 'through medicine you shall kill with impunity' (*cum medicina eos impunem occidetis*). Diego de Simancas went further, however, and recounted the story of an unnamed *judeoconverso* burnt at the stake by the Inquisition 'many years ago'. Simancas told his readers that the murderous *judeoconverso* doctor would return home 'after murdering many Christians' and be greeted by his children with the following phrase: 'Welcome, avenger'. To this greeting, he would reply 'Come to [greet] the avenger', as if he were stating 'Welcome, here is the avenger'. Without providing any specific information or references to support this claim, Simancas claimed that the Inquisition had unmasked six hundred cases of murder committed by *judeoconverso* doctors.[26]

The story of the *judeconverso* doctor who took himself to be 'the avenger' of the Jewish people in Iberia was repeated by other authors, usually with little or no alteration. Ignacio del Villar Maldonado, a respected jurist who reproduced (in vernacular Spanish) the forged letters in his influential work *Sylua responsorum iuris: In duos libros diuisa*, first printed in Madrid in 1614, presented the same cautionary tale to his readers (but in vernacular Spanish):

> We know it for certain, and it has been verified, that a doctor who was descended from Jews was condemned as a heretic by the inquisitors of a certain district of Spain. It was confirmed that the man, whilst residing in a certain village whose name has not been revealed in order to avoid offending anyone (as this is not my intention), murdered more than three hundred people through his false and adulterated medicine as well as his poisons. Every time that he returned home from visiting the sick, or perhaps it should be said after procuring their death, his wife who was also one of that race of Jews would say: 'our avenger is welcome'. To this, her Jewish husband would lift and wave his right arm whilst clenching his fist as a sign of victory and utter 'he is come and will wreak his revenge'.[27]

This tale became a standard element of any antisemitic work produced in Spain and Portugal during the seventeenth and eighteenth centuries, to such an extent that it can be described as part of the antisemitic canon in the early modern Iberian world. Among those who repeated it were the Portuguese author Vicente da Costa Mattos in his 1622 *Breve discurso contra a heretica perfidia do judaismo*; Francisco de Torrejoncillo in his 1674 *Centinela contra Judíos*; Fray Félix Alamín in his 1727 *Impugnacion contra el Talmud de los judios, al Coran de Mahoma, y contra los hereges*; and Antonio de Contreras in his 1736 *Mayor fiscal contra Judíos.*

In both Spain and Portugal, the fears generated by the myth of medical murder by *judeoconversos* reached new paroxysms of fantasy during the seventeenth century. Far from being confined to uneducated peasants and the populace, it found ready acceptance within the ecclesiastical hierarchy. As we have already seen, as early as the 1580s bishop Andrés de Noronha felt moved to petition King Philip II urgently to take action against the *judioconversos*. Similarly, a gathering of Portuguese bishops and ecclesiastics, assembled in the town of Tomar in the spring and summer of 1629 to find a 'remedy against Judaism' (*remédio do Judaismo*), alluded to the danger posed by *judeoconverso* doctors in their report to the Crown. They demanded an immediate ban on the practice of medicine and pharmacy by *judeoconversos*.[28]

Another recurring element of the legend is the widely reported notion that *judeoconverso* doctors followed a careful modus operandi and murdered one in every five, ten or twelve of their Old Christian patients. Although the origins of this aspect of the myth are unclear, it probably evolved to lend verisimilitude to it as well as to counter or deflect criticism that doctors who systematically murdered their patients would rapidly go out of business or come to the attention of the authorities. Vicente da Costa Mattos, as the opening quote produced at the start of this chapter demonstrates, favoured the notion that New Christian doctors murdered one out of every twelve patients in his 1622 polemic. Likewise, Fray Francisco de Torrejoncillo perpetuated the same claim in his 1674 *Centinela contra Judíos*:

> Of yet another, whom it is claimed was burnt in Lisbon, it is told that he killed many monks, men of the Church, and members of the nobility. He killed one out of every dozen of his patients.[29]

The Inquisition appears to have played a major role in stimulating the rise of fears about Jewish medical murder. It is worth remembering that Bishop Andrés de Noronha received his manuscript copy of the forged letters implicating the Iberian *judeoconversos* in a worldwide Jewish conspiracy (of which the medical conspiracy was just one element) from an inquisitor. Many inquisitors themselves appear to have lent credence to claims of a medical conspiracy. Particularly striking in this respect is a letter sent in 1619 by an inquisitor of the tribunal based in Coimbra to the General Council of the Portuguese Inquisition in Lisbon. The inquisitor informed the Council that his tribunal had arrested a number of *judeo-*

converso doctors and pointed its attention to an old case of religiously motivated murders perpetrated by a *judeoconverso* doctor, although, characteristically, he did not name the man or offer any specific details about the case:

> A [*judeoconverso*] doctor confessed to the Holy Office (after confessing his Judaism) that he killed many Old Christians using purgatives and other drugs that did not cure the illnesses from which they were suffering. If he treated some [Old Christian patients] with the appropriate drugs, it was to preserve his standing and reputation. [He acted in this way because], had he killed all of his patients, nobody would have wanted to be treated by him and he would thus not have been able to earn a living through his profession.[30]

Perhaps inevitably, the public hysteria about medical murder was stoked by incendiary manuscript pamphlets listing the names and places of residence of individual medical practitioners accused of murdering their patients. By way of illustration, the Portuguese pamphlet *Treatise in which it is proved that the New Christians of the* [Hebrew] *Nation who dwell in Portugal are secret Jews and in which the evils that they inflict upon Old Christians are pointed out*, circulating in the 1630s, enumerated the names of fifty-one New Christian physicians, surgeons and apothecaries working in Portugal and Spain convicted by the Inquisition of crypto-Jewish beliefs and, in some cases, even of mass murder. The most conspicuous listees were a physician named Garcia Lopes of Portalegre, who was accused of having poisoned no less than one hundred and fifty Old Christian patients, including twenty-five *fildalgos* (members of the lower nobility), and a certain Pero Lopes of Goa, who had allegedly killed seventy Old Christian patients. Yet another was the apothecary Gabriel Pinto, a resident of Coimbra burnt at the stake in 1600, who had 'confessed to killing many Old Christians, including churchmen and nuns'.[31] The fact that the list featured the names of numerous genuine individuals who had been prosecuted by the Inquisition is interesting, for it indicates that the anonymous author wished to make his claims appear real. Nevertheless, the recent research of José Alberto Rodrigues da Silva Tavim, who has examined the extant inquisitorial trial records in the National Portuguese Archives of those accused of being murderous doctors, has not found any trace of such accusations in their actual trials.[32]

The Myth of Medical Murder and the *Morisco* Medical Practitioners

In comparison with *judeoconversos*, medical practitioners from the *morisco* minority largely escaped suspicions that they were orchestrating a campaign of secret medical murder. While the statutes of *limpieza* regulating admittance to many university colleges as well as colleges of surgeons and apothecaries also applied to them, no lasting conspiracy theory of medical murder by *moriscos* materialised. The reason why this was the case is not entirely clear. The research of Luis García Ballester has demonstrated that *morisco* medical practitioners tended to practise their skills not only within their communities, but also among a wide variety of Old Christian patients as unlicensed 'healers' (known as *sanadores*), and even King Philip II had recourse to *morisco* doctors to attend to the illnesses of his sons.[33] Various factors may have mitigated the fear that Old Christian patients might unknowingly be entrusting their lives to crypto-Muslim doctors who had evil intentions. Firstly, the *morisco* populations residing in those parts of Spain where *morisco* settlement was densest—Granada (until 1571) and Valencia—remained quite culturally distinct (and thus visually recognisable) from the Old Christian population after their forced conversion, speaking an Arabic dialect and wearing distinctive clothing. This distinctiveness doubtless helped to alleviate the fears of Old Christian patients that they could be receiving treatment from a *morisco* doctor or healer without realising it. It would therefore seem that the fear of the 'secret identity' of the *judeoconverso* doctor did not affect attitudes towards the much more 'visible' *moriscos* in a similar manner. Secondly, in stark contrast to the Jews and *judeoconversos*, there did not exist an established medieval tradition of suspicion relating to Muslim medical practitioners, any myths comparable to that of Rabbi Meir Alguadex or conspiratorial documents such as the forged letter from Constantinople. Thirdly, and finally, the fact that there were few licensed or university-trained *morisco* doctors, and thus few reasons for professional rivalry between *moriscos* and licensed Old Christian doctors, may also have played a role.

Anti-Muslim sentiment among Christians did not spare *morisco* medical practitioners. Most of them were not university trained and were part of what Luis García Ballester has described as a 'medical subculture' (*subcultura médica*). Working as unlicensed 'healers', the *moriscos* were the object of suspicion and their medical skills were the subject of accusations of sorcery and demonic links. Yet such accusations of demonic sorcery were used to account for the medical knowledge of the *moriscos* (and occasionally their seemingly inexplicable success) and did not translate into claims of a medical plot against Old Christian patients. This was the case even for a famous *morisco* healer such as Jerónimo Pachet, who was summoned by no less than King Philip II of Spain to help cure the illnesses of his sons when they were desperately ailing, but who was also prosecuted by the Inquisition twice (in 1567 and 1580).[34] It is interesting to note that Christian authors could simultaneously express admiration for and suspicion of *morisco* medical knowledge. Citing the very example of Jerónimo Pachet, the noted historian Gaspar Escolano typified such an ambivalent attitude when discussing *morisco* medical knowledge in a work printed in 1610:

> The moorish doctors who live amongst us are favoured by the skills that they are known to possess and have benefitted us by bringing these to our knowledge, by means of which they concoct incredible cures, as it has been witnessed in a certain [doctor] named Pachet who was penanced by the Holy Office [of the Inquisition] because he had a [demonic] familiar and the Devil assisted him as his herbalist.[35]

Ironically, and in stark contrast to *judeoconverso* doctors and apothecaries, it would seem that *morisco* medical practitioners were often accused of sorcery and supernatural medical powers because their cures kept Old Christian patients *alive* when other licensed and university-trained doctors had abandoned all hope of a recovery.

While Christian authors could ascribe *morisco* medical knowledge and skills to demonic agency, it is striking that they were not as ready to make the leap to subscribing to conspiracist beliefs about the murder of Old Christian patients. There are a very few instances in which *morisco* physicians were accused of conspiring to murder Old Christian patients, and these must be examined very carefully. A rare reference to an alleged plot by *morisco* doctors can be found in a September 1607 meeting of the parliament of Castile, when the parliamentarian Pedro de Vesga called for *moriscos* to be prohibited from studying at medical faculties and even from attending public medical lectures. His fiery rhetoric matched his extreme views and he began straightaway by asking why, if *moriscos* were not allowed to bear arms because of the fear that they might rebel and aid Ottoman or North African raids on the Spanish coast, they were not also banned from medical practice, since 'the ability to cure is the greatest weapon'. Vesga claimed that *moriscos* were using medical knowledge to 'kill more [Catholic Christians] of this kingdom than the Turks and English' and secretly cause pregnant Old Christian women to suffer abortions. Moreover, he argued that a *morisco* doctor in Madrid named 'the Avenger' had apparently murdered three thousand of his patients with a 'poisonous ointment', while another *morisco* had used his skills to mutilate his patients in order to stop them from being able to use weapons.

Pedro de Vesga's assertions must, however, both be examined in their immediate historical context and compared with the claims of a *judeoconverso* medical plot. Vesga certainly expressed his claims in an atmosphere of heightened anti-*morisco* popular anxiety and in the years immediately preceding the royal decision to decree the expulsion of the *moriscos* from Spain in 1609. The decades preceding the expulsion decree were characterised by increasingly strident and alarmist claims targeting the *moriscos*: from the allegation accusing them of systematically conspiring with Spain's Muslim adversaries in the Mediterranean to that of seeking to take over the kingdoms through their allegedly explosive birth rate.[36] Moreover, the allegations of Pedro de Vesga were clearly inspired by the conspiracist beliefs about the 'medical plot' of *judeoconverso* doctors. The reference to a doctor known as 'the Avenger' was a manifest and unambiguous (even clumsy) recycling of the myth of the murderous 'avenging' *judeoconverso* doctor that featured in the work of Diego de Simancas. Vesga's argument was therefore nothing more than a crude attempt to usurp the antisemitic legend of Jewish medical murder and apply it to serve and justify anti-*morisco* aims.[37]

In addition to Vesga's claims, the Dominican chronicler Fray Jaime Bleda also claimed that *morisco* medical practitioners murdered Christian patients in his work *Corónica de los Moros de España, en ocho libros*, printed in Valencia in 1618. Fray Bleda accused 'some [*morisco*] doctors' in Valencia of poisoning 'many Christians', and claimed than one *morisco* doctor named Castellano had admitted to another *morisco* that he systematically killed 'at least' one out of every ten Christian patients. Bleda's only source was an anonymous Arabic-speaking Christian who had allegedly eavesdropped on the conversation between the two *moriscos* and had personally related it to Bleda.

This claim of a *morisco* medical conspiracy is nevertheless just as derived from the anti-Jewish tradition as that made by Pedro de Vesga in the previous decade. Beyond the highly suspicious flimsiness of his 'evidence', an overheard conversation reported by a single anonymous individual, the accusation of the systematic decimation of Old Christian patients is extremely reminiscent of the claims made against *judeoconverso* doctors. Bleda was a propagandist seeking to justify post facto the expulsion of the *moriscos* from Spain, and he was correspondingly ready to hurl any libel against the *moriscos* in his *Corónica de los Moros de España*. Earlier, in 1610, he had already printed a polemical work attacking the *moriscos* and including a chapter entitled '*Moriscos* should be forbidden from practising the art of medicine as well as that of theology and other studies'. Bleda's principal accusation against the *morisco* medical practitioners in the Latin *Defensio fidei* was that they used their prestige and authority to spread Muslim beliefs within *morisco* communities.[38] Moreover, Bleda was extremely familiar with the claims of medical murder made against Jews and *judeoconversos*. In both the *Defensio fidei* and the *Corónica de los Moros de España*, he discussed Jewish medical murder, including that allegedly committed by Meir Alguadex, in far more detail. For Bleda, the notion of a conspiracy of medical murder by *moriscos* was certainly associated with that of the *judeoconverso* medical plot, and this association is revealed in his concluding remark to the passage: 'the injuries committed against the [Old] Christians by Muslim and Jewish doctors are wondrous indeed'.[39]

Conclusion

When the question of whether antisemitism and Islamophobia may have had a 'shared history' is examined in relation to attitudes towards Jewish and Muslim minorities in the historical context of medieval and early modern Europe, it rapidly becomes apparent that the notion of a 'shared' experience should not imply an 'identical' experience. Jews and Muslims—or rather, *judeoconversos* and *moriscos* in an early modern Iberian context—were the subject of hostile conspiracy theories, but these did not always target both groups evenly. The conspiracy theory of medical murder stands as a case in point. A comparison of the fears of medical murder by crypto-Jews and crypto-Muslims that existed in early modern Spain and Portugal unambiguously demonstrates the extent to which the anxieties relating to doctors and medical knowledge can generate a potent conspiracy theory when they are combined with xenophobic suspicions about the place and role of religious/ethnic minorities. Yet it also reveals the limits of the notion of a 'shared history', since it was focused almost exclusively on the *judeoconversos* rather than the *moriscos*.

The exceptional 'success' of the myth of medical murder targeting Jews and their converted descendants—if this 'success' can be measured by its repetition in printed works and longevity across the centuries—is in no small part due to factors that were particular to the situation of Jews/*judeoconversos* in the early modern Iberian world. The reasons why the *judeoconversos* were targeted by such a medical conspiracy theory are certainly numerous: (1) the success of Jewish medical practitioners in the Iberian Christian kingdoms during the medieval period; (2) the existence of a medieval myth of Jewish medical murder and the Catholic Church's support for laws that (unsuccessfully) sought to prevent Christians from interacting with Jewish doctors; (3) the fears generated by the invisibility of alleged judaisers who were no longer forced to wear distinguishing clothing; (4) the appearance of the forged letters between the *judeoconversos* of Toledo and Jews of Constantinople; and (5) the credibility conferred on the conspiracy theory by the officially sanctioned statutes of racial purity, introduced by numerous medical faculties and schools.

Although they were frequently accused of demonic links, *morisco* healers and medical practitioners were not accused of a conspiracy of medical murder. The best efforts of Pedro de Vesga and Jaime Bleda in the early seventeenth century to claim that a Muslim medical conspiracy existed reveal themselves, on careful examination, to be nothing more than a desperate ploy to exploit the anti-*judeoconverso* legend to suit their anti-*morisco* agenda.

Given its ancient origins, the concept of Jewish medical murder has unsurprisingly proven itself to be an enduring and powerful element in antisemitic thought and conspiracy theories. Even one of the foremost figures of the Spanish Enlightenment in the second half of the eighteenth century—the Benedictine monk Benito Jerónimo Feijóo y Montenegro—could not bring himself to reject the libel in its entirety. His reaction to such tales is worth reading in full:

> Firstly, there is no doctor whatsoever who does not treasure his own personal interest and reputation more than the ruin of others. For this reason, he will seek to cure his patients, upon which his reputation depends and therefore also his personal interest. The only exceptions will be one or two individual cases of [*judeoconverso*] doctors who hope not to be detected [as a result of their high patient death rates]. There can be no doubt that a doctor at whose hands so many patients died would lose his reputation. Secondly, even if some were to succeed in their malevolent intent, then within two or three months everyone would flee from such a mortiferous doctor, even if they only attributed his patient death rates to his ignorance or bad luck. ...

> What I, therefore, will only choose to believe is that a few of that rabble may cause the death by homicide of Christian [patients] in spite of the difficulty of doing this. Apart from a few patients that they decide to eliminate due to some private hatred, they will specifically target those persons whom they consider to be useful to the Church or the most zealous in the true faith. This is reason enough to flee from Jewish doctors and to loathe them.[40]

While Feijóo therefore discounted the idea that there was a systematic campaign of medical murder, he simply could not bring himself to discount its existence and was prepared to endorse its practice by individual *judeoconverso* doctors.

This state of affairs and difference in treatment between Jews and Muslims has survived into the modern era. Jewish doctors have continued to be suspected of medical machinations against Gentiles. In Germany, the Nazi propaganda outlet *Der Stürmer* accused Jewish doctors of experimenting on Christian patients, with fatal consequences for the latter, and of practising abortions on German-Gentile women in order to 'destroy the German *Volk*'.[41] In a similar vein, the French collaborationist and antisemite Doctor Fernand Querrioux published a violent pamphlet entitled *La médecine et les juifs* (Paris, 1940), claiming that, with the aid of the socialist *Front populaire* and freemasonry, medicine in France had become 'a kingdom ruled by Jews'. Among the accusations hurled at the *médecins-juifs* was that, as covetous Jews, they wilfully neglected impecunious patients and allowed them to die.[42] Even closer to the early modern Iberian example was the (in)famous 'Doctors' Plot' in the USSR in 1952–53, which began with an article in the state newspaper *Pravda* denouncing a conspiracy of Jewish 'saboteur-doctors' and 'poisoner-doctors' with Zionist and imperialist sympathies, whose goal was 'shortening the lives of leaders of the Soviet Union by means of medical sabotage'. The antisemitic medical purge in Russia that led to hundreds of arrests only came to a halt with the death of Stalin on 5 March 1953.[43] Even today, antisemitic polemicists claim, both explicitly and implicitly, that there are sinister motives behind the demographic overrepresentation of Jews in medicine and other scientific fields. Moreover, neo-Nazi, white supremacist and extremist Catholic/Protestant propaganda produced in the United States and Europe continues to present the Jews as the driving force behind the legalisation of abortion, which is presented as the medical murder of unborn children and as an integral part of a wider anti-Gentile plot.[44]

Notes

1. V. da Costa Mattos, *Breve Discurso contra a heretica perfidia do judaismo* (Lisbon, 1623), fols 56v–58r (da Costa Mattos 1623).
2. Pliny the Elder, *Natural History*, Book XXIX, ch. 7.
3. É. Lévi-Provençal, *Séville musulmane au début du XIIe siècle. Le traité d'Ibn ʿAbdun sur la vie urbaine et les corps de métiers* (Paris: Maisonneuve & Larose, 2001), p. 91 (Lévi-Provençal 2001).
4. J. Shatzmiller, *Jews, Medicine, and Medieval Society* (Berkeley: University of California Press, 1994), pp. 104–8 and L. García-Ballester, *Medicine in a Multicultural Society: Christian, Jewish and Muslim Practitioners in the Spanish Kingdoms, 1220–1610* (Aldershot: Ashgate, 2001), III, pp. 34–5 (Shatzmiller 1994; García-Ballester 2001).
5. J. Trachtenberg, *The Devil and the Jews: The Medieval Conception of the Jew and its Relation to Modern Antisemitism* (Philadelphia: The Jewish Publication Society, 1983) (Trachtenberg 1983).
6. *Decretum Gratiani, Pars Secunda, Causa XXVIII, question I*, c. XIII.
7. See chapter five in Shatzmiller, *Jews, Medicine, and Medieval Society*.
8. D.E. Carpenter, *Alfonso X and the Jews: Edition of and Commentary on Siete Partidas 7.24 "De los judíos"* (Berkeley: University of California Press, 1986), pp. 87–8 (Carpenter 1986).
9. Translation by Shatzmiller, *Jews, Medicine, and Medieval Society*, pp. 87–8.
10. F. Diago, *Historia de la vida y milagros, muerte y discípulos* (Barcelona, 1600), pp. 134–5 (Diago 1600).
11. M.V. Amasuno, 'The *Converso* Physician in the anti-Jewish Controversy in Fourteenth-Fifteenth Century Castile', in *Medicine and Medical Ethics in Medieval and Early Modern Spain: An Intercultural Approach*, eds. Samuel S. Kottek and Luís García-Ballester (Jerusalem: Magnes Press, 1996), pp. 92–118 (Amasuno 1996).
12. Translation by Amasuno, 'The *Converso* Physician', pp. 103–5.
13. B. Netanyahu, *The Origins of the Inquisition in Fifteenth Century Spain* (New York: New York Review Books, 1995), pp. 177–82 (Netanyahu 1995).
14. A. de Espina, *Fortalitium fidei* (Nürnberg, 1494), fols 172v–173r (de Espina 1494).

15. da Costa Mattos, *Breve Discurso*, fol. 66v; Fray Francisco de Torrejoncillo, *Centinela contra Judíos* (Madrid, 1674), pp. 129–30; A. de Contreras, *Mayor Fiscal contra Judíos* (Madrid, 1736), p. 121 (Francisco de Torrejoncillo 1674; de Contreras 1736).
16. D. de Colmenares, *Historía de la insigne ciudad de Segovia y Conpendio de la historias de Castilla* (Segovia, 1637), p. 324; P. de San Nicola, *Siglos Geronimianos. Historia general eclesiastica, monastica y secular* (Madrid, 1744), XIX, p. 433; M. de Torrecilla, *Encyclopedia canonica, civil, moral, regular y orthodoxa* (Madrid, 1721), p. 495 (de Colmenares 1637; de San Nicola 1744; de Torrecilla 1721).
17. F. Máximo de Moya Torres, *Manifiesto universal de los males envejecidos que España padece, y de las causas de que nacen, y remedios que à cada uno en su clase corresponde* (Madrid, s.d.), pp. 121–6 (Máximo de Moya Torres s.d.).
18. I. Loeb, 'La correspondance des Juifs d'Espagne avec ceux de Constantinople', *Revue des Études Juives*, XV (1888), 262–76 (Loeb 1888).
19. D.C. Goodman, *Power and Penury. Government, Technology and Science in Philip II's Spain* (Cambridge: Cambridge University Press, 1988), p. 219 (Goodman 1988).
20. A.H.N., *Sección Inquisición, libro* 497, fols 22–3.
21. Goodman, *Power and Penury*, pp. 219–20.
22. A. Fernández Luzón, *La Universidad de Barcelona en el Siglo XVI* (Barcelona: Edicions Universitat Barcelona, 2005), pp. 256–8; Fray Juan B. Guardiola, *Tratado de Nobleza, y de los Títulos y Ditados que oy dia tienen los varones claros y grandes de España* (Madrid, 1591), p. 10 (Fernández Luzón 2005; Juan B. Guardiola 1591).
23. Goodman, *Power and Penury*, p. 220; Teófilo Braga, *Historia Da Universidade De Coimbra Nas Suas Relações com a Instucção Publica Portugueza. Vol II.1555 a 1700* (Lisbon, 1895), pp. 779–83 (Braga 1895).
24. J. Valdemar Guerra, 'Judeus e Cristãos-novos na Madeira (1461–1650)', *Arquivo Histórico da Madeira. Série de Transcrições Documentais. Transcrições Documentais 1* (Funchal, 2003), pp. 163–4, n. 331 (Valdemar Guerra 2003).
25. C. Heinrich Freiesleben, *Corpus iuris canonici* (Basel, 1757), II, cols 125–6 (Heinrich Freiesleben 1757).

26. D. Velázquez, *Defensio Statuti Toletani a Sede Apostolica saepe confirmati* (Antwerp, 1575), pp. 18–20 (Velázquez 1575).
27. Ignacio del Villar Maldonado, *Sylua responsorum iuris: in duos libros diuisa* (Madrid, 1614), fol. 133r (del Villar Maldonado 1614).
28. M.A. Cohen, *The Canonization of a Myth. Portugal's "Jewish Problem" and the Assembly of Tomar 1629* (Cincinnati: Hebrew Union College, 2002), p. 48 (Cohen 2002).
29. de Torrejoncillo, *Centinela contra Judíos*, p. 87.
30. J. Lúcio de Azevedo, *História dos cristãos novos portugueses* (Lisbon: Clássica Editora, 1921), pp. 465–9, doc. 12 (Lúcio de Azevedo 1921).
31. A.N.T.T., *Conselho Geral do Santo Ofício*, book 301, fols. 66r–97v, *Tratado em que se prova serem christãos fingidos os da nação, que viuem em Portugal, apontando os males que fazem aos christãos velhos.*
32. J.A. Rodrigues da Silva Tavim, '"Murdering Doctors" in Portugal (XVI–XVII Centuries): The Accusation of a Revenge', in *El Prezente. Studies in Sephardic Culture. Magic and Folk Medicine*, V, eds. Tamar Alexander, Yaakov Bentolila and Eliezer Shaul (2011), pp. 81–98 (Rodrigues da Silva Tavim 2011).
33. L. García Ballester, *Los Moriscos y la medicina. Un capítulo de la medicina y la ciencia marginadas en la España del siglo XVI* (Barcelona: Labor Universitaria, 1984), pp. 99–118 (García Ballester 1984).
34. See García Ballester, *Los Moriscos y la medicina.*
35. G. Escolano, *Decada primera de la historia de la insigne y coronada ciudad y reyno de Valencia* (Valencia, 1610), cols 686–7 (Escolano 1610).
36. For a concise analysis of anti-*morisco* fears and conspiracy theories, see Francois Soyer, 'Faith, Culture and Fear: Comparing Islamophobia in Early Modern Spain and Twenty-first-century Europe', *Ethnic and Racial Studies*, XXXVI, Issue 3 (2013), 399–416 (Soyer 2013).
37. *Actas Cortes de Castilla*, XXIII (Madrid), pp. 583–7.
38. Fray Jaime Bleda, *Defensio fidei in causa neophytorum siue Morischorum Regni Valentiae totiusque Hispaniae* (Valencia, 1610), pp. 365–74 (Jaime Bleda 1610).

39. Fray Jaime Bleda, *Corónica de los Moros de España, en ocho libros* (Valencia, 1618), pp. 546 and 861 (Jaime Bleda 1618).
40. Benito Jerónimo Feijóo y Montenegro, *Teatro crítico universal, ó, Discursos varios en todo género de materias, para desengaño de errores comunes* (Madrid, 1773), V, pp. 110–11 (Feijóo y Montenegro 1773).
41. R. Keysers, *L'intoxication nazie de la jeunesse allemande* (Paris: L'Harmattan, 2011), pp. 102–3 (Keysers 2011).
42. F. Querrioux, *La médecine et les juifs* (Paris, 1940), pp. 7–8 (Querrioux 1940).
43. B. Pinkus and J. Frankel, *The Soviet Government and the Jews 1948–1967: A Documentary History* (Cambridge: Cambridge University Press, 1984), pp. 198–201 (Pinkus and Frankel 1984).
44. M. Durham, *White Rage: The Extreme Right and American Politics* (New York: Taylor & Francis, 2007), pp. 85–6 (Durham 2007).

References

Amasuno, M.V. 1996. The *Converso* Physician in the anti-Jewish Controversy in Fourteenth-Fifteenth Century Castile. In *Medicine and Medical Ethics in Medieval and Early Modern Spain: An Intercultural Approach*, ed. Samuel S. Kottek and Luís García-Ballester. Jerusalem: Magnes Press.

Braga, Teófilo. 1895. *Historia Da Universidade De Coimbra Nas Suas Relações com a Instucçâo Publica Portugueza. Vol II.1555 a 1700.* Lisbon.

Carpenter, D.E. 1986. *Alfonso X and the Jews: Edition of and Commentary on Siete Partidas 7.24 "De los judíos"*. Berkeley: University of California Press.

Cohen, M.A. 2002. *The Canonization of a Myth. Portugal's "Jewish Problem" and the Assembly of Tomar 1629.* Cincinnati: Hebrew Union College.

da Costa Mattos, V. 1623. *Breve Discurso contra a heretica perfidia do judaismo.* Lisbon.

de Colmenares, D. 1637. *Historía de la insigne ciudad de Segovia y Conpendio de la historias de Castilla.* Segovia.

de Contreras, A. 1736. *Mayor Fiscal contra Judíos.* Madrid.

de Espina, A. 1494. *Fortalitium fidei.* Nürnberg.

de San Nicola, P. 1744. *Siglos Geronimianos. Historia general eclesiastica, monastica y secular*, XIX. Madrid.

de Torrecilla, M. 1721. *Encyclopedia canonica, civil, moral, regular y orthodoxa.* Madrid.

del Villar Maldonado, Ignacio. 1614. *Sylua responsorum iuris: in duos libros diuisa.* Madrid.

Diago, F. 1600. *Historia de la vida y milagros, muerte y discípulos*. Barcelona.
Durham, M. 2007. *White Rage: The Extreme Right and American Politics*. New York: Taylor & Francis.
Escolano, G. 1610. *Decada primera de la historia de la insigne y coronada ciudad y reyno de Valencia*. Valencia.
Feijóo y Montenegro, Benito Jerónimo. 1773. *Teatro crítico universal, ó, Discursos varios en todo género de materias, para desengaño de errores comunes*, V. Madrid.
Fernández Luzón, A. 2005. *La Universidad de Barcelona en el Siglo XVI*. Barcelona: Edicions Universitat Barcelona.
Francisco de Torrejoncillo, Fray. 1674. *Centinela contra Judíos*. Madrid.
García Ballester, L. 1984. *Los Moriscos y la medicina. Un capítulo de la medicina y la ciencia marginadas en la España del siglo XVI*. Barcelona: Labor Universitaria.
García-Ballester, L. 2001. *Medicine in a Multicultural Society: Christian, Jewish and Muslim Practitioners in the Spanish Kingdoms, 1220–1610*, III. Aldershot: Ashgate.
Goodman, D.C. 1988. *Power and Penury. Government, Technology and Science in Philip II's Spain*. Cambridge: Cambridge University Press.
Heinrich Freiesleben, C. 1757. *Corpus iuris canonici*. Basel.
Jaime Bleda, Fray. 1610. *Defensio fidei in causa neophytorum siue Morischorum Regni Valentiae totiusque Hispaniae*. Valencia.
———. 1618. *Corónica de los Moros de España, en ocho libros*. Valencia.
Juan B. Guardiola, Fray. 1591. *Tratado de Nobleza, y de los Títulos y Ditados que oy dia tienen los varones claros y grandes de España*. Madrid.
Keysers, R. 2011. *L'intoxication nazie de la jeunesse allemande*. Paris: L'Harmattan.
Lévi-Provençal, É. 2001. *Séville musulmane au début du XIIe siècle. Le traité d'Ibn 'Abdun sur la vie urbaine et les corps de métiers*. Paris: Maisonneuve & Larose.
Loeb, I. 1888. La correspondance des Juifs d'Espagne avec ceux de Constantinople. *Revue des Études Juives* XV: 262–276.
Lúcio de Azevedo, J. 1921. *História dos cristãos novos portugueses*. Lisbon: Clássica Editora.
Máximo de Moya Torres, F. s.d. *Manifiesto universal de los males envejecidos que España padece, y de las causas de que nacen, y remedios que à cada uno en su clase corresponde*. Madrid.
Netanyahu, B. 1995. *The Origins of the Inquisition in Fifteenth Century Spain*. New York: New York Review Books.
Pinkus, B., and J. Frankel. 1984. *The Soviet Government and the Jews 1948–1967: A Documentary History*. Cambridge: Cambridge University Press.
Querrioux, F. 1940. *La médecine et les juifs*. Paris.
Rodrigues da Silva Tavim, J.A. 2011. "Murdering Doctors" in Portugal (XVI–XVII Centuries): The Accusation of a Revenge. In *El Prezente. Studies in Sephardic Culture. Magic and Folk Medicine*, V, ed. Tamar Alexander, Yaakov Bentolila, and Eliezer Shaul. Beer-Sheva:Ben-Gurion University Of The Negev.

Shatzmiller, J. 1994. *Jews, Medicine, and Medieval Society*. Berkeley: University of California Press.

Soyer, Francois. 2013. Faith, Culture and Fear: Comparing Islamophobia in Early Modern Spain and Twenty-first-century Europe. *Ethnic and Racial Studies* XXXVI, Issue 3: 399–416.

Trachtenberg, J. 1983. *The Devil and the Jews: The Medieval Conception of the Jew and its Relation to Modern Antisemitism*. Philadelphia: The Jewish Publication Society.

Valdemar Guerra, J. 2003. Judeus e Cristãos-novos na Madeira (1461–1650). *Arquivo Histórico da Madeira. Série de Transcrições Documentais. Transcrições Documentais 1*. Funchal.

Velázquez, D. 1575. *Defensio Statuti Toletani a Sede Apostolica saepe confirmati*. Antwerp.

PART II

Empire

CHAPTER 4

Fear and Loathing in the Russian Empire

Robert D. Crews

The Russian empire was remarkably fertile ground for hostility towards Jews and Muslims. The Russian tsars had ruled Muslims since the fifteenth and sixteenth centuries and Jews from the late eighteenth century. However, tsarist elites only came to see both groups as major challenges to the imperial order in the late nineteenth and early twentieth centuries, when Muslims and Jews became stock figures in an ominous pantheon of ethnic and religious groups that seemed to pose a threat to the imperial political order. The shared story of how they came to inhabit analogous, yet unique, positions in the anxious imaginations of Russian elites is the focus of this chapter.

In the Russian imperial setting of the late nineteenth and early twentieth centuries, anti-Jewish thinking had a great deal in common with anti-Muslim prejudice. Stereotypes cut across Russia's increasingly polarised political scene. In the terrifying world conjured by right-wing thinkers, Jews and Muslims shared odious traits. They were cunning, frighteningly clever and keen to outsmart and exploit the good-hearted Russian peasant. They disdained Christianity and Christians and mocked the faith of the Russian tsar. Not only was their religion abominable, they worshipped profit above all else, a trait exemplified by their wealthy merchants and

R.D. Crews (✉)
Stanford University, Stanford, CA, USA

J. Renton, B. Gidley (eds.), *Antisemitism and Islamophobia in Europe*, DOI 10.1057/978-1-137-41302-4_4

entrepreneurs. At the same time, though, Jews and Muslims, particularly the poor among them, appeared to be mired in backwardness; their social isolation could not be explained by geography or poverty alone, but also by a haughty and religiously inspired exclusivity, which caused them to shun their Christian neighbours and would-be brothers in the family of empire. They were even in cahoots with foreigners.

To those inclined to see the problem through a more scientific lens, on the other hand, it was also clear that Muslims and Jews shared the burden of being stamped with immutable physical and moral deficiencies. If none of this were bad enough, they were, each in their own way, sexual deviants and, with their large families and rapid population growth, a demographic threat to boot. They were, in short, dangerous, a threat to the solidity of the autocracy, to the empire, and to the pre-eminent position of the Russian people and their Orthodox Christian faith within it.[1]

For all of these commonalities, though, there were important differences between anti-Jewish and anti-Muslim dispositions. Jews haunted the imaginations of Russian thinkers on the right and the left who were gripped by anxieties about Russia's changing place in the globe and about the novel forces—capitalism, industrialisation, urbanisation and the explosion of commercial culture—that were transforming the empire. On the right, critics imagined Jews to be the source of the revolutionary movements that haunted the empire from the 1870s; on the left, they were to blame for the failure of various radical campaigns.

Russian intellectuals and policy-makers brought a somewhat different lens to bear on Muslims, however. By the early twentieth century, they had come to imagine a 'Muslim Question' alongside the 'Jewish', 'Polish', 'German' and other 'Questions' about the loyalty of non-Russian and non-Orthodox groups.[2] But until the end of the nineteenth century they tended to concentrate on different populations in specific locales within the empire or along its frontiers, acknowledging ethnic, linguistic and social distinctions among these groups. In fact, some of these communities had been subjects of the tsar for centuries. Others had been the object of more recent—and brutal—tsarist campaigns to subjugate populations in the North Caucasus and Central Asia. Thus the charge of 'fanaticism'—itself the creature of a secularising critique of European Christianity and, simultaneously, a reaction to anticolonial movements around the globe—was one that Russian elites levelled against these diverse Muslim populations in very distinct ways.

Tsarist policies and attitudes towards both groups were shifting and dynamic. The tsarist regime consistently privileged the dominant position of Orthodox Christianity in the empire. Yet, from the era of Catherine the Great (r. 1762–96), the state selectively granted religious toleration to non-Orthodox groups. In this setting, toleration never meant a laissez-faire policy of non-interference, however. In fact, the regime continually worked with non-Orthodox elites to shape the empire's confessional groups into entities that would undergird the strength and integrity of the empire. Tsarist officials might have preferred that all of their subjects join the tsar's Orthodox faith. But they came to see non-Orthodox personnel and even religious doctrine as potentially useful to the empire. Beginning with Alexander I (r. 1801–25) and Nicholas I (r. 1825–1855) in particular, the tsarist government created institutions to integrate non-Orthodox clerics (and, in some instances, create social groups that resembled a Christian clergy). The Russian state backed certain clerical powers and religious norms among Jews, Muslims, Protestants, Catholics and Buddhists. In return, clerical elites preached obedience to the tsarist order and regulated morality and the family. Until the collapse of the Romanov system in 1917, tsarist officials tended to value religion—even in forms they judged completely wrongheaded or inferior—as a constellation of moralising prohibitions and controls that constrained their subjects' behaviour and safeguarded the status quo. They may have disdained Islam and Judaism, much as they did Catholicism and Buddhism, but such antipathy did not preclude taking an instrumental view of the matter. As a seemingly universal force that contributed to a conservative disposition, religion appeared to be an indispensable foundation of the empire.[3]

The Orthodox Church maintained its own, dissenting point of view, of course, preferring to launch missions to convert such 'heathens' and appealing to the government to repress them. Several prominent officials sympathised with the Church hierarchy's critique of imperial policy on this score. Nonetheless, in the case of Muslims and Jews, their non-Orthodox institutions survived the late nineteenth-century shift towards the scapegoating of these groups, a phenomenon that, in any event, never wholly consumed tsarist policy-makers. What, then, accounts for the appearance among tsarist elites in the late nineteenth and early twentieth centuries of a heightened sense that Muslims and Jews—each in their distinctive way—posed such a pernicious danger to Russian society and the imperial order writ large?

The fact that Russian elites came to see Muslims and Jews through very similar lenses in the late imperial era appears all the more remarkable when we consider the very different historical trajectories of these communities before the second half of the nineteenth century. One crucial factor was the antiquity of Russian contacts with Muslims, first in the Russians' role as subjects of one of the successor states to the Golden Horde (whose elites had converted to Islam), and then as rulers of Muslim subjects. Traditional religious antipathy was one part of the story. Reviewing Russian historical chronicles of the medieval principalities that would become Rus' and then Muscovy, we frequently confront hostile representations of the eastern and southern steppe peoples and their religion. As the institution responsible for such imagery, the Orthodox Church presented itself as a bulwark against what its clergy viewed as the abomination of Islam.[4]

Yet with contempt came a degree of familiarity, even intimacy. The Muscovites frequently pursued pragmatic relations with the Golden Horde and its successors, even intermarrying with the Muslim elites of Kazan.[5] From the fifteenth century, as Muscovy expanded, it incorporated territory inhabited by Muslims. Muscovites and Muslim Tatars clashed on the battlefield, and the Orthodox portrayed their conquest in 1552 of the seat of their former overlords at Kazan as a victory for Christendom. Frequent wars with the Ottoman empire, raids from the southern and eastern steppes and the threat of being abducted and lost to the slave markets of Central Asia kept this image of the menacing 'infidel' in the minds of the common folk as well. Yet all of this also meant that the Muscovite state had to come to terms with substantial Muslim populations and, whenever possible, seek to co-opt Muslim elites into the imperial ruling class. Further territorial expansion between the seventeenth and nineteenth centuries would bring more and more Muslim subjects under the tsar's authority. By the twentieth century, they were the largest non-Orthodox group in the empire, with a population of some twenty million.

Their demographic weight, then, was another key variable that distinguished the experience of Russia's Muslims from that of the Jews, who numbered just over five million at the turn of the century. The size and geographic distribution of Russia's Muslims meant that they were of political interest not only for the tsarist government. In the context of inter-imperial rivalries, they loomed in the minds of the rivals of the tsarist empire, including the Ottomans, as a potential liability for the Russian state. The perception of a potential Muslim challenge to the borders of the empire was a perennial anxiety. Indeed, the southern frontier of the tsarist realm, stretching from the Black Sea to China, was populated by heterogeneous Muslim populations inhabiting both sides of the borders that Russia shared with Muslim powers. This geopolitical sensitivity at times worked to heighten more radical officials' antipathy towards Muslims, and at others strengthened the hand of authorities who called for a more cautious and accommodating approach to these communities. Clear lines of division frequently appeared between representatives of the Ministries of the Interior and War, as institutions responsible for domestic social control, and those of the Foreign Ministry, who tended to be more attentive to how tsarist policies would be perceived abroad by enemies and foes alike. Many officials in St. Petersburg, as well as in tsarist consular outposts throughout the Middle East, even calculated that Russian Muslims who travelled abroad as "merchants" and pilgrims could enhance the reach of tsarist power abroad.[6] And particularly during the reign of the Ottoman ruler Abdülhamid II (r. 1876–1909), when the sultan projected an image of himself as the defender of Muslims everywhere, competition for Muslim loyalties became an international affair.[7]

In stark contrast to the much longer history of interactions between Russian rulers and Muslim subjects, Jews were a rarity—and often more an abstraction—for lengthy periods in the history of the empire. The late fifteenth-century Church had been divided by a controversy that drew attention to a group that supposedly wanted to 'Judaise' the faith. However, the Blood Libel—a trope that exercised the imagination of so many in Europe—appears not to have circulated as widely in Muscovy. Still, the tsars banned actual Jews from settling there and periodically blocked visiting Jewish merchants as well. For instance, Empress Elizabeth (r. 1741–62) ordered the expulsion of a small community of Jews from Moscow. This stance would not change until the late eighteenth century, when imperial expansion to the west and south and, under Catherine the Great, the partitions of Poland in 1772, 1792, and 1795 made half a million Jews subjects of the Russian empire.[8]

Olga Litvak has characterised tsarist policy from the period of these partitions through the reign of Nicholas I as wavering 'between a vision of radical reform and the reality of social and administrative conservatism heavily laced with Judaeophobia'.[9] Catherine sought to integrate the Jews into imperial social hierarchies and to break down, over time, the old institutions of communal autonomy that had developed under Polish rule. In practice, though, the state relied on these very institutions to tax and discipline Jewish merchants, artisans and petty traders. Later, Nicholas's strategy for incorporating the Jews was to rely on the military as a kind of 'experimental school of citizenship'.[10]

Whereas most state authorities had largely given up in the eighteenth century on making Muslim conversion to Christianity an official priority, they continued to hold out hope for the conversion of the Jews. Thus Nicholas saw military service as a gateway to the tsar's faith. In 1827, he introduced conscription for Jewish communities. Unlike Muslim and other non-Orthodox conscripts, these inductees into the Russian army were expected to become Orthodox Christians (Jews were also denied promotion to the rank of officer). Paradoxically, it was the Russian military, Litvak shows, that 'fostered an acute awareness of Jewish individuality at the expense of both the tsarist vision of Jewish difference erased in the ranks and Jewish communal authority derived from the tsarist regime which itself supported Jewish confessional discipline'.[11] Beyond the military, despite the privileged position of Orthodoxy, tsarist law also permitted Jewish conversion to other Christian denominations.[12]

The international political scene was yet another factor that distinguished the Jews from Muslims and other tsarist subjects. In the second half of the nineteenth century, the plight of Jews in the Russian empire became a matter of concern for international Jewish organisations such as the Alliance Israélite Universelle established in 1860 in Paris. And at key moments, some Russian Jewish leaders received a sympathetic, if often ambivalent, hearing in Britain, the United States and elsewhere. Most notably, anti-Jewish pogroms in 1881–82 in the southwestern provinces of the Russian empire provoked international outrage. They tarnished Russia's reputation and repulsed foreign investors. More than two million Jews would emigrate. Nevertheless, foreign powers were not willing to risk more forceful action on behalf of Russia's Jews: until 1915, no army stood on the borders of the empire to march to the Jews' defence.[13]

Yet, despite these important differences in the circumstances of Russia's Muslims and Jews and in the tsarist approach towards each of these communities, over the course of the nineteenth century Russian officials adopted measures that brought Muslims and Jews, together with other religious groups, into the same regulatory framework. Under Nicholas I, the state placed Islam and Judaism within a single legislative system intended to administer the non-Orthodox Christian confessions. The Russian *Digest of Laws* defined the rights and obligations of clerics in each religious community in the empire and spelled out the terms of their subordination to the imperial ministries. Law-makers pursued a degree of standardisation, with the Orthodox clerical hierarchy and parish as an implicit model. For Muslims and Jews, in particular, these new legal distinctions introduced novel ways of conceiving of religious authority. For the Jews, however, this official endorsement of particular aspects of clerical and communal life coincided with a renewed drive for conversion.

Other policies pursued similarly contradictory ends. In 1835, a residential restriction consolidated earlier curbs to limit Jews formally to former Polish and Ottoman territories, in an era that came to be known as the 'Pale of Settlement', and in 1844, the government dissolved the Jewish communal government structure of the *kahal*. At the same time, the state sponsored primary schools and agricultural colonies designed to integrate Jews more deeply into imperial life. Meanwhile, for some Muslim populations, for example the Bashkirs in the Ural Mountains region, a state-imposed system of military administration brought these communities into the fold of imperial service.

Even more pervasive was the influence of an official hierarchy of Islamic authorities, established in 1788, that oversaw mosques throughout the empire (separate institutions performed similar functions for Muslims in the Crimea and, later, in the Caucasus). Tsarist police backed the authority of Muslim clerics in these communities, particularly in matters that seemed to shore up the state, for instance by preaching loyalty to the emperor, respect for public morality, and by assisting in expanding the bureaucratic reach of the state. Muslim clerics were more than mere servants of the state, though, and sometimes lay men and women could appeal to the regime to discipline wayward prayer leaders and scholars. Indeed, it was the tsarist government's defence of particular Islamic legal claims that made it so central to disputes about Islam among these communities.[14]

Attitudes towards Muslims and Jews proved malleable, though, and as a focus of the anxieties of Russian elites, they were hardly alone. From the 1860s, wide-ranging transformations swept across the empire. Tsarist defeat in the Crimean War (1853–56) prompted the autocracy to seek new ways to enhance its legitimacy and to strengthen the empire. A social and economic order grounded in serfdom was partially unravelled by tsarist decree. Investing in railroad construction and rapid industrialisation, the regime unleashed urbanisation and mass migration. Russia was on the move like never before. Throughout European Russia, the ranks of imperial urban centres swelled. Official efforts to narrow the gap separating Russia from its more formidable European rivals included policies that at once acknowledged the centrality of Jewish merchants and financial intermediaries in the tsarist economy, and selectively threw up impediments to Jewish commercial activities in the name of 'protecting' Russian peasants and other subjects from their supposed exploitation at the hands of the Jews. For instance, Jewish merchants who had grown wealthy as tax farmers petitioned Alexander II (r. 1855–81) to loosen residential restrictions for 'useful' Jews. In 1859, Alexander II permitted merchants of the first (that is, the most elite) guild to live and work outside the Pale. Similarly, tsarist legislation expanded the rights of particular Jewish social groups, including Jewish university graduates in 1861, some Jewish artisans in 1865, Jewish veterans in 1867 and post-secondary graduates in 1879. On the other hand, in 1880 the regime introduced bans on Jewish military doctors (similar measures applied to Poles) and in 1887 a quota (the '*numerus clausus*') limited Jewish enrolment in secondary and higher education.[15]

Meanwhile, the frontiers of the empire stretched towards the east and the south. The period of the 'Great Reforms' of the 1860s–1870s coincided with the Russian subjugation of the mountainous North Caucasus region and the conquest of Central Asia.[16] In the mountains of Daghestan and Chechnya, indigenous Muslim mountaineers declared *jihad* against imperial forces, amplifying the challenge of incorporating these vast and diverse Muslim populations. Tsarist retribution could be fierce. Beginning with the Crimean War, tsarist military forces had forcibly removed Muslim communities, spurring emigration to Ottoman lands. Tsarist troops unleashed violence across the Muslim highland communities in the Caucasus as well. Tsarist reprisals forced several hundred thousand Muslims from their homes. Yet, away from the Russo-Ottoman borderlands, the state continued to accommodate Muslims and Islamic institutions, while new questions began to haunt the imaginations of tsarist elites.

From the perspective of Russian officials and many other members of educated society, this moment was all the more precarious because a novel and dynamic form of politics was changing Europe, a phenomenon most visible—and disconcerting—in the German states, which by 1871 had been forged together, led by Prussia, as a single, powerful state on the principle of German national unity. Russia, by contrast, was still behind Europe in the metrics that mattered to Russian educated society. From the early nineteenth century, the Romanov dynasty had sought to recast itself as the embodiment of the Russian nation, but this agenda failed to banish questions about the viability of the tsarist imperial project in an era of industry, urban life and mass politics in a national key.[17]

The monarchy could look to a solid political base of a few thousand Russian noble families with enormous landed wealth. The Orthodox Church was another pillar of the tsarist system. But what about the political loyalties of non-Orthodox and non-Russian subjects? The Germans had long been a source of cultural anxiety, one only intensified by the foundation of the German empire. Polish nationalists revived memories of political independence. Their priests looked to Rome—and they rebelled. Polish uprisings in 1830, 1863 and 1905 took significant deployments of tsarist military force to quell them. At key moments in the nineteenth century, anti-Catholicism dominated official thinking. Muslims and Jews, then, were hardly the sole objects of suspicion and hostility in the last decades of the tsarist empire. Thus, like other non-Orthodox groups, including varied sectarian groups who had broken from the tsar's church, they stood outside the fold of ethnic Russianness and Orthodoxy that increasingly became the markers of political loyalty.

The extent to which official anxieties about the status of Jews had spread beyond elite circles in the 1870s exploded to the surface in 1881 when, following the assassination of Alexander II by populist revolutionaries, rioters targeted Jews in dozens of locales across the empire. Fuelled by rumours about Jewish culpability as well as by alcohol, these pogroms continued into 1882, claiming hundreds of lives and wrecking Jewish property and businesses. The autocratic state was slow to restore control, leading some contemporaries to conclude that the government had instigated the violence or at least condoned it. Scholars have since revised this view, emphasising the regime's discomfort with popular mobilisation, especially when it led to a breakdown of public order. Still, government authorities were quick to cast blame: on the victims. If Russian peasants and townspeople had turned on their Jewish neighbours, these officials concluded, then it was because these simple people had been forced to resort to violence by the Jews who exploited them. Much of Russian public opinion agreed with this judgement. In Kiev, what the historian Simon Dubnow dubbed an 'inferno of Russian Israel' as a result of the pogroms there, accusations against the Jews came from the political left and right. The South Russian Workers' Union warned workers 'one should not beat the Jew [*'zhid'*] because he is a Jew . . . one should beat him because he is robbing the people, he is sucking the blood of the working man'. In a similar vein, the right-wing newspaper *Kievlianin* rationalised the violence as the result of the 'insolence, impertinence, and exploitation' of the Jew.[18] These attitudes prompted the government to introduce legislation to further restrict Jewish residency and economic activities to 'protect' Russian populations. In 1891, the government expelled some fifteen thousand Jews from Moscow. Some Jews responded by emigrating, and intellectuals turned to socialism and Zionism. Nevertheless, popular anti-Jewish violence persisted. The Kishinev pogrom of 1903 became a global event, provoking sharp critiques of the tsarist government and eliciting Jewish calls for self-defence. During the revolutionary events of 1905–07, rioters turned on Jews once again, but on a scale not yet seen in the tsarist empire.

Muslims confronted violence, too, but under quite different circumstances. Outbreaks of popular violence against Muslims were quite rare. In Kazan Province, for instance, when Russian nationalist agitators of the right-wing, monarchist political movement known as 'the Black Hundreds' voiced their grievances against non-Russians in the region during the revolution of 1905–07, they focused their attentions primarily on the Jews, who were a relatively small minority in the region, not the far more demographically significant Muslim Tatars. These Muslim communities had long been at the receiving end of the hostility of the local Orthodox Church bishops and some of their allies in provincial administration. While like-minded Black Hundreds leaders at times campaigned against the supposed threat posed by Muslims, many of them also held out hope of forging a union with Muslims in support of the autocracy and in opposition to the revolutionary movement.[19] Indeed, it was really only in 1905 in Baku, the epicentre of Russian oil production on the Caspian Sea, when Christian crowds turned on Muslims. Such violence was highly localised, a product of growing tensions among communities at the margins of the late imperial oil industry boom. These attacks differed from the majority of anti-Jewish pogroms too in that they quickly descended into a chaotic scene of vicious intercommunal fighting, in this case mostly between Muslims and Armenians.[20]

Anxieties about the character of the Russian nation and the solidity of the empire rippled through Russian educated society, including the world of artists and writers, who played a crucial role in framing how a wider Russian public came to adopt new ways of seeing Jews, Muslims and other non-Russians. For writers and artists searching for the key to Russian national character, portrayals of the heterogeneous peoples of the empire in literature and art presented an opportunity to define Russianness. For early nineteenth-century Russian painters looking for Russian national themes of glory and heroism, Catholic Poles and Muslim Tatars had made the most appealing villains. Yet Jews also became part of this world, especially for artists who tried to connect with the lower orders of Russia's starkly hierarchical society. Provincial theatre was one such venue that bridged the empire's social divides, and theatre troupes sometimes staged plays with antisemitic and anti-Jewish jokes.[21]

For many of these artists, Jewishness had become much more than a religious identity; indeed it was an indelible aspect, and a negative one at that, of the character of Jews and of their offspring. Viewed through this antisemitic lens, it became more difficult for conversion to erase one's Jewishness, and some Jewish converts to Christianity never entirely escaped their roots. The arts world is revealing in this respect as well. The painter Ivan Kramskoi (1837–87), for instance, derided his boss, the convert Yakov Danilevsky, as having 'the character of a Jew [*zhid*]'.[22] Similarly, the virtuoso pianist Anton Rubenstein, who came from a Jewish family that converted in a mass baptism in Berdichev in 1831, found patrons at the imperial court but faced abuse from the likes of the composer Mikhail Glinka, who dismissed Rubenstein as 'an impudent Jew [*zhid*]'.[23] In artist circles feverishly committed to finding a 'national style', a Rubenstein became suspect. In 1861, Vladimir Stasov asserted that Rubinstein was 'a foreigner, with nothing in common either with our national character or our art'. Another composer with a partly Jewish background, Alexander Serov, was similarly biting in his antisemitic remarks about Rubinstein and the St Petersburg Conservatory, a school open to Jews and other groups, which Serov condemned as the 'Piano Synagogue'. By contrast, rival composers whose family names betrayed Tatar (and thus Muslim) origins, including Alexander Borodin and Mily Balakirev, were associated with a more thoroughly authentic Russianness, in both their character and their music.[24] More recent Muslim converts such as Mirza Aleksander Kazem-Bek (1802–70), a scholar and government expert on Islamic matters, also fared better than Jewish converts. He enjoyed a distinguished career as a leading Orientalist at Kazan University and adviser to government institutions that included the Ministry of the Interior.

Questions of identity were also taken up by tsarist policy-makers, who weighed in on whether conversion could erase Jewishness or whether Jews could otherwise become 'Russian' by adopting new names. In 1877, the High Commission for Review of Legislation Pertaining to the Jews of Russia resolved to ban name changes for Jews (in 1850 this ban had been established for baptised Jews). In 1893, Nicholas II himself directed the Chancellery of Petitions not to deviate from the prohibition. However, second-generation converts often received approval. Yakov Brafman, a Jew who had converted to Christianity and who served as an authority on all things Jewish for the Russian bureaucracy, warned against the practice. In several provinces, governors heeded his call, making Jewish shopkeepers and traders display their names prominently. In an effort to forestall all efforts to shed the social stigma and legal prohibitions that came with a Jewish name in this setting, the tsarist bureaucracy even created an archive of Jewish names to facilitate scrutiny of various documents.[25]

Subjects marked by Jewishness could thus be dangerous whether they wanted to remain Jews or not. By the 1880s, moreover, the tsarist police had firmly established a connection between the Jews and the revolutionary movement that had taken up violence against the regime. As the Moscow chief of police put it, 'The very people who resist a transition to the peaceful program are the Jews, who recently have been quietly attempting to grasp the initiative of the revolutionary movement into their hands.'[26]

As elsewhere in Europe, Russian antisemites at the turn of the century associated Jews with all kinds of threats to the nation. Critics portrayed them as simultaneously too insular and too cosmopolitan. They were capitalists and leftist revolutionaries. Sexually depraved, they preyed on Gentile girls by masterminding the practice of 'white slavery'. Purveyors of alcohol, they exploited the simple Russian peasants' love of drink—and were therefore responsible for inciting them to anti-Jewish violence in the pogrom. 'Not only were the Russian Jews associated with the unleashing of popular passions', writes Laura Engelstein, 'but their own desire was a source of suspicion and fear'.[27] Antisemites imagined Russians locked in a demographic war with Jewish enemies who, by peddling pornography, contraception and abortion, sought to unleash sexual desire while reducing the Russian birthrate.[28]

Over time the tsarist secret police, in particular, would associate both Jewish and Muslim populations with conspiracy in the early twentieth century, but they imagined their paths to the destruction of the empire to be quite different. Thus Muslims, too, faced scrutiny for ostensibly posing a threat to the Russian nation. But here critics tended to place the accent on Islam as an impediment to their integration. Moreover, in places such as the eastern province of Ufa, local church authorities interpreted the Muslim majority as a mortal danger to the Orthodox Christians in the region. In fact, in 1891 the bishop of Ufa cautioned that the more than one million Muslims (along with hundreds of thousands of other non-Orthodox) were 'all enemies of Orthodox Christians and Christ's Church'. Christians in the area were, in his eyes, like 'the martyrs of the first centuries of Christianity'.[29]

Whereas such critiques of tsarist Muslim subjects voiced dire expectations about Muslims committing violence against Christians, actual rebellions, though relatively few and far between and quite localised, seemed to Russian commentators to be part of a much wider and geographically expansive pan-Islamist campaign centred in Istanbul (and by some accounts in Kabul or elsewhere). Once backed by the Ministry of the Interior and local police, Muslim clerics, in particular, came in for acute criticism by the end of the nineteenth century. As the purveyors of a supposedly 'fanatical' ideology, they repeatedly faced the charge, particularly from Orthodox churchmen, that they were undermining converts to Christianity (especially those from families that had once been Muslim or animist) and that they were stalling the march of Russian-language schools and other mechanisms of integration and acculturation. For their part, the police pointed to conspiracies, often supposedly led by itinerant preachers and mystics, whose aim was to promote the Ottoman cause by stoking Muslim violence against the tsarist order. For the Russian revolutionaries on the far left, though, insofar as they thought about Muslims as political actors, they eventually folded Muslims into the wider body of the 'oppressed' whose supposed zeal could be unleashed in a broader anticolonial struggle against the tsars and imperial rulers everywhere, and whose faith might be reinterpreted as a prelude to socialism. Despite such rhetoric, Muslims remained integrated into the tsarist institutional landscape, retaining important positions in the official Islamic clerical hierarchies, the military, and in various provincial and local administrative and commercial bodies.[30]

At the same time, and much as in our own era, antisemitism and hostility towards Islam and Muslims were ideologies that transcended borders. Russia gave the world the imagery of the pogrom and the tragic figure of the Jewish conscript. Yet Europe played a key role in shaping how Russians formed antipathies towards Jews and Muslims alike. From the scholarly to the sensationalist, books and articles by Western European authors on topics ranging from the 'Blood Libel' and 'white slavery' to the biography of the Prophet Muhammad and *jihad* circulated in Russian reading circles. The Russian reading public consumed numerous antisemitic texts read throughout Europe, Otto Weininger's *Sex and Character* (1903) among them. Nevertheless, some antisemitic tropes—for instance, about their innate criminality or their ubiquitous exposure to syphilis—did not find adherents in Russia.[31] For Russian scholars of Islam, French and British writing about North Africa and the Indo-Afghan frontier were of particular interest. European descriptions of 'wild mountaineers' and 'mad mullahs' resonated with Russian audiences. Yet Russian readers generally, and Russian officialdom in particular, remained confident that the legacy of experience and lengthy contact with Muslims had endowed tsarist authorities with a hand that was more deft in dealing with the 'Muhammadans' than their European counterparts and rivals. Thus it would be simplifying matters to say that Russian clerics, scholars, journalists or officials passively absorbed Western writing about Jews and Muslims. Russian thinkers drew inspiration from, but also engaged with and reworked, such material. Russia produced its own university-based Orientalist scholars, skilled in philology and the translation of ancient texts, alongside race- and sex-obsessed popularisers of anti-Jewish and anti-Muslim stereotypes in the popular press.[32]

Russia parted ways with Europe again on the eve of the First World War and, more dramatically, during the first months of the conflict. In 1911, the Ministry of Justice charged Mendel Beilis, a Jewish artisan, with ritual blood murder, a move that strengthened the antisemitic credentials of the monarchy and its backers who had mobilised in the wake of 1905 in the name of a militant brand of Russian nationalism. While the Beilis trial further alienated many critics of the regime, it found support in influential circles. Indeed, in 1911–12 a body of the landed gentry, the Congress of the United Nobility, along with many senior military figures and other officials, wanted to expel Jews from the army. By this time, the Ministry of War too had become an institutional centre of antisemitism.[33] When the war broke out, the authorities charged numerous Jews with espionage and executed at least one hundred of them early in the war. 'From the kikes take everything' was among the instructions sent by the Fourth Army to units assigned with requisitioning.[34] During the war, they bore the brunt of deportations from the frontlines. The state removed roughly a quarter of a million Jews, and many more fled. Spontaneous anti-Jewish violence broke out as troops were mobilised.[35] In Grodno, Volynia and Minsk provinces, attackers raped Jewish women, desecrated synagogues and cemeteries, and ransacked homes and property. Arson was another of their tactics. During the occupation of Galicia, the Russian army also attacked local Jews. During the civil war that followed the collapse of the Romanov dynasty in February 1917 and the Bolshevik seizure of power in October 1917, perhaps as many as two hundred thousand Jews suffered at the hands of men who joined in the pogroms against them.

To the east, in tsarist Central Asia, the First World War also strained relations between Muslims and the state. When some Muslim communities rebelled against the authorities' attempts to draft them into labour battalions in 1916, massive reprisals followed. As in the western borderlands, the civil war would unleash violence that was informed by what were now decades of dehumanising rhetoric about Muslims and Islam.

Rooted in the anxieties of nationhood and European modernity, antisemitism and Islamophobia in the tsarist empire were key components of Russian nationalist ideology. Born at a moment of dramatic transformation that upended the status quo throughout the empire, they grew out of many of the same uncertainties among Russia's cultural and political elites about social change and the coherence of the Russian nation. Though these ideas emerged as part of Russian thinkers' contact with contemporary European ideas, the Russian context was crucial to the elaboration of such politics in the empire. What made them so lethal, though, was the breakdown of all kinds of social and political constraints on the mobilisation of violence, by diverse actors who ranged across the social spectrum, against people whom elites in Russian society had cast as enemies.

Notes

1. On shifting images of Jews and Muslims in the Russian empire and the transformations brought about by imperial rule, see Yuri Slezkine, *The Jewish Century* (Princeton, NJ and Oxford: Princeton University Press, 2004); and Robert D. Crews, *For Prophet and Tsar: Islam and Empire in Russia and Central Asia* (Cambridge, MA: Harvard University Press, 2006) (Slezkine 2004; Crews 2006).
2. Elena I. Campbell, *The Muslim Question and Russian Imperial Governance* (Bloomington: Indiana University Press, 2015) (Campbell 2015).
3. Crews, *For Prophet and Tsar*; *idem.*, 'Empire and the Confessional State: Islam and Religious Politics in Nineteenth-Century Russia', *American Historical Review*, 108, no. 3 (February 2003), 50–83; and Paul Werth, *The Tsar's Foreign Faiths: Toleration and the Fate*

of Religious Freedom in Imperial Russia (Oxford: Oxford University Press, 2014) (Crews 2003; Werth 2014).

4. See, for example, 'The Tale of Events beyond the Don' and other texts in Serge A. Zenkovsky, ed. and trans., *Medieval Russia's Epics, Chronicles, and Tales* (New York: E. P. Dutton and Co., 1963) (Zenkovsky 1963).
5. Andreas Kappeler, *The Russian Empire: A Multiethnic History*, trans. Alfred Clayton (Harrow, UK: Longman, 2001), pp. 21–4 (Kappeler 2001).
6. Robert D. Crews, 'Muslim Networks, Imperial Power, and the Local Politics of Qajar Iran', in *Asiatic Russia: Imperial Power in Regional and International Contexts*, ed. Uyama Tomohiko (London: Routledge, 2012), pp. 174–88; and Eileen Kane, *Russian Hajj: Empire and the Pilgrimage to Mecca* (Ithaca, NY: Cornell University Press, 2015) (Crews 2012; Kane 2015).
7. Crews, *For Prophet and Tsar*, chapters 5–6. See also James H. Meyer, *Turks across Empires: Marketing Muslim identity in the Russian-Ottoman Borderlands, 1856–1914* (Oxford: Oxford University Press, 2014); and Michael Reynolds, *Shattering Empires: The Clash and Collapse of the Ottoman and Russian Empires, 1908–1918* (Cambridge: Cambridge University Press, 2011) (Meyer 2014; Reynolds 2011).
8. See the very valuable survey in Benjamin Nathans, 'Jews', in *The Cambridge History of Russia*, vol. II: *Imperial Russia, 1689–1917*, ed. Dominic Lieven (Cambridge: Cambridge University Press, 2006), pp. 184–201 (Nathans 2006).
9. Olga Litvak, *Conscription and the Search for Modern Russian Jewry* (Bloomington and Indianapolis: Indiana University Press, 2006), pp. 15–16 (Litvak 2006).
10. Ibid., p. 18.
11. Ibid., p. 21.
12. See Ellie Schainker, 'Jewish Conversion in an Imperial Context: Confessional Choice and Multiple Baptisms in Nineteenth-Century Russia', *Jewish Social Studies*, 20, no. 1 (Fall 2013), 1–31 (Schainker 2013).
13. See John Klier, *Imperial Russia's Jewish Question, 1855–1881* (Cambridge: Cambridge University Press, 1995) (Klier 1995).
14. See Crews, *For Prophet and Tsar*, chapters 2–3.
15. Nathans, 'Jews', p. 194.

16. On state mediation of relations between indigenous Jews and Muslims in Central Asia, see Robert D. Crews, 'Islamic Law, Imperial Order: Muslims, Jews, and the Russian State', *Ab Imperio*, no. 3 (2004), 467–90 (Crews 2004).
17. See Richard S. Wortman, *Scenarios of Power: Myth and Ceremony in Russian Monarchy*, 2 vols. (Princeton, NJ: Princeton University Press, 1995) (Wortman 1995).
18. Michael F. Hamm, *Kiev: A Portrait* (Princeton, NJ: Princeton University Press, 1993), pp. 126–34 (Hamm 1993).
19. I.E. Alekseev, *Chernaia sotnia v Kazanskoi gubernii* (Kazan: Izdatel'stvo 'DAS', 2001), pp. 138–76 (Alekseev 2001).
20. Ronald Grigor Suny, *The Baku Commune, 1917–1918: Class and Nationality in the Russian Revolution* (Princeton, NJ: Princeton University Press, 1972) (Suny 1972).
21. Richard Stites, *Serfdom, Society and the Arts in Imperial Russia: The Pleasure and the Power* (New Haven, CT: Yale University Press, 2005), pp. 259, 278, 302 (Stites 2005).
22. Ibid., p. 378.
23. Ibid., pp. 123–5.
24. Ibid., pp. 389–96.
25. Eugene M. Avrutin, *Jews and the Imperial State: Identification Politics in Tsarist Russia* (Ithaca, NY and London: Cornell University Press, 2010), pp. 147–79 (Avrutin 2010).
26. Ibid., pp. 163–4.
27. Laura Engelstein, *The Keys to Happiness: Sex and the Search for Modernity in Fin-de-Siècle Russia* (Ithaca, NY: Cornell University Press, 1992), pp. 299–304 (Engelstein 1992).
28. Ibid., pp. 305–6.
29. *Ufimskie eparkhial'nye vedomosti*, 1 January 1891, pp. 17 and 19.
30. Charles Steinwedel, *Threads of Empire: Loyalty and Tsarist Authority in Bashkiria, 1552–1917* (Bloomington: Indiana University Press, 2016); and Robert D. Crews, 'The Russian Worlds of Islam', in *Islam and the European Empires*, ed. David Motadel (Oxford: Oxford University Press, 2014), pp. 35–52 (Steinwedel 2016; Crews 2014).
31. Engelstein, *The Keys to Happiness*, pp. 306–7.
32. See David Schimmelpennick van der Oye, *Russian Orientalism: Asia in the Russian Mind from Peter the Great to the Emigration* (New Haven, CT and London: Yale University Press, 2010);

Austin Jersild, *Orientalism and Empire: North Caucasus Mountain Peoples and the Georgian Frontier, 1845–1917* (Montreal and Kingston: McGill-Queen's University Press, 2002); Robert P. Geraci, *Window on the East: National and Imperial Identities in Late Tsarist Russia* (Ithaca, NY: Cornell University Press, 2001); Yuri Slezkine, *Arctic Mirrors: Russia and the Small Peoples of the North* (Ithaca, NY: Cornell University Press, 1994); and Jeffrey Brooks, *When Russia Learned to Read: Literacy and Popular Literature, 1861–1917* (Princeton, NJ: Princeton University Press, 1985) (Schimmelpennick van der Oye 2010; Jersild 2002; Geraci 2001; Slezkine 1994; Brooks 1985).

33. Yohanan Petrovsky-Shtern, *Jews in the Russian Army, 1827–1917: Drafted into Modernity* (Cambridge: Cambridge University Press, 2009), chapters 5 and 7 (Petrovsky-Shtern 2009).
34. Oleg Budnitskii, 'Shots in the Back: On the Origin of the Anti-Jewish Pogroms of 1918–1921', in *Jews in the East European Borderlands: Essays in Honor of John D. Klier*, ed. Eugene M. Avrutin and Harriet Murav (Boston, MA: Academic Studies Press, 2012), pp. 187–201, here, 193 (Budnitskii 2012).
35. Eric Lohr, *Nationalizing the Russian Empire: The Campaign against Enemy Aliens during World War I* (Cambridge, MA: Harvard University Press, 2003) (Lohr 2003).

References

Alekseev, I.E. 2001. *Chernaia sotnia v Kazanskoi gubernii*. Kazan: Izdatel'stvo 'DAS'.

Avrutin, Eugene M. 2010. *Jews and the Imperial State: Identification Politics in Tsarist Russia*. Ithaca, NY and London: Cornell University Press.

Brooks, Jeffrey. 1985. *When Russia Learned to Read: Literacy and Popular Literature, 1861–1917*. Princeton, NJ: Princeton University Press.

Budnitskii, Oleg. 2012. Shots in the Back: On the Origin of the Anti-Jewish Pogroms of 1918–1921. In *Jews in the East European Borderlands: Essays in Honor of John D. Klier*, ed. Eugene M. Avrutin and Harriet Murav. Boston, MA: Academic Studies Press.

Campbell, Elena I. 2015. *The Muslim Question and Russian Imperial Governance*. Bloomington: Indiana University Press.

Crews, Robert D. 2003. Empire and the Confessional State: Islam and Religious Politics in Nineteenth-Century Russia. *American Historical Review* 108 (3), February: 50–83.

———. 2004. Islamic Law, Imperial Order: Muslims, Jews, and the Russian State. *Ab Imperio* (3): 467–490.

———. 2006. *For Prophet and Tsar: Islam and Empire in Russia and Central Asia*. Cambridge, MA: Harvard University Press.

———. 2012. Muslim Networks, Imperial Power, and the Local Politics of Qajar Iran. In *Asiatic Russia: Imperial Power in Regional and International Contexts*, ed. Uyama Tomohiko. London: Routledge.

———. 2014. The Russian Worlds of Islam. In *Islam and the European Empires*, ed. David Motadel. Oxford: Oxford University Press.

Engelstein, Laura. 1992. *The Keys to Happiness: Sex and the Search for Modernity in Fin-de-Siècle Russia*. Ithaca, NY: Cornell University Press.

Geraci, Robert P. 2001. *Window on the East: National and Imperial Identities in Late Tsarist Russia*. Ithaca, NY: Cornell University Press.

Hamm, Michael F. 1993. *Kiev: A Portrait*. Princeton, NJ: Princeton University Press.

Jersild, Austin. 2002. *Orientalism and Empire: North Caucasus Mountain Peoples and the Georgian Frontier, 1845–1917*. Montreal and Kingston: McGill-Queen's University Press.

Kane, Eileen. 2015. *Russian Hajj: Empire and the Pilgrimage to Mecca*. Ithaca, NY: Cornell University Press.

Kappeler, Andreas. 2001. *The Russian Empire: A Multiethnic History*. Trans. Alfred Clayton. Harrow, UK: Longman.

Klier, John. 1995. *Imperial Russia's Jewish Question, 1855–1881*. Cambridge: Cambridge University Press.

Litvak, Olga. 2006. *Conscription and the Search for Modern Russian Jewry*. Bloomington and Indianapolis: Indiana University Press.

Lohr, Eric. 2003. *Nationalizing the Russian Empire: The Campaign against Enemy Aliens during World War I*. Cambridge, MA: Harvard University Press.

Meyer, James H. 2014. *Turks across Empires: Marketing Muslim identity in the Russian-Ottoman Borderlands, 1856–1914*. Oxford: Oxford University Press.

Nathans, Benjamin. 2006. Jews. In *The Cambridge History of Russia, Vol. II: Imperial Russia, 1689–1917*, ed. Dominic Lieven. Cambridge: Cambridge University Press.

Petrovsky-Shtern, Yohanan. 2009. *Jews in the Russian Army, 1827–1917: Drafted into Modernity*. Cambridge: Cambridge University Press.

Reynolds, Michael. 2011. *Shattering Empires: The Clash and Collapse of the Ottoman and Russian Empires, 1908–1918*. Cambridge: Cambridge University Press.

Schainker, Ellie. 2013. Jewish Conversion in an Imperial Context: Confessional Choice and Multiple Baptisms in Nineteenth-Century Russia. *Jewish Social Studies* 20 (1), Fall: 1–31.

Schimmelpennick van der Oye, David. 2010. *Russian Orientalism: Asia in the Russian Mind from Peter the Great to the Emigration*. New Haven, CT and London: Yale University Press.
Slezkine, Yuri. 1994. *Arctic Mirrors: Russia and the Small Peoples of the North*. Ithaca, NY: Cornell University Press.
———. 2004. *The Jewish Century*. Princeton, NJ and Oxford: Princeton University Press.
Steinwedel, Charles. 2016. *Threads of Empire: Loyalty and Tsarist Authority in Bashkiria, 1552–1917*. Bloomington: Indiana University Press.
Stites, Richard. 2005. *Serfdom, Society and the Arts in Imperial Russia: The Pleasure and the Power*. New Haven, CT: Yale University Press.
Suny, Ronald Grigor. 1972. *The Baku Commune, 1917–1918: Class and Nationality in the Russian Revolution*. Princeton, NJ: Princeton University Press.
Werth, Paul. 2014. *The Tsar's Foreign Faiths: Toleration and the Fate of Religious Freedom in Imperial Russia*. Oxford: Oxford University Press.
Wortman, Richard S. 1995. *Scenarios of Power: Myth and Ceremony in Russian Monarchy*. Princeton, NJ: Princeton University Press.
Zenkovsky, Serge A., ed. and trans. 1963. *Medieval Russia's Epics, Chronicles, and Tales*. New York: E. P. Dutton and Co.

CHAPTER 5

The End of the Semites

James Renton

[E]verything was against us in our secret partnership; time itself—for this could not go on forever.

—*Joseph Conrad, 'The Secret Sharer' (1910)*

By the end of the nineteenth century, the idea of the Semites had become the principal manifestation in Western European thought of the Christian tradition of linking Judaism and Islam, the Jew and the Muslim. This concept posited that both religions belonged to a single race, which was bound by its own family of languages and the product of a unique geographical space: Western Asia. Since Edward Said described 'the Islamic branch' of Orientalism as the 'strange, secret sharer of Western anti-Semitism' in 1978, a significant body of scholarship has been produced on the idea of the Semites—though it pales in comparison to the explosion of writing on Orientalism more broadly.[1] Yet even within the specialised field of post-Semitic studies, as we might call it, few have examined precisely when and why the idea ceased to be common currency in Western thought—for fallen from grace it surely has.[2] The end of the explicit use and naming of the Semitic category, however, is of enormous significance

J. Renton (✉)
Edge Hill University, Ormskirk, UK

J. Renton, B. Gidley (eds.), *Antisemitism and Islamophobia in Europe*, DOI 10.1057/978-1-137-41302-4_5

for understanding the trajectory of the relationship between Islamophobia and antisemitism; it is the start of the story in which the European ideas of the Jew and the Muslim splintered into two separate sides of global politics: the West versus the Islamic East.

This chapter will argue that the pinpointing of when the Semite disappeared and the context of that precise moment, which has been the focus of previous scholarship, explains very little. Instead, it will be contended that the possibility of the collapse of the Semitic category was present from its very beginnings at the end of the eighteenth century, and we need to attend closely to that point of departure. The idea of the Semites was very much of the Enlightenment. The quest for origins, scientific classification and rigid boundaries led to this coalescence into a singular category of what often had been only implicit epistemological connections between Christian conceptions of the two religions. However, for all of the scientific impulses that demanded and shaped the form of the Semite idea, it will be argued that its intellectual content derived from a medieval Christian cosmology, rooted in the Bible. For this reason, the Jew and Judaism were much more significant in the Semitic concept than the Arab and Islam, and were perceived very differently. While Islam was, of course, one of the Abrahamic monotheisms and, unlike Judaism, a universalist religion that competed with Christianity, at least in theory, for adherents, it did not feature in Christian theology. Conversely, Judaism and the Jew were integral to Christian conceptions of the past, present and future of the world. This asymmetry made the Semitic idea vulnerable if at any point European understandings of the Jew and the Arab were pulled in different directions. That moment arrived, the chapter will argue, after the First World War when the British and French empires, particularly the former, attempted to move the Semitic idea from the intellectual to the political realm. Policy-makers deployed the Semite as the conceptual basis for a new *political* geography for Western Asia in place of the Ottoman empire. As with much of post-1918 colonial Asia, this European attempt at geopolitical re-engineering led to violent conflict—in this case, between Arab and Jew. These developments began the process that saw the disappearance of the Semite as a mainstream notion within 25 years. The end of the Semites belongs in part, therefore, to the history of the end of the European empires, and this chapter is intended as a contribution to the incorporation of anti-Jewish and anti-Muslim racisms into our understanding of that subject.[3] To a much greater extent, though, this analysis seeks to add to the growing body of work that places religion and Christianity at the centre of the history of European Orientalism and antisemitism,[4] and, more widely, the scholarship that is challenging the secularism of post-Enlightenment Europe.[5]

This analysis of the end of the Semites is by necessity transnational and transcontinental, bridging Europe and Western Asia. Before its demise, the concept was as established in parts of the western littoral of Western Asia and Egypt as it was in France, the German-speaking lands and Britain. One of the striking features of the process by which the Semitic idea fell out of circulation was the degree of its influence in these spaces, and more, right up until its sudden decline. How could such a prominent and well-established aspect of global historical, linguistic and scientific thought slip away so quickly and comprehensively? The answer is to be found in the hidden fissure at the heart of the idea's foundations, a structural fragility that derived from the nature of its inception.

Origins

The idea of the Semites was based on the Bible's genealogy of the peoples of the world, which begins with the aftermath of the Flood and its surviving human remnant, Noah and his family, as set out in the book of Genesis, chapters 9 to 11. 'And the sons of Noah that went forth of the ark', the Bible states, 'were Shem, and Ham, and Japheth ... These *are* the three sons of Noah: and of them was the whole earth overspread.'[6] The chronicle explains that Noah cursed Canaan, son of Ham, and blessed 'the LORD God of Shem'. He went on to prophesize that 'God shall enlarge Japheth, and he shall dwell in the tents of Shem', suggesting a special bond between their descendants.[7] The line of the sons of Shem, who were thus the peoples of God, led, the chronicle tells us, to Abraham, the patriarch of the children of Israel and the descendants of Ishmael.[8] Evidently, this account of human origins is primarily a genealogical schema that places at its centre the notion of lineage, and a biological division of the world's population into familial groups: 'These *are* the families of the sons of Noah, after their generations, in their peoples [*begoyehem*]: and by these were the peoples [*ha-goyim*] divided in the earth after the flood.'[9] However, the Bible story also territorialises its ontology of ethnological separation—it frames geography as an integral, though secondary, defining feature of a people's essence. As part of this picture, we are told that the descendants of Shem lived from 'Mesha, as thou goest toward Sephar, unto the mountain of the east'.[10]

The third aspect of the Bible's paradigm for understanding human origins and separation, alongside history and geography, is language. This element is introduced in the penultimate verse of chapter 10, in which the three components of peoplehood are identified together: 'These are the sons of Shem, after their families, after their languages [*lilshonotam*], in their lands, after their peoples.'[11] Then, chapter 11 steps back in time, to before the descendants of Shem possessed their own languages, and tells the story of the tower of Babel. 'And the whole earth was of one language, and of one speech', it begins. Yet in punishment for humanity's attempt to construct a city with a tower that would reach into heaven, in order to 'make us a name', God 'did there confound' their language and 'scattered' them across the earth. Significantly, the Babel story finishes a third of the way through chapter 11, which then goes back to the lineage of Shem; it is at this point that the narrative details the genealogical link between him and Abraham.[12] The Bible thus places language and its evolution at the centre of the history of human origins and the antecedents of the patriarch of the chosen people.

The biblical narrative of human origins had a critical influence on medieval European world-views, and the interest in, and conception of, Western Asia. Neither Jewish nor Christian scholars knew of the location of the landmarks of the zone inhabited by the peoples of Shem, as specified in Genesis. Nevertheless, European Christian thought took on the general notion of a territorially delimited and defined people of Shem, along with the Bible's general schema of the evolution of humanity. In the sixth century, the Spanish monk Isidore of Seville translated the Bible framework into a global cartography, in which the different branches of humanity descended from the sons of Noah were located in separate geographical zones: Japhet in Europe, Ham in Africa and Sem (Shem transliterated into Latin) in Asia.[13] For the most part, however, medieval thinkers interested in Asia referred to Western Asia as the Orient, and judged there to be a unity of what St Jerome called Oriental languages.[14] Hence, scholars of Asia were known as Orientalists, not Semiticists.

The pre-modern Orientalist was chiefly interested in the study of ancient languages, with the goal of coming closer to the truth of the Old Testament. Accepting the notion of an original language before Babel, St Augustine established the Christian orthodoxy that it was Hebrew,[15] an idea accepted as self-evident fact more than a thousand years later by the English explorer Sir Walter Raleigh, among many others.[16] The Reformation and the ensuing religious conflict that engulfed Western and Central Europe led to the flowering of Orientalist scholarship: a competition to decipher the original word of God. Early modern Orientalism was, in short, primarily an exercise in philological exegesis.[17]

The concept of the Semite was born when this tradition collided with the Enlightenment; specifically, it was the product of the development of comparative philology. Crucially, however, the aim and the context remained the Bible. From the middle of the eighteenth century, scholars began increasingly to study Arabic and cognate languages in order to further their understanding of Hebrew. Protestant universities in German-speaking lands became the most important sites of this scholarship. In particular, Professor of Oriental Studies at the University of Göttingen, Johann David Michaelis, and his students played an important role.[18] And it was one of Michaelis' former students, August Ludwig von Schlözer, who in 1781 was the first to use, and define, the term 'Semitic' as a label for the language of a singular people in Western Asia that included 'Arabs' and 'Hebrews'. In his multi-volume world history, von Schlözer had already written in 1771 of 'the Semites' possessing one language, in the context of his discussion of the Phoenicians. And his colleague at Göttingen, Johann Christoph Gatterer, referred to 'the Semites', 'Semitic lands' and 'Semitic tribes' in his own 'universal history' published in the same year.[19] However, it was in von Schlözer's article, 'Von den Chaldäern', published a decade later, that he explained the 'Semitic':

> When the world was young (until Cyrus) there were not many languages, and even fewer peoples, or vice versa.
>
> From the Mediterranean to the Euphrates, and from Mesopotamia down through Arabia, only one language prevailed, as is generally known. So the Syrians, Babylonians, Hebrews, and Arabs, were one people. Also the Phoenicians (Hamites) spoke this language, which I would like to call Semitic; they had, however, only learnt it at their frontiers.[20]

Von Schlözer's first suggestion of the Semitic stemmed from the long-standing Christian epistemology of human origins, which was underscored by the fact that he went on to cite Genesis chapter 10, verse 11, in his discussion. The notion of one people descended from Shem, connected by language and inhabiting a discrete geographical zone, were all familiar elements from Genesis. In addition, von Schlözer refers to Hamites on the border of Shem, and goes on to discuss a people to the north of the Semites, which, citing Moses (!) and the pioneering early Enlightenment scholar Gottfried Wilhelm Leibniz, he calls Japhetic. As intimated by the reference to Leibniz, von Schlözer's Semitic was not just the result of a medieval biblical tradition of thought; the Enlightenment innovations are also striking. The Semitic label evidently belonged to the global intellectual project of identifying fixed, inter-linked taxonomic categories that were applied to entire geographical zones across time, and unveiling uniform characteristics of populations and their evolution in these spaces. However, for our purposes, the biblical substance of the Semitic concept—the theological content of the Enlightenment form—is particularly important for understanding its inner fragility.

While the thrust of Orientalist comparative philology was to look outwards from the language and 'chosen people' of the Old Testament, that text remained the central concern. Hence, the journal in which von Schlözer published his article, edited by another former student of Michaelis, Johann Gottfried Eichhorn, was entitled *Repertorium für biblische und morgenländische [Oriental] Litteratur*. In addition, the second major piece of scholarship to use the term Semitic was a study not of West Asian languages or history, but of Hebrew poetry, published in 1782. *The Spirit of Hebrew Poetry* was penned by Johann Gottfried Herder, the hugely influential historian, Bible scholar and Lutheran clergyman.[21] Herder did not consider Hebrew to be humanity's first language, the primordial *Ursprache*, and thus diverged from Augustinian tradition. However, he did judge Hebrew to be one of its ancient offshoots and, critically, the paradigm of all poetic expression.[22] It is of no small significance that Herder deployed the concept of the Semitic in the service of this very particular biblical preoccupation. He might have argued that Hebrew poetry—the Hebrew Bible—was ultimately the product of human agency,[23] but he also maintained that behind the human lay the divine creator. And, crucially, like the Semitic idea itself, Herder's fascination with the Old Testament as a basis for human culture spoke of a marked continuity with the theological universe of pre-Enlightenment Europe.

The Christian theological centre of the Semite and the Semitic is most strikingly apparent in the work of Ernest Renan, one-time Chair of Hebrew at the Collège de France, whose immensely successful career stretched from the 1850s until his death in 1892. Many scholars have argued that Renan did more than any other figure to develop and popularise in Europe the notion of the Semite as a racialised category that was essentially apart from its opposite, the Aryan, or, its synonym in Renan's lexicon, the linguistic category Indo-European.[24] In the literature on Renan's racial ideas, particular attention has been paid to his first major work, *Histoire générale et système comparé des langues sémitiques*. Published in 1855, this book set out Renan's argument that the Semites possessed a linguistic unity and, critically, distinctive characteristics, or rather a 'Semitic spirit', which separated them from Indo-Europeans. Renan was clear that he was concerned with making 'judgements about races'.[25] Yet as Robert D. Priest has argued, Renan's understanding of race was much more nuanced and difficult to pin down than many scholars have given credit.[26] In his *Histoire générale*, Renan argued for the limitations of the influence of race, and pointed to the example of contemporary 'Israelites' who had surpassed their Semitic heritage thanks to the effects of 'civilisation':

> [T]he primordial influence of race, as immense a part that it plays in the dynamics of human affairs, is offset by a crowd of other influences, which sometimes seem to overcome or even smother [*étouffer*] entirely that of blood. How many Israelites today, who are descended directly from the ancient inhabitants of Palestine, have nothing of the Semitic character, and are only modern men, swept along and assimilated by this great force superior to race that we call civilisation![27]

'All assertions on the Semites', Renan continued, 'entail similar reservations.'[28] Language, he argued, was more important than 'race' in shaping identity. And as a philologist, Renan was convinced that the study of language held the key to unlocking the inner secrets of culture and history. Language, in his assessment, ultimately shaped society. No Indo-Europeans or Semites, he argued, could come from, or descend to, a state of savagery with their level of linguistic, and therefore cultural, sophistication. Renan wrote that religion came second to language in separating Aryans and Semites—they were, after all, linguistic categories.[29] However, it was religion that lay at the heart of Renan's conception of the Semite, and that sustained his fascination with the subject. It would be a profound error to judge that he had a secular approach to his research due to his abandonment of his early studies for the Catholic priesthood, and his disenchantment with the Catholic Church. In the preface to the first volume of his *Histoire du peuple d'Israël*, published in 1887, he wrote: 'The true God of the universe, the unique God, the one that we adore … is there for eternity. It is the certainty of having served, in my way … this excellent cause, which inspires in me an absolute confidence in the divine good.' As to his assessment of this magnum opus on ancient Israel, which spanned from the beginnings of the Semitic tribes until the eve of Jesus, he judged: 'I am sure that overall I have understood well the unique achievement that the Breath of God, that is to say the soul of the world, was realised through Israel.'[30] This history of Judaism, the final volume of which was published posthumously, was the last step in Renan's career-spanning preoccupation with deciphering the origins of Christianity, which included a seven-volume history that began with the publishing sensation *Vie de Jésus* in 1863.[31] Christianity, he argued, 'is the realisation … [,] the goal, the final cause of Judaism … the masterpiece of Judaism, its glory, the summation of its evolution'. Indeed, Renan judged that it was 'by Christianity that Judaism truly conquered the world'.[32] There is evidence to suggest that he thought of a Semitic-Aryan/Jewish-Christian symbiosis. He did use the term 'Judeo-Christianity'.[33] And he contended that Indo-Europeans—the racial basis of white, Christian Europe—sprang from the same space as the Semites: the region of the 'Belourtag' or Hindu Kush mountains;[34] Semites and Aryans were thus like twins.[35] They were *both* white, and together constituted the highest level of racial civilisation. In their reciprocal relationship, they gave the world morality and religion: Aryans developed philosophy and science, and Semites produced religion—that is to say, true religion in the shape of Abrahamic monotheism.[36] Nevertheless, Renan did not view Semites and Aryans as equals, or as inhabiting the same civilisational time frame. He possessed a diachronic and supersessionist conception of racial civilisational development: the Semitic moment ended with the coming of Christianity, which, in the hands of the Aryans, or Indo-Europeans, took up the baton of progress.[37] Certainly, Renan asserted in the *Histoire générale* in 1855 that Islam was the crown of the Semitic oeuvre, as a simplification of the human spirit.[38] However, he judged that this phase of Semitic evolution paled in comparison to the foundational achievements of ancient Israel, at least as he saw them by the end of his life, and the revolution of Christianity.

Renan argued in the *Histoire du people d'Israël* that what Greece had been for intellectual culture, Rome was for politics, and the nomadic Semites were for religion.[39] In the *Histoire générale*, he stated that the Semites were 'the people of God and the people of religions'.[40] He meant by this assertion that the Semites alone had produced, and only they were capable of producing, the Abrahamic monotheism that led to Christianity. The principal developments in this story, by his account, were the establishment by a cadre and culture of prophets, or what he called 'prophetism', of the concept of a single universal God, and morality. In this process, ancient Israel was the central actor. What he called 'the Arab form or Islamism' (Arab was synonymous with Islam; it was a religious category) was, in his analysis, a pure manifestation of the 'Semitic spirit', alongside the 'Hebraic' or 'Mosaism'.[41] However, as he put it, Judaism undertook all of the work of religion for humanity: Christianity and Islam were mere offshoots.[42] On the opening page of the *Histoire du people d'Israël*, Renan argued that for those interested in origins, only three histories were of principal interest: those of Greece, Israel and Rome.[43] Although he displayed a similar level of interest in Arab and Jew in his landmark comparative study of Semitic languages in 1855, by the time of his final work the Semite was largely synonymous with ancient Israel and Judaism. And Renan did not stand alone. As Maurice Olender has observed, nineteenth-century Orientalists commonly applied the characteristics that they saw in the ancient Hebrews to all Semites, and the Hebrew language, which was identified with the monotheistic religion of the Israelites, often shaped the questions asked of the Semitic.[44]

Renan did not like the term Semitic, which he attributed to Eichhorn. He argued that some of the peoples who had spoken 'Semitic' languages, such as Phoenicians and certain Arab tribes, were not descendants of Sem, as set out in Genesis chapter 10. And others who did belong to that lineage spoke non-'Semitic' languages, he claimed. Instead of Semitic, Renan preferred the appellation '*syro-arabes*', but he used Semitic nonetheless because, for all of his reservations, it was a simple and conventional term. Semitic had become by the middle of the nineteenth century part of the terminological furniture of Orientalism. Yet despite his scepticism about the label of Sem, his analysis was deeply influenced by the Genesis framework—'this precious document'. For Renan, the sense of that text and the name Sem was ultimately geographical: it referred to 'the middle zone of the earth, without distinction of race'.[45] Geography was absolutely central to his explanation of the Semitic religious achievement of Abrahamic monotheism. The isolated nomadic tribal life of the tent, of the desert, in Western Asia produced, he argued, the culture of simplicity and purification that enabled the birth of Judaism and then Islam.[46]

Antisemitism

The biblical basis of the Semitic idea, whether expressed in terms of geography, language or lineage, not only gave Judaism and Israel a privileged position in Orientalist studies of the ancient past such as Renan's; it also resulted in a focus on Jews in the application of the Semitic label to the contemporary world. As mentioned earlier, Renan did not judge nineteenth-century European Jewry to be Semitic—they had, he believed, transcended Semitic civilisation. Yet this opinion was not shared, of course, by those who developed the racial pseudo-sciences of the second half of the century, or the agitators of the pan-European popular politics of Judeophobia in the same period.[47] The invoking of the term 'antisemitism' as a banner for an anti-Jewish politics from the end of the 1870s applied what had been an academic Orientalist concept for understanding ancient human origins to an exercise in internal European racial colonialism. For 'antisemites', the Semitic label functioned as a means of Orientalising contemporary European Jewry as a foreign body that was produced in Western Asia, could not co-exist with the Aryan and had to be removed. This notion was contemporaneous with increasingly rigid racial boundaries and preoccupations with purity, corruption and spatial separation in the European colonial world.[48] In addition, the biblical origins of the term meant that it provided a shorthand that linked the Jew to theological Christian anti-Jewish mythology.

Said's suggestion that 'Islamic Orientalism' was a 'strange, secret sharer' of 'Western antisemitism' misses the specificity of these developments.[49] European Orientalists included both Jew and Arab in the Semitic category. However, only Jews were, or could be, the subject of political antisemitism, and not just because of their location within Europe. Antisemitism in fin-de-siècle Europe could not be a secret; it had to be performed and sated, publicly—it was a mass political movement. We can certainly see powerful connections between European notions of the Jew and Arab/Muslim, without which the Semitic category could not have made sense, as Said's work helped us to understand, along with important scholarship that has followed. Yet Judaism and the idea of the Jew possessed a primacy in European Christian society that Arabs and Islam did not, due their central place in Christian Europe's politico-theology. And this ideational significance and difference were reflected within the Semitic idea itself because, at base, it was a Christian theological construct.

The Reach of the Semitic

The imbalance within the idea of the Semites meant that it possessed an inherent fragility. If European notions of Jews and Arabs were pulled in different directions, they could be prised apart. Before the First World War, however, the Semitic held fast as the frame for understanding the origins and substance of Western Asia. Although the European antisemitism movement had no concern with the Arab/Muslim, the link between the Jew and the Orient served a useful function, as already mentioned. Hence, antisemitic and Orientalist ideas of the Semite could and did co-exist. Prior to 1914, European societies did not evince a significant active interest in the contemporary Arab/Muslim that might problematise the relationship with the Jew. For all of the interest of Orientalist scholars in the Arab/Muslim by the end of the nineteenth century, many of whom had moved far away from the *explicitly* theological concerns of Michaelis and Herder to an ethnographic interest, most saw a figure frozen in time, outside of contemporary history.[50] This conception was shared in wider Western European intellectual culture, and in society at large, where, in sharp contrast to the Jew, the Western Asian Arab hardly registered as a contemporary subject.[51]

It would be difficult to overestimate the currency of the Semitic by 1914: it was the dominant frame for thinking about Western Asia in Western Europe. As testimony to this framing, the frontispiece of one of the most influential works in Britain on the historical geography of the Holy Land, written by George Adam Smith and first published in 1894, was a map of 'The Semitic World' (Fig. 5.1); the map of Palestine was tucked away at the back of the book.[52] When General Allenby was appointed head of the British military campaign for Jerusalem in 1917, the year after Smith was knighted, Prime Minister David Lloyd George gave him a copy of this work. Allenby is said to have read the book, which was in its sixteenth edition by 1910, almost daily alongside the Bible.[53] At the end of the war, when the world order had been turned upside down and the Ottoman empire destroyed, Lloyd George and his colleagues turned to Smith's *Atlas of the Historical Geography of the Holy Land*, published in 1915, as their reference work for consideration of the borders for the new state of Palestine.[54] The first map in the atlas was 'The Semitic World', which included a submap of 'Jewish Babylonia', reflecting the Jewish kernel of the Semitic at that moment—though it is worth noting that the submap in *The Historical Geography* was 'The Babylonian Empire'. In the

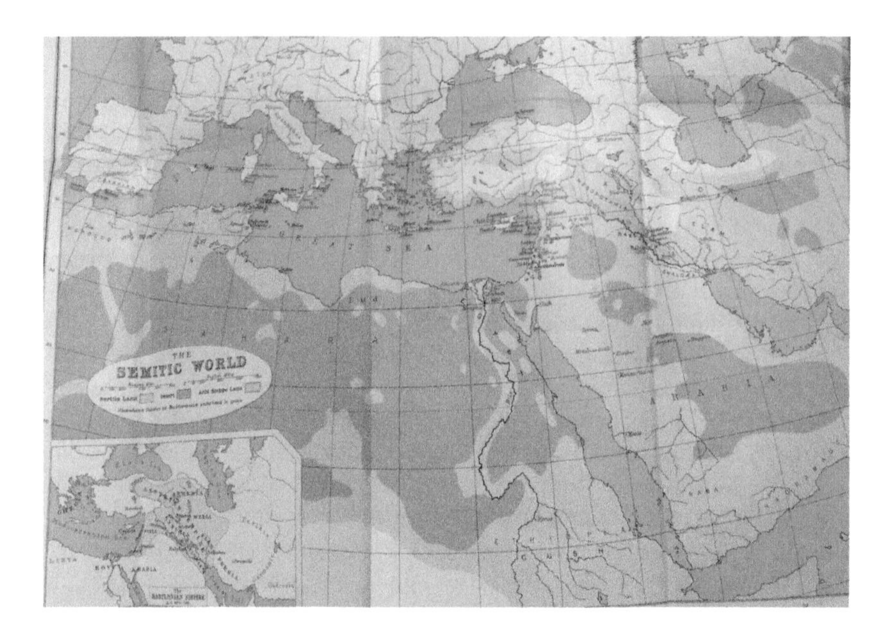

Fig. 5.1 Frontispiece of George Adam Smith, *The Historical Geography of the Holy Land* (London: Hodder and Stoughton, 16th edn, 1910)

explanatory notes in the *Atlas*, Smith commented that Israel belonged to 'the Semitic race', for whom the Arabian 'peninsula and the deserts obtruding from it upon Syria have been from time immemorial their breeding ground and proper home'.[55] In map 6 and its accompanying note, he set out the biblical basis for the Semitic category and its geography: 'The World and its Races According to the Old Testament' (Fig. 5.2).[56]

Renan's oeuvre remained essential reading. In his memoirs, Ronald Storrs, the British Orientalist who was appointed military governor of Jerusalem in December 1917, recalled that, aside from some professional and personal dealings with Jews prior to his position in Palestine, his sum 'knowledge of Jewry' was confined to 'the Old Testament (Psalms almost by heart) and Renan's *Histoire du Peuple d'Israel* [sic]'. Storrs read Renan again in Jerusalem. He recorded in his diary: 'a little out of date, but very stimulating: not very popular with the Jews, who dislike (for instance) Abimelech being described (rightly) as a worshipper of Moloch. Renan

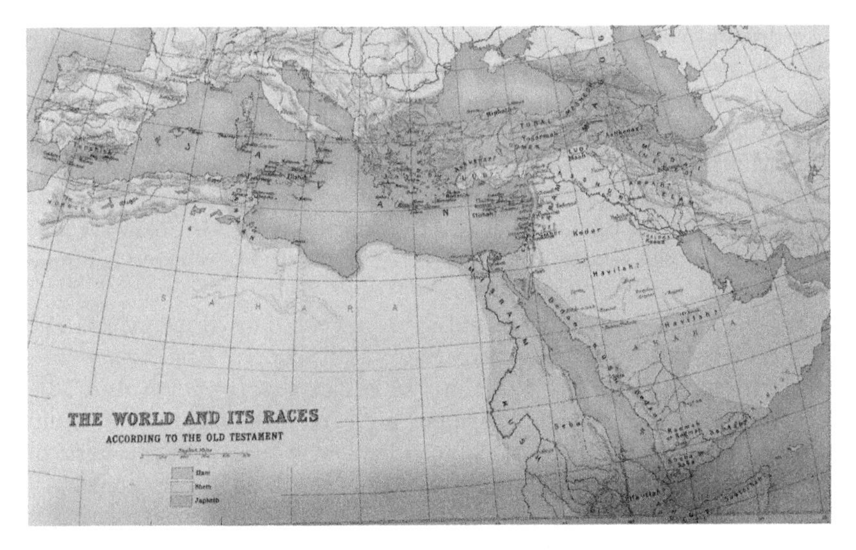

Fig. 5.2 George Adam Smith, *Atlas of the Historical Geography of the Holy Land* (London: Hodder and Stoughton, 1915), p. 6

himself venerates the Patriarchs and the Prophets, but appears to dislike all between them.'[57]

The Political Semite

It was the First World War, however, that witnessed the zenith of the idea of the Semitic, as it passed from the intellectual fields of historical geography, linguistics and ethnography to that of the political: state-building and governance. The Semite for the first time became a *political* actor and category, not merely an object of study or domination. In 1916, the British and French empires agreed to dismantle the Ottoman Empire after the war. The aim was to secure their imperial interests in the region while also fulfilling a pledge to support Arab national independence in exchange for an anti-Ottoman Arab revolt, led by Sharif Hussein of Mecca.[58] By the beginning of 1917, British and French planners had also begun to consider inserting a Jewish nation alongside the Arab area as the basis for a constitutionally and cartographically re-engineered Western Asia.[59]

The British Conservative MP and influential Middle East adviser Sir Mark Sykes was at the forefront of this re-thinking of the future for the region, along with François Georges Picot, the former French consul-general in Beirut and agent of the French colonial party. Sykes, who became Britain's main Middle East liaison with the French government, did not aim simply to re-draw borders and establish zones of influence. Rather, he looked to fill a conceptual vacuum that would result from the Ottoman absence (if left to his own devices, Georges Picot would have preferred a simple European imperial replacement[60]).

Prior to 1914, the British and French governing elites viewed the Ottoman empire as a composite state, and Western Asia as a political unity that belonged to that imperial edifice—*la Turquie d'Asie*.[61] By 1917, Sykes, in particular, felt the need for a complete scheme that would replace that state: a vision that hung together in tune with the spirit of the time—the principle of nationality. As per much of British intellectual culture in this period, Sykes was captivated by racial nationalist thought. He believed that the nation constituted the historical basis for stable and civilised political organisation.[62] Sykes judged that the war had changed the norms of global politics, away from colonialism and imperialism towards self-determination. Nationality, he believed, had to be the new basis for a Western Asia without the Ottomans. Sykes did not think that the region was ready for complete national independence; like the majority of his colleagues in Britain and France, he assumed that Western Asia needed European, specifically British and French, supervision due to its stage of racial development. But Sykes was emphatic that the nation had to be the building block for a re-constituted 'Asiatic Turkey'.[63]

In addition to their considerations as to which political ideas and structures would best suit the region, Sykes and Georges Picot had to pay close attention to how they would sell their post-Ottoman future—to the peoples of the region, and to the world. The imperial elites of the First World War saw and projected their conflict as a struggle over ideas, and they were as preoccupied with global public opinion, and its shaping, as they were with the battlefield. The post-Ottoman future had to make sense and inspire as a concept, if it was to be accepted by its populations and the West. From this perspective, the rise of the notion of self-determination from 1917 became the dominant concern. Armed with a narrative of inherent Ottoman despotism and brutality, the Entente presented Arabs and Jews as sharing a history of national oppression, which the British and French governments claimed to be bringing to an end. They were, their policy-makers contended, waging a war of liberation. Hence, Sykes worked to associate the revival of the Jewish and Arab nations with that of the Armenians to the north in Anatolia, whose slaughter in 1915 at the hands of the Ottoman state was used by the Entente as their pre-eminent icon of Ottoman barbarism. After the Balfour Declaration of 2 November 1917, which pledged British governmental support for a national home in Palestine for the Jewish people, Whitehall policy-makers publicly spoke of Armenian, Arab and Jewish national emancipation as a war aim. The government exploited its global propaganda apparatus and collaborated with its nationalist partners and allies to project the story of a new epoch for Western Asia.[64] The tripartite alliance was not only rhetoric. In preparation for the building of the new national political architecture, and the tripartite alliance that would be its foundation, Sykes strove to bring together Armenian, Jewish and Arab representatives in Britain, France and Egypt, and established a joint committee in London to maintain 'common harmony' and work for the 'common cause'.[65]

For Sykes, the concept of the Semites provided a ready-made racial notion that rationalised a post-Ottoman sub-Anatolian Western Asia based on both Zionist and Arab nationalisms. The Semitic held the potential to be the racial thread that bound the post-Ottoman tapestry, as designed in London and acceded to in Paris. Soon after Sykes decided to include Zionism in his plans for the region in February 1917, he sought to promote the Semitic bond. In a meeting with Herbert Sidebotham, a journalist who belonged to the pro-Zionist British Palestine Committee, Sykes pointed to the 'important service' that its journal *Palestine* 'might render by emphasising the cultural connection between Jews and Arabs in the East'.[66] The Zionist leader Chaim Weizmann reported that Sykes was going to distribute the publication in Mecca.[67] At a public meeting in Manchester to celebrate the Balfour Declaration, Sykes said of the Arabs, to a chiefly Jewish audience, 'You know the Semite sleeps but never dies.' The pamphlet in which the speech was published, printed by the Zionist organisation and circulated covertly by the British government across the Jewish diaspora,[68] claimed that this statement was followed by 'Loud cheers'. Sykes went on to say that it 'was the destiny of the Jews to be closely connected with the Arab revival', emphasised the need for 'co-operation and goodwill from the first' and 'warned the Jews to look through Arab glasses'. 'We will, we will!' the audience were quoted as saying in response.[69] Just over a week later, the day after British imperial troops occupied Jerusalem, this speech was cabled to Cairo and published in Arabic press.[70] In Sharif Hussein's paper, *al-Qibla*, produced in Mecca and disseminated by British agents, the comment about the undying Semite was printed in large font.[71]

Zionist leaders in Britain understood that they were expected to promote the Arab–Zionist connection, as well as the alliance with Armenians. Writing to Sykes, Zionist executive member Nahum Sokolow remarked with obedient enthusiasm: 'your idea of an Arab-Armenian-Zionist Entente is excellent indeed'.[72] And at the Manchester meeting, Sokolow dutifully articulated the Semitic bond and its importance for the future of Western Asia:

> Our membership of the Semitic race, our title to a place in the civilisation of the world and to influence the world and take our share in the development of civilisation, have always been emphasised. If racial kinship really counts, if great associations exist which must serve as a foundation for the future, these associations exist between us and the Arabs. I believe in the logic of these facts.[73]

Earlier in the year, Sokolow held meetings with Syrians and Armenians in Paris to further the alliance project,[74] and in August 1917, Weizmann explained to a US colleague that the development of close relations with Arab and Armenian leaders was a present point of interest.[75] After the Balfour Declaration, Sokolow and Weizmann, acting jointly, pushed to develop ties in the United States and Egypt, and Weizmann became the Zionist representative on Sykes' London committee.[76] Following the British occupation of Jerusalem in December 1917, the government sent a Zionist Commission to the Holy Land led by Weizmann, a principal aim of which was to ameliorate relations with the Arab population. And in June 1918, Weizmann performed the apogee of the narrative of the political Semite when he travelled for approximately two weeks from Palestine to Aqaba, via the Red Sea, and then to Waheida near Ma'an in present-day Jordan, to meet with the head of the Arab Northern Army and third son of Sharif Hussein, Prince Faisal.[77] This meeting, which took place in the desert headquarters of the Arab army, produced an image that visually articulated Sykes' dream of the political Semite: Weizmann and Faisal standing next to each other both wearing a *keffiyeh*, a traditional headdress and quintessential Orientalist icon of Arabness (Fig. 5.3).

Yet despite the Zionist leadership's willingness to satisfy Sykes and his colleagues, they did not uniformly adopt the notion of the political Semite,

Fig. 5.3 Chaim Weizmann and Emir Faisal, June 1918. Courtesy of Yad Chaim Weizmann, Weizmann Archives, Rehovot, Israel

and behind closed doors expressed major reservations. Weizmann was particularly sceptical. In his speech to an Arab audience in Jerusalem in April 1918, he did not make any reference to the Semitic or even a prospective partnership. He did say that Jews, Arabs and Armenians shared 'the highest claim to a life of their own'. And he claimed that Zionists 'watched with deepest sympathy and profound interest the struggle for freedom which the ancient Arab race was now waging'. But he did not speak of a racial bond.[78] In his private correspondence, Weizmann referred to 'Arabs and Syrians' in the British Palestine administration as 'enemies', talked of Palestinian Arabs as 'the local baystrucks [bastards]' and told Foreign Secretary Arthur J. Balfour that 'there is a fundamental qualitative difference between Jew and Arab'.[79] After his speech in Jerusalem, Weizmann wrote to his wife, 'I feel that I do not need to concern myself with the Arabs any more; we have done everything that was required of us, we have explained our point of view publicly and openly: *c'est à prendre ou à laisser* [take it or leave it].'[80] Certainly, Weizmann saw the Arab national project outside of Palestine differently. He described Faisal as 'the first real Arab nationalist I have met', 'a leader!' and 'handsome as a picture!'[81] Faisal also had the benefit of being 'contemptuous of the Palestinian Arabs whom he doesn't even regard as Arabs!'[82] But Zionist ambivalence and even stark opposition in regard to Hussein and his family sat alongside such positive assessments. Israel Sieff, a young member of the British Zionist leadership who was Weizmann's secretary in Palestine in 1918, argued the previous year that Sykes had 'in short "sold us" to the Arabs, whose support is of much greater importance to him than that of the Zionists ... We cannot, and dare not, sell our just rights to make matters easier for the consummation of an Arab agreement ....' It was, Sieff argued, 'our holy duty to combat an Arab Palestine'. And if Zionist press articles about the future borders of Palestine 'harm the Arab kingdom', he argued, 'that is no concern of ours'.[83]

Among the leading Semitic partners designated for the Zionists by the British government—the Hussein family of the Hijaz—the position on the Arab–Zionist alliance was similarly ambivalent. Before the British promoted the idea of the political Semite in 1917, *al-Qibla* published an article criticising Zionism. Sykes took action via his colleagues in Cairo to put a stop to this opposition, and the newspaper later published a piece that praised Jewish settlement.[84] After the war, Faisal, who is said to have seen Zionism as a potentially important source of economic and political support for the Arab cause,[85] acceded to the invocation of the political Semite. On 3 January 1919, he signed an agreement with Weizmann for the implementation of the Balfour Declaration and Zionist assistance for the economic development of the expected Arab state. The first line of the text pointed to the Semitic as the starting point for their collaboration. As representatives of the Arab kingdom of Hijaz and the Zionist organisation, they were, the document declared, 'mindful of the racial kinship and ancient bonds existing between the Arabs and the Jewish people'. Faisal added, however, a handwritten note that made the pact entirely provisional on the implementation of his own demands for Arab independence in the peace settlement—demands that were not to be realised.[86]

Zionists and Arab nationalists were not the only ones to be unconvinced by the concept of political Semitic unity; British and French Orientalists working for their respective government's war machine also expressed profound concerns. Brigadier-General Gilbert Clayton, the Chief Political Officer in the Egypt Expeditionary Force, wrote to Sykes:

> It is an attempt to change ... the traditional sentiment of centuries. ... [A]s regards the Jew the Bedouin despises him and will never do anything else, while the sedentary Arab hates the Jew and fears his superior com-

> mercial and economic ability ... Whatever protestation Jews like Sokolow and Weismann [sic] may make and whatever Arabs, whom we may put up as delegates, may say, the fact remains that an Arab-Jewish entente can only be brought about by very gradual and cautious action. The Arab does not believe that the Jew with whom he has to do will act up to the high-flown sentiments which may be expressed at Committee meetings. In practice he finds that the Jew with whom he comes in contact is a far better business man than himself and prone to extract his pound of flesh. This is a root fact which no amount of public declarations can get over.[87]

Two months before Clayton's letter, the French Orientalist Louis Massignon, a second lieutenant working with the Franco-British–sponsored Arab Legion, wrote about the thirty-two 'Israelite Arabs' who had been recruited for that army in Baghdad. Such terminology could imply that Massignon conceived of an Arab–Jewish symbiosis. However, this was not the case. Instead, he doubted that it was possible to recruit 'patriotic Israelite Arabs' and that, as Zionism was opposed to the expansion of King Hussein into Syria, the movement was, 'in this capacity, an agent of moderation in the sense of the inter-confessional balance' in that country advocated by France. In other words, Massignon saw Zionism as a useful element to undermine Arab nationalism to further French imperial aims, rather than as its Semitic partner.[88] In Cairo, the French Minister, Albert Defrance, was entirely opposed to Allied backing for a Zionist–Arab partnership, which he thought would lead to serious problems for France in the Arab world.[89] Even Sykes' diplomatic partner, Georges Picot, wished to minimise France's role in publicly supporting Zionism, despite the imagined benefits of winning 'Jewish power' around the world.[90] The French government-funded Arabic newspaper *al-Mustaqbal* went so far as to publish articles that attacked Zionism.[91]

Unlike their British imperial colleagues, French Middle East policymakers had a precedent for a significant and actual 'Semitic' colonial relationship, as opposed to a fantasy, in North Africa. Following the Crémieux decree of 1870 that gave Jewish Algerians full French citizenship, the ethno-political division of Jew and Arab became colonial practice. It is perhaps in part for this reason that Georges Picot and other members of the French Middle East establishment were so averse to the whole project. In addition, the simple fact that the French imperial state governed a vast stretch of Arab lands meant that Orientalist scholarship was not their first reference point for imagining the future for Western Asia. It was not a coincidence that some of the staunchest British criticism of the concept of the political Semite worked in Egypt, the British empire's principal Arab territory.

If we probe further into Sykes' thinking about the project of an Arab–Jewish Middle East, we can see that his idea itself was based on divergent conceptions of the two 'Semites' that rendered it vulnerable. Sykes did not consider there to be parity between the Jew and the Arab in terms of their civilisational development, or inner character. He is reported to have said in March 1917: 'Although they [Arabs] were Nomads they were all civilized as he put it, at the back of their heads.' As per the Orientalist convention of medieval Arab sophistication and modern decline, he argued that 'the race had the habit, as it were, of efflorescence into high urban civilization which however, was rarely lasting'. The Arabs, he asserted, 'were not stayers, the Jews were'. The following month, he confessed to the head of the new civil administration in Baghdad, Sir Percy Cox, that the '[t]he idea of Arab nationalism may be absurd'. Yet, he continued, the British case at the post-war peace conference 'will be good if we can say we are helping to develop a race on nationalist lines under our protection'.[92]

In contrast, Sykes and other figures in the British government saw Zionism as a well-established movement that would help to bring civilisation to the backward Holy Land.[93] Moreover, the concept of the Jewish Return was deep-rooted in British Christian culture; it resonated. The usually dry Sir Edward Grey is said to have remarked to the Jewish Minister, Sir Herbert Samuel, in November 1914 that he held a 'sentimental attraction' to the idea of a Jewish state: the 'historical appeal was very strong'.[94] There was no equivalent historical Christian aspiration for Arab restoration.

Sykes considered the differences between the contemporary Jew and Arab to be so marked that persuasion and stage management were central to his effort to construct the political Semite. To Arabs, he stressed the political benefits of gaining international Zionism as a friend, and to Jews, he spoke of the strength and authenticity of the 'Arab revival'.[95] Sykes pleaded for 'co-operation and goodwill', or, he warned, ultimate disaster would overtake both Jew and Arab'.[96] Even for its chief proponent, therefore, the creation of the political Semite required tremendous work and faced many challenges; Sykes did not consider it to be the inevitable outcome of a simple racial connection and similarity. Nor should we expect him to have thought along such lines. As mentioned earlier, even Renan judged contemporary Jewry to be far removed from the Semitic Arab.

Given the major differences between the Arab and Jewish nations in the minds of British and French policy-makers alike, it should not be surprising that the turn to the Semitic as a political idea was not a reflexive response to the possible disappearance of the Ottoman imperial state. After the British correspondence with Hussein in late 1915, Sykes and Georges Picot signed in May 1916 what became their infamous plan for the region in the event of an Ottoman defeat. The Sykes–Picot agreement (Fig. 5.4) divided Western Asia into a blue area under French authority; a red space in British hands; zones A and B, which were to be independent

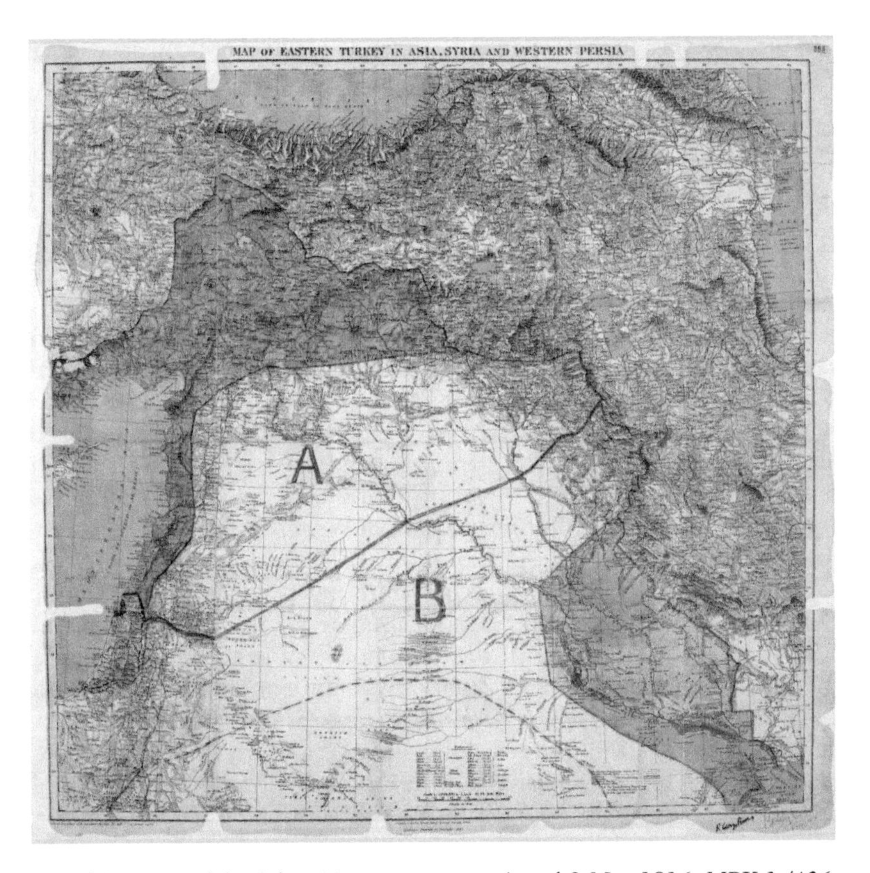

Fig. 5.4 Map of the Sykes–Picot agreement, signed 8 May 1916, MPK 1/426, The National Archives, Kew, UK

Arab areas 'protected' by France and Britain, respectively; and an internationally administered yellow territory (referred to as brown in British documentation) in the Holy Land. This cartography was predicated on a specifically *Arab* racial exclusivity—the only Western Asian population named in the agreement; the Semite was nowhere to be seen.[97]

The map followed the stipulation in the McMahon–Hussein correspondence that zones in Western Asia that were not 'purely Arab' would remain outside of the Arab independent area.[98] Sykes and Georges Picot almost included Zionism in their plan, but only as an afterthought in response to discussions in London aimed at gathering supposed Jewish power behind the Allies. In the event, French premier Aristide Briand would not accede to a pro-Zionist policy. Among other concerns, he was not convinced that Zionism would be compatible with the promised Arab state.[99] As a result, Jews, Zionism and the Semitic were kept out of the new cartography. This was a case of absence, not of erasure. The invention of the Semitic as a political category was counter-intuitive, and required the undoing of the Semitic idea's intellectual foundations. Let us recall that the Semitic emerged as a category to further the study of the politico-theological history of an ancient people—the Jews—whose genius, and agency, were thought to have been superseded by Christianity centuries past.

However, the inclusion of Zionism in British and French plans from early 1917, as the perceived urgency of enlisting Jewish power outweighed concerns about the Jewish place in the future of Western Asia, scuppered the racial unity of the original scheme. Sykes needed a new racial frame to explain the region's post-Ottoman future and make it work. This would not have been so much of an issue if the Entente only intended to give Jewish nationalism a small, fringe corner of Western Asia. Geographically speaking, this was the case. And constitutionally, the role allotted to Jews, as sketchy as it was, did not even feature independent statehood; the Balfour Declaration specified 'a national home for the Jewish people' in Palestine, but the meaning of this terminology was not defined during the war. The precise future of Zionism was not a priority for the British government, who saw Jewish nationalism principally as a weapon of warfare.[100] And yet, ideationally, the British government envisaged a major place for the Jews in the post-Ottoman future—well beyond any political and geographical limits—alongside the Arab nation. In large part, this prominence resulted from the conviction among Entente policy-makers that Jews wielded global political power. In British propaganda, much was to be made of the Jewish place in the political transformation of Asiatic

Turkey.[101] Hence, the Jews could not be sidelined. They had to be incorporated into the racial ontology of the region.

Despite all of the doubts and dissonances in the concept of a new Western Asia based on an Arab–Jewish partnership, British propagandists attempted to present a positive image of Zionist aims in their widely distributed Arabic media, including in occupied Palestine, in 1918.[102] For example, the Cairene paper *al-Kawkab*, secretly funded and distributed by British authorities, published the article 'Palestine and the Jews—by an Arab Son of Shem'. The author, who made reference to the 'Semitic Arab' and the 'Semitic Jew', argued that a Jewish state was incompatible with the Balfour Declaration, that the whole population of Palestine could benefit from Jewish agriculture, and advocated a joint commission of Jews on the one side, and Christians and Muslims on the other, to make decisions to secure the public interest.[103]

The government's investment in this kind of material, however, was small in comparison to its separate Zionist and Arab nationalist propaganda operations, which spoke of a new era of redemption for their respective movements.[104] Britain's Arab project included the creation of an Arab national flag that featured the colours of past Islamic Arab kingdoms, which was held aloft by Faisal's troops in celebration of Arab, not Semitic, liberation. British-sponsored Zionist ceremony included the laying of the foundation stones of the Hebrew University in Jerusalem, and processions of uniformed nationalist youth under banners adorned by the Star of David.[105] And when the British government secured the most compelling evidence of the beginning of the Semitic future—the meeting between Faisal and Weizmann—they stepped back. 'No Publicity', Sykes wrote on the file cover of Clayton's report.[106] The dream of the political Semite could be promoted as an aspiration, but its realisation was to be kept hidden. Sykes did not provide an explanation. Perhaps he did not consider that the Jewish and Arab worlds were ready, that the collaboration required for the success of the political Semite was such a departure that more time was needed.

Smiting the Semite

Despite this caution, the notion of the political Semite did not survive the transition from the idea factory of Whitehall to the reality of twentieth-century Palestine. Sykes and his British Orientalist colleagues had not viewed the Palestinian Arab as the Semitic partner for the Jews in the post-Ottoman future. The true blood of the Arab was to be found elsewhere, in the desert—the authentic Semitic space—in the shape of Faisal and the Hashemite royal family. British Orientalists saw the Holy Land as a place that could not be inhabited or possessed by authentic Arabs. The Christian conception of Palestine meant that only the Israelites and their descendants truly belonged in this landscape. Arabs were necessarily foreigners, and could not have a racial purity that came from a rooted attachment to the land. Hence, British government Middle East specialists, and many Western European Orientalists, commonly judged that much of the Palestinian population was, to quote a report by the Zionist Commission's Political Officer, 'of the most mixed race'.[107] In the Christian Orientalist mind, the Palestinian Arabs were a non-people, simply 'non-Jewish communities', as per the words of the Balfour Declaration and, in turn, the League of Nations Mandate for Palestine.[108]

This view of Palestinian Arab society was, however, incorrect. Inspired by the Allied message of Arab national freedom, a specifically Palestinian nationalist movement erupted in the first year of British colonial rule in the Holy Land. This political movement drew on the powerful currents of Palestinian Arab national identity that existed before the war.[109] By the end of 1918, the Palestinian elite had rallied around the newly formed Muslim-Christian Association that started in Jaffa and, alongside young intellectual groups such as the Arab Club and the Harvest of Studies in Jerusalem, demanded and expected Arab independence for *Filastin*. This statist nationalist political ideology mirrored the development of Zionism among the majority of its adherents, who, by 1918, expected to obtain Jewish national sovereignty over all of *Eretz Israel*.[110]

By the end of 1918, a political conflict had begun for national sovereignty between Palestinian Arabs and Zionists, which started the process of the undoing of the idea of the Semite. The conflict exposed the stark differences between European conceptions of the Jew and the Arab/Muslim. An early Palestinian Arab challenge to the notion of the political Semite towards the end of 1918 is instructive regarding its incompatibility with the conflict, but also the profound influence of the Semitic category up until that point. Towards the end of 1918, a Zionist pamphlet published in Jaffa deployed the Semitic argument as part of a case for Zionist colonisation, which also cited the backwardness and stagnation of the Arab population. The author, Elie Eberlin, a lawyer from Paris, argued that the Jews and Arabs were 'sister nations' who were 'two vigorous branches of the big Semitic tree, the shade of which had covered the ancient world'.[111]

British censors reported a Palestinian Arab response to Eberlin (they did not specify who, unfortunately), which declared: 'We don't deny that the Arabs and the Jews are two branches issuing from this tree and that they are united by kinship and language, but we do deny that there is anything else in common.'[112]

Deeply interested in European scholarship, members of the Palestinian Arab elite had long embraced the idea of the Semite.[113] The decision not to abandon or ignore the concept at this juncture, but to minimise its significance, indicates the extent of its currency at that time, as does the transcontinental field across which this exchange took place: Palestinian Arabs arguing with a Parisian Jew in Jaffa about the Semitic, reported, with different degrees of emphasis, by French imperial officials to Paris (focused on Eberlin's pamphlet)[114] and British military intelligence (concentrating on the Palestinian Arab reaction), whose account in a briefing on Palestine was sent on to London.[115]

The strategy of retaining but minimising the Semitic frame was still in evidence almost two decades later in Storrs' account of the conflict. Central European Jews were, he wrote:

> foreigners ... to the Arabs of Palestine, despite the oft-quoted Semitic bond of language—foreigners in all the essentials of civilization, and mainly Western both in their qualities and their defects. *Identity* of language is a bond: a common linguistic origin of several thousand years ago is no more than an academic fact.[116]

An alternative approach to minimising or displacing the Semitic was to re-configure its meaning. We can see this response in T.E. Lawrence's account of the Arab revolt against the Ottoman empire, *Seven Pillars of Wisdom*, which he completed in 1922 and revised for publication in 1926. In that book, Lawrence retained Renan's emphasis on the intrinsic relationship between the Semite and the environment of Western Asia.[117] He also used the term Semite synonymously with his main subject of interest. But unlike Renan, particularly at the end of his career, Lawrence's principal Semite was the Arab. Indeed, he almost removed the Jew entirely from the picture. Even for Lawrence, this conception marked a big shift from his earlier understanding of the Jew and the Orient. In 1909, he wrote from Beirut to his mother concerning a month that he had just spent in 'northern Palestine'. It is 'such a comfort', he told her, 'to *know* that the country was not a bit like this in the time of Our Lord. The

Renaissance painters were right, who drew him and his disciples feasting in a pillared hall, or sunning themselves on marble staircases.' Under Roman rule, Palestine was 'a decent country then, and could so easily be made so again'. Lawrence's answer was Zionism: 'The sooner the Jews farm it all the better: their colonies are bright spots in the desert.'[118] However, after the explosion of the Zionist–Palestinian conflict at the end of the war, Lawrence produced his narration of the Semitic in *Seven Pillars* with little reference to Jews except, for the most part, to mark them out as essentially different or even foreign. In contrast to dynamic Christianity and Islam, Jewry was, he wrote, 'unchanging',[119] and, unlike the vast majority of Semites, could exist outside of Western Asia. In Palestine itself, Jewish colonists 'had introduced strange manners, and strange crops, and European houses'.[120] Lawrence also criticised Sykes, the architect of the Jewish–Arab entente: 'He saw the odd in everything, and missed the even.'[121] Without explaining or even mentioning Sykes' plans, Lawrence wrote cryptically of his former colleague's 'awful realization of the true shape of his dreams' just before he died at the beginning of 1919, and his effort in Paris 'to say gallantly, 'I was wrong: here is the truth'.'[122] In Lawrence's account of the British government's promises about the future of Western Asia, he would not mention the Balfour Declaration by name, and described it disparagingly as a document 'to Lord Rothschild, a new power, whose race was promised something equivocal in Palestine'.[123] At one point, *Seven Pillars* seems to suggest that Zionists had ceased entirely to be Semites. In Renan's original account of the Semites in 1855, he specified that they lacked 'almost completely the capacity to laugh'.[124] Perhaps in tribute to Renan, Lawrence used the latter's Semitic criteria to debunk the existence of the Jewish Semite; damningly, he wrote of the 'watchmen at Zion's gate who drank beer and laughed in Zion because they were Zionists'.[125]

Two political developments in the 1930s signalled, however, that the Semite was not to be re-written or sidelined, but dropped as a concept from European intellectual culture, and the global political public sphere of which Palestine had become a part, following the creation of the League of Nations in 1919. Perhaps the longest-lasting testament to the splintering of the Semite was the 'two-state solution' to the Zionist–Palestinian conflict, which first emerged in international politics in 1937. Following the outbreak of the Palestinian uprising in 1936, the British Cabinet despatched a Royal Commission to examine the causes of the conflict. In its report, published the following year, the Commission recommended that the goal of co-existence should be abandoned. Jews and Arabs were fundamentally divided, racially: the former was of Europe, the latter of Asia. Though 'it is linked with ancient Jewish tradition', the Jewish national home of Mandate Palestine was, the commissioners argued, 'predominantly a culture of the West'. The Asian Palestinian Arabs, therefore, possessed 'little kinship' with the Zionist community.[126] The Semitic family was no more. 'There is', the report stated, 'no common ground between them'. The nationalism of the two communities was the greatest barrier to peace, but the separation went further: 'Their cultural and social life, their ways of thought and conduct, are as incompatible as their national aspirations.'[127] Indeed, the authors argued, 'It has long been obvious that the notion of a cultural "assimilation" between Arab and Jew is a phantasy.'[128] The Commission talked of Jews and Palestinians as two separate races, and contended that the only possibility for peace lay in territorial partition and racial separation, with a transfer of populations to solve the 'cardinal problem' of minorities living in what were to be racially defined states.[129]

The other major development that signalled the end of the Semite was in Nazi Germany. The Nazi state's geostrategic aim of gaining influence in the Arab world made the term 'antisemitism' extremely problematic. The Nazis' response, just two years after they had come to power, was for the Propaganda Ministry to tell the press to avoid using 'antisemitic' and 'antisemitism'. Once the Second World War began, the Nazi Office of Racial Politics itself backed the abolition of Semitic terminology. 'Anti-Judaism', Adolf Eichmann remonstrated on trial in Jerusalem in 1961, was the correct term.[130]

The ease with which Europe's racial state abandoned the Semitic laid bare the fragility of the idea that had existed since its inception at the end of the eighteenth century, and the fissure at its core. The Christian theological basis of the concept meant that for all of the apparent commonalities between the Semites, as conceived in Western European thought, the couplets of the Jew and Judaism and the Arab and Islam possessed very different meanings and levels of cultural importance. Up until the First World War, European intellectual culture was not exposed to circumstances that might apply pressure to the Jewish–Arab connection—though we can see a very different picture in the French colonial state due to the situation in Algeria after 1870. In Europe before 1914, the Semite was an abstraction, not a political reality, which made sense because of its Christian religious architecture. But as soon as Sykes attempted to transform the Semitic into a political category, the tension at the heart of the idea—the sharp division between Christian Europe's relationships with Judaism and Islam—started to come to the fore. From the start, the political Semite was bedevilled with opposition: among Zionists, Arabs and agents of the British and French imperial states. And once the new global public sphere, embodied in the League of Nations, was presented with the figure of the anti-Zionist Arab, the Semitic category ceased to make sense and began to unravel.

It would be a mistake to conclude that the conflict alone could have put an end to the Semites. This explanation fails to account for the ease with which the Semitic fell from favour and what followed—the splintering of an idea that was deeply embedded in European thought into two separate categories, Jew/Judaism and Arab/Islam, belonging to two opposing camps in the European understanding of the world: the West and the Orient. This history is only intelligible if we acknowledge that the Christian foundation of the Semitic idea created an imbalance in its apparent unity from the moment that the name was articulated by von Schlözer: the figures of the Jew and the Arab had been connected, but they had never been the same. It is for this reason that Renan could write of 'Judeo-Christianity'—not in opposition to the Semitic, but as part of his theological understanding of, and interest in, its history. This was the paradox of the Semites: the idea possessed tremendous cultural capital that derived from centuries of European thinking about human origins, but the source of that tradition and its power—the Christian Bible story—also held the seeds of the idea's destruction.

The Semite did not disappear without trace once Europeans stopped uttering the word. The concept derived from texts, traditions of interpretation and beliefs at the centre of a theological culture that arguably runs through the foundations of much of Europe's intellectual universe. By the mid-nineteenth century, the notion of the Semites traversed the national borders of Western Europe, and was as recognised in London as it was in Paris and Berlin, well beyond the confines of Orientalist scholarship. By 1914, the Semitic had moved into the intellectual culture of the eastern Mediterranean itself. It should not be surprising, therefore, that, as Gil Anidjar has observed, the shadow of the Semite, an unnamed category,[131] has continued to influence European and wider Western political thought about Jews, Judaism, Arabs and Islam—though this is an area that requires further research. The neat epistemological divide in post-Semitic European thought between a Western Jewry and an Asian Islamic Arab world is more apparent than real. When the Palestine Royal Commission reported that the Zionists of the Holy Land were predominantly European in 'outlook and equipment', they added the qualification, 'if not in race'.[132] In the time of the Semitic, Jews and Judaism were more significant for Christian European culture than Arabs and Islam, but it did not follow that either were, or could be, synonymous with the West. In addition, while the idea of the Jew was not the same as the concept of the Arab/Muslim, they were connected; otherwise the Semitic would have served no purpose, and possessed no influence. An acknowledgement of the inherent divisions within, and paradox of, the Semitic should not detract from the intellectual influence held by that category in the past, nor since its demise, even as an unnamed shadow.

Acknowledgements The research for this chapter was made possible by funding from the UK's Arts and Humanities Research Council and Edge Hill University. For research assistance, I would like to thank Harith Bin Ramli. For their help, I am also most grateful to Monica Gonzalez-Correa and Elke Weissmann.

Notes

1. Edward W. Said, *Orientalism: Western Conceptions of the Orient-Reprinted with a New Afterword* (London: Penguin, 1995), pp. 27–8, 286; Maurice Olender, *Les Langues du Paradis—Aryens et Sémites: un couple providential, Préface de Jean-Pierre Vernant* (Paris: Gallimard/Le Seuil, 1989); James Pasto, 'Islam's "Strange Secret Sharer": Orientalism, Judaism, and the Jewish Question', *Comparative Studies in Society and History*, 40, no. 3 (July 1998), 437–74; Gil Anidjar, *Semites: Race, Religion, Literature* (Stanford: Stanford University Press, 2008); Ivan Davidson Kalmar, 'Anti-Semitism and Islamophobia: The Formation of a Secret', *Human Architecture: Journal of the Sociology of Self-Knowledge*, 7, no. 2 (Spring 2009), 135–44; idem, 'Arabizing the Bible: Racial Supersessionism in Nineteenth Century Christian Art and Biblical Scholarship', in *Orientalism Revisited: Art, Land and Voyage*, ed. Ian Netton (London: Routledge, 2013), pp. 176–86; Abdellali Hajjat and Marwan Mohammed, *Islamophobie: Comment les élites françaises fabriquent le "problème musulman"* (Paris: La Découverte, 2013), pp. 184–90; Andrew N. Rubin, 'Orientalism and the History of Western Anti-Semitism: The Coming End of an American Taboo', *History of the Present*, 5, no. 1 (Spring

2015), 95–108 (Said 1995; Olender 1989; Pasto 1998; Anidjar 2008; Kalmar 2009, 2013; Hajjat and Mohammed 2013; Rubin 2015).

2. See, in particular, Anidjar, *Semites*, pp. 20–1, ch. 1.
3. On Islamophobia, see Martin Thomas, *Empires of Intelligence: Security Services and Colonial Disorder after 1914* (Berkeley: University of California Press, 2007); Priya Satia, *Spies in Arabia: The Great War and the Cultural Foundations of Britain's Covert Empire in the Middle East* (Oxford: Oxford University Press, 2008). On antisemitism and European empire, see Hannah Arendt, *The Origins of Totalitarianism* (Oregon: Harcourt, c. 1976), Part II (Thomas 2007; Satia 2008; Arendt 1976).
4. Suzanne L. Marchand, *German Orientalism in the Age of Empire: Religion, Race, and Scholarship* (Cambridge: Cambridge University Press, 2009); Ivan Kalmar, *Early Orientalism: Imagined Islam and the Notion of Sublime Power* (London: Routledge, 2012); Susannah Heschel, *The Aryan Jesus: Christian Theologians and the Bible in Nazi Germany* (Princeton, NJ: Princeton University Press, 2008) (Marchand 2009; Kalmar 2012; Heschel 2008).
5. For a recent discussion in this debate, see the contributions by Talal Asad and Gil Anidjar in *Critical Inquiry*, 41, no. 2 (Winter 2015).
6. Genesis 9: 18–19. All Bible quotations here are from the King James Version, with the exception of some words that I have translated differently, each of which are followed by a transliteration of the Hebrew original.
7. Ibid., 9: 25–27.
8. Ibid., 10: 21–31; 11: 10–27.
9. Ibid., 10: 32.
10. Ibid., 10: 30.
11. Ibid., 10: 31.
12. Ibid., 11: 10–32.
13. Kalmar, *Early Orientalism*, pp. 34–5.
14. Ernest Renan, *Histoire générale et système comparé des langues sémitiques* (Paris: Imprimerie impériale, 1855), p. 1, incl. n. 2 (Renan 1855).
15. Olender, *Les Langues du Paradis*, p. 13.

16. Marchand, *German Orientalism*, p. 12; David Nirenberg, *Anti-Judaism: The Western Tradition* (New York: W.W. Norton, 2013), p. 321 (Nirenberg 2013).
17. Marchand, *German Orientalism*, pp. 1–14.
18. See, in particular, ibid., ch. 1, and Jonathan M. Hess, 'Johann David Michaelis and the Colonial Imaginary: Orientalism and the Emergence of Racial Antisemitism in Eighteenth-Century Germany', *Jewish Social Studies*, 6, no. 2 (Winter 2000), 56–101 (Hess 2000).
19. Martin F.J. Baasten, 'A Note on the History of "Semitic"', in *Hamlet on a Hill: Semitic and Greek Studies Presented to Professor T. Muraoka on the Occasion of His Sixty-Fifth Birthday*, eds. M.F.J Baasten and W. Th. van Peursen (Leuven: Peeters, 2003), pp. 65–7, n. 35 (Baasten 2003).
20. August Ludwig Schlözer, 'Von den Chaldäern', *Repertorium für biblische und morgenländische Litteratur*, no. 8 (1781), 161 (Schlözer 1781).
21. Olender, *Les Langues du Paradis*, pp. 35–6, n. 70.
22. Ibid., pp. 68–77.
23. Ibid., p. 75.
24. Robert D. Priest, 'Ernest Renan's Race Problem', *The Historical Journal*, 58, no. 1 (2015), pp. 318–20 (Priest 2015).
25. Renan, *Histoire générale*, p. vii.
26. Priest, 'Renan's Race Problem', pp. 312–15, 322, 329–30.
27. Renan, *Histoire générale*, pp. vii–viii.
28. Ibid., p. viii.
29. Ibid., p. 468.
30. Ernest Renan, *Histoire du peuple d'Israël. Tome premier* (Paris: Calmann-Lévy, 1887), pp. xxviii–xxix (Renan 1887).
31. See Henriette Psichari (ed.), *Oeuvres complètes de Ernest Renan* (Paris: Calmann-Lévy, 1947) (Psichari 1947).
32. Ernest Renan, *Histoire du peuple d'Israël. Tome cinquième* (Paris: Calmann-Lévy, 1893), pp. 414–15 (Renan 1893).
33. Ibid., p. 412.
34. Renan, *Histoire générale*, pp. 465, 475.
35. Ibid., pp. 465, 470.
36. Ibid., pp. 463–4.

37. On the ideas of the Semite and the Aryan as a racial expression of Christian supersessionist thought, see Kalmar, 'Arabizing the Bible'.
38. Renan, *Histoire générale*, p. 475.
39. Idem, *Histoire du peuple d'Israël. Tome Premier*, p. 26.
40. Idem, *Histoire générale*, pp. 3–4.
41. Ibid., p. 13.
42. Ernest Renan, *Histoire du peuple d'Israël. Tome deuxième* (Paris: Calmann-Lévy, 1889), pp. ii–iii (Renan 1889).
43. Idem, *Histoire du peuple d'Israël. Tome premier*, p. i.
44. Olender, *Les Langues du Paradis*, pp. 36–7.
45. Renan, *Histoire générale*, p. 2.
46. See esp. ibid., pp. 470–1; idem, *Histoire du peuple d'Israël. Tome premier*, pp. 13–26, 42–3, 61–2.
47. See Arendt, *Origins of Totalitarianism*, Pt 1; George L. Mosse, *Toward the Final Solution: A History of European Racism* (New York: H. Fertig, 1978) (Mosse 1978).
48. Tony Ballantyne and Antoinette Burton, *Empires and the Reach of the Global, 1870–1945* (Cambridge, MA: Harvard University Press, 2012), pp. 29–40 (Ballantyne and Burton 2012).
49. Said, *Orientalism*, pp. 27–8.
50. Ibid., chs. 1–2, p. 234.
51. See, for example, Nicholas Tromans (ed.), *The Lure of the East: British Orientalist Painting* (London: Tate, 2008); Lynne Thornton, *Les Orientalistes: peintres voyageurs, 1828–1908* (Paris: ACR edition/Pochecouleur, 1993) (Tromans 2008; Thornton 1993).
52. George Adam Smith, *The Historical Geography of the Holy Land*, 16th edn. (London: Hodder and Stoughton, 1910) (Smith 1910).
53. Roza M.L. El-Eini, *Mandated Landscape: British Imperial Rule in Palestine, 1929–1948* (London: Routledge, 2004), p. 38, n. 160; John D. Grainger, *The Battle for Palestine 1917* (Woodbridge: Boydell Press, 2006), p. 83 (El-Eini 2004; Grainger 2006).
54. Henry Laurens, *La Question de Palestine. Tome premier, 1799–1922, L'invention de la Terre sainte* (Paris: Fayard, 1999), p. 487 (Laurens 1999).

55. George Adam Smith, *Atlas of the Historical Geography of the Holy Land* (London: Hodder and Stoughton, 1915), p. xi (Smith 1915).
56. Ibid., pp. xii–xiii.
57. Ronald Storrs, *Orientations* (London: I. Nicholson & Watson, 1937), p. 359 (Storrs 1937).
58. Vincent Cloarec, *La France et la question de Syrie: 1914–1918* (Paris: CNRS, 2010), ch. 3 (Cloarec 2010).
59. James Renton, *The Zionist Masquerade: The Birth of the Anglo-Zionist Alliance, 1914–1918* (Basingstoke: Palgrave Macmillan, 2007), chs. 3–4 (Renton 2007a).
60. Christopher M. Andrew and A.S. Kanya-Forstner, *France Overseas: The Great War and the Climax of French Imperial Expansion* (London: Thames and Hudson, 1981), pp. 89–90 (Andrew and Kanya-Forstner 1981).
61. Cloarec, *La France et la question de Syrie*, chs. 1–2; James Renton, 'Changing Languages of Empire and the Orient: Britain and the Invention of the Middle East, 1917–1918', *The Historical Journal*, 50, no. 3 (2007), 646 (Renton 2007b).
62. Renton, *Zionist Masquerade*, pp. 15–16, 28–9.
63. Sykes, 'Memorandum on the Asia-Minor Agreement', 14 August 1917, Sykes Collection, Middle East Centre Archive, Oxford (MECA) (Sykes 1917).
64. Renton, 'Changing Languages'.
65. Shmuel Tolkowsky diary, 7 August 1917, Central Zionist Archives, Jerusalem, Israel (CZA) A248/2; Sykes to Clayton, 14 November 1917, and draft of Sykes to Clayton, 16 November 1917, Sykes Collection, MECA; Chaim Weizmann to Aaron Aaronsohn, 16 January 1918, no. 66, in Dvorah Barzilay and Barnet Litvinoff (eds.), *The Letters and Papers of Chaim Weizmann*, Vol. VIII, Series A (New Brunswick, NJ: Transaction, 1977), p. 55, incl. n. 7 (Tolkowsky diary 1917; Sykes to Clayton 1917; Weizmann to Aaronsohn 1918).
66. Memorandum by Herbert Sidebotham, 9 March 1917, Weizmann Archives, Rehovot, Israel (WA) (Sidebotham 1917).
67. Tolkowsky diary, 7 March 1917, CZA A248/2.
68. *Great Britain, Palestine and the Jews: Jewry's Celebration of its National Charter* (London: Zionist Organisation, 1918), no. 952, Wellington House Schedule, Wellington House Collection,

Department of Printed Books, Imperial War Museum; 'Report of Meeting of Propaganda Commitee, 14 December–30 June 1918', CZA Z4/243 (Zionist Organisation 1918; Zionist propaganda committee (London) 1918).
69. *Great Britain, Palestine and the Jews*, pp. 39–40.
70. Sokolow and Weizmann to Jacques Mosseri, 11 December 1917, no. 22, in Barzilay and Litvinoff, *Letters and Papers*, p. 20 (Sokolow and Weizmann to Mosseri 1917).
71. 'Resolutions are Measured Against Those Who Make Them', *al-Qibla*, Issue 151, 18 Shawwal 1336/31 January, 1918, pp. 1–2 (Resolutions 1918).
72. Sokolow to Sykes, 28 March 1917, Sledmere Papers, copy, WA (Sokolow to Sykes 1917).
73. *Great Britain, Palestine and the Jews*, p. 51.
74. Report on Sokolow and J.A. Malcolm mission to Paris, Samuel Landman papers, CZA A226/30/1 (Sokolow and Malcolm report n.d.).
75. Weizmann to Felix Frankfurter, 17 August 1917, no. 479, in Leonard Stein (ed.), *The Letters and Papers of Chaim Weizmann, Vol. VII, Series A* (Jerusalem: Israel Universities Press, 1975), p. 489 (Weizmann to Frankfurter 1917).
76. Weizmann and Sokolow to Aaron Aaronsohn, 16 November 1917, no. 8 and to Jacques Mosseri, 11 December 1917, no. 22, in Barzilay and Litvinoff, *Letters and Papers*, pp. 7, 20, incl. ns. 2–3; Sykes, on behalf of the joint committee, to the Syria Welfare Committee, Cairo, 15 February 1918, The National Archives, Kew, UK (TNA) Foreign Office records (FO) 371/3398/27647 (Weizmann and Sokolow to Aaronsohn 1917; Sykes to the Syria Welfare Committee 1918).
77. Weizmann to Vera Weizmann, 26 May 1918, no. 207, Barzilay and Litvinoff, *Letters and Papers*, p. 197 and n. 6; and 17 June 1918, no. 213, p. 211.
78. 'Jewish National Movement. The Commission in Palestine. Jews, Arabs, and Armenians. Statements by Dr. Weizmann', *Jewish Chronicle*, 10 May 1918, p. 11 (Jewish National Movement 1918).
79. Weizmann to Vera Weizmann, 20 May 1918, no. 200, and to Arthur J. Balfour, 30 May 1918, no. 208, in Barzilay and Litvinoff, *Letters and Papers*, pp. 190, 191, 202 (Weizmann to Vera Weizmann, 20 May 1918; Weizmann to Balfour, 30 May 1918).

80. Weizmann to Vera Weizmann, 30 April 1918, no. 181, in ibid., p. 171 (Weizmann to Vera Weizmann, 30 April 1918).
81. Weizmann to Vera Weizmann, 17 June 1918, no. 213, in ibid., p. 210 (Weizmann to Vera Weizmann, 17 June 1918).
82. Ibid.
83. Sieff to Weizmann, 19 February 1917, WA (Sieff to Weizmann 1917).
84. A.L. Tibawi, *Anglo-Arab Relations and the Question of Palestine, 1914–1921* (London: Luzac, 1978), p. 162; 'Wisdom is the Legacy of the Believer, He Has a Right to It Wherever He Finds It', *al-Qibla*, 13 Sha'ban 1336/23 May 1918, p. 1 (Tibawi 1978; Wisdom is the Legacy of the Believer 1918).
85. Clayton to the Secretary of State for Foreign Affairs, 1 July 1918, F.R. Wingate Papers, Sudan Archive, Durham University Library, UK (SAD.) 149/1/3 (Clayton to the Secretary of State for Foreign Affairs 1918).
86. Faisal-Weizmann agreement, 3 January 1919, London Bureau Papers, CZA Z4/40065 (Faisal-Weizmann agreement 1919).
87. Clayton to Sykes, 15 December 1917, F.R. Wingate Papers, SAD. 147/3/92 (Clayton to Sykes 1917).
88. Massignon, Louis, to G. Maugras: Report on the state of French propaganda among the Arab Legion, 26 October 1917, le Centre des Archives diplomatiques de Nantes, France, 294PO/B/149 Jerusalem Consulat General, Beyrouth: 2367- Correspondence of Georges Picot, 1917–1919, sub-folder 10. (Massignon to Maugras 1917).
89. See, for example, Defrance, Albert, to Ministry of Foreign Affairs (MFA), 17 January 1918, Le Centre des Archives diplomatiques de La Courneuve (CADC), Guerre 1914–1918, Sionisme V, Vol. 1200, January–February 1918, sub-folder 2. (Defrance to Ministry of Foreign Affairs 1918).
90. Georges Picot, François, to MFA, 27 November 1917, CADC, Guerre 1914–1918, Sionisme, Vol. 1199, folder 2, sub-folder 1. (Georges Picot to MFA 1917).
91. See, for example, 'Ali al-Ghayati, 'Here and There', *al-Mustaqbal*, no. 90, 30 December 1917, p. 2 (al-Ghayati 1917).
92. Sykes to Sir Percy Cox, 23 May 1917, Sykes collection MECA (Sykes to Cox 1917).

93. David Vital, *Zionism: The Crucial Phase* (Oxford: Clarendon, 1987), p. 292 (Vital 1987).
94. Sir Herbert Samuel note, 9 November 1914, Herbert Samuel papers, copies, Parliamentary Archives, UK (Samuel note 1914).
95. Sykes to Syria Welfare Committee, 15 February 1918, TNA FO 371/3398/27647; memorandum on Sykes meeting with Zionists, 7 February 1917, p. 15, CZA A226/30/1 (Sykes meeting with Zionists 1917).
96. *Great Britain, Palestine and the Jews*, p. 40.
97. Paul Cambon to Sir Edward Grey, 9 May 1916, and translation, 'Arab Proposals. Amended Version', TNA FO 371/2777/88317 (Cambon to Grey 1916).
98. Cloarec, *La France et la question de Syrie*, p. 234.
99. Ibid., p. 270; Isaiah Friedman, *The Question of Palestine, 1914–1918: British-Jewish-Arab Relations* (New Brunswick, NJ: Transaction, 1992), pp. 60–1 (Friedman 1992).
100. James Renton, 'Flawed Foundations: The Balfour Declaration and the Palestine Mandate', in *Britain, Palestine and Empire: The Mandate Years*, ed. Rory Miller (Farnham, 2010), pp. 15–37 (Renton 2010).
101. Idem, *Zionist Masquerade*, chs. 4–5.
102. Clayton to Foreign Office, 7 June 1918, TNA FO 371/3396/102624 (Clayton to Foreign Office 1918).
103. 'Palestine and the Jews—by an Arab Son of Shem' (translation sent to London and read by Sykes), *al- Kawkab*, no. 84, 5 March 1918, TNA FO 371/3398/57971 (Palestine and the Jews 1918).
104. Renton, 'Changing Languages'.
105. For examples, see photos of the Anglo-Arab entrance into Aleppo and Zionist ceremony in Jerusalem: Q51258 and Q13191, Photograph Archive, Imperial War Museum. On the Arab flag, see Renton, 'Changing Languages', p. 655.
106. Sykes minute, c. 13 June 1918, TNA FO 371/3398/105824. (Sykes minute 1918).
107. William Ormsby-Gore report, 22 August 1918, p. 5, TNA FO 371/3395/152266 (Ormsby-Gore report 1918).
108. Rashid Khalidi, *The Iron Cage: The Story of the Palestinian Struggle for Statehood* (Oxford: Oneworld, 2007), pp. 32–3 (Khalidi 2007).

109. On pre-war Palestinian national identity, see, in particular, Rashid Khalidi, *Palestinian Identity: The Construction of Modern National Consciousness* (New York: Columbia University Press, 1997), chs. 2–6; and Haim Gerber, *Remembering and Imagining Palestine: Identity and Nationalism from the Crusades to the Present* (Basingstoke: Palgrave Macmillan, 2008), ch. 2 (Khalidi 1997; Gerber 2008).
110. James Renton, 'The Age of Nationality and the Origins of the Zionist-Palestinian Conflict', *The International History Review*, 35, no. 3 (2013), 576–99 (Renton 2013).
111. E. Eberlin, 'Aux Arabes: Lettre Ouverte d'un Sioniste', n.d., enclosed in Durieux to Georges Picot, 4 December 1918, CADC, Levant 1918–1929, Palestine Vol. 11, E312 4 Sionisme, October–December 1918, sub-folder 3. (Eberlin 1918).
112. 'Palestine Censorship. A Summary of Political and Economic Information obtained from the Censorship of Civil Correspondence in Palestine during the period April, 1918, to January 1919', TNA FO 371/4229/83514 (Palestine Censorship 1919).
113. Jonathan Marc Gribetz, *Defining Neighbors: Religion, Race, and the Early Zionist-Arab Encounter* (Princeton, NJ: Princeton University Press, 2014), pp. 139–50 (Gribetz 2014).
114. Georges Picot to Stephen Pichon, 15 December 1918, CADC, Levant 1918–1929, Palestine Vol. 11, E312 4 Sionisme, October–December 1918, sub-folder 3. (Georges Picot to Pichon 1918).
115. 'Palestine Censorship', TNA FO 371/4229/83514.
116. Storrs, *Orientations*, p. 352.
117. T.E. Lawrence, *Seven Pillars of Wisdom* (London: Wordsworth, 1997), p. 16 (Lawrence 1997).
118. Lawrence to his mother, 2 August 1909, in David Garnett (ed.), *Selected Letters of T.E. Lawrence*, 2nd edn. (London: World Books, 1941), pp. 26–7 (Lawrence to his mother 1909).
119. Lawrence, *Seven Pillars*, p. 347.
120. Ibid., p. 321.
121. Ibid., p. 40.
122. Ibid., pp. 40–1.
123. Ibid., pp. 551–2.
124. Renan, *Histoire générale*, p. 11.
125. Lawrence, *Seven Pillars*, p. 24.

126. *Palestine Royal Commission Report*, Cmd. 5479 (London: H.M Stationery Office, 1937), p. 117 (Palestine Royal Commission Report 1937).
127. Ibid., p. 370.
128. Ibid., p. 120.
129. Ibid., chs. 20–22, p. 379.
130. David Motadel, *Islam and Nazi Germany's War* (Cambridge, MA: Harvard University Press, 2014), pp. 58–9 (Motadel 2014).
131. Anidjar, *Semites*, p. 38.
132. *Palestine Royal Commission*, p. 122.

References

Andrew, Christopher M., and A.S. Kanya-Forstner. 1981. *France Overseas: The Great War and the Climax of French Imperial Expansion*. London: Thames and Hudson.

Anidjar, Gil. 2008. *Semites: Race, Religion, Literature*. Stanford: Stanford University Press.

Arendt, Hannah. c.1976. *The Origins of Totalitarianism*. Oregon: Harcourt.

Baasten, Martin F.J. 2003. A Note on the History of "Semitic". In *Hamlet on a Hill: Semitic and Greek Studies Presented to Professor T. Muraoka on the Occasion of His Sixty-Fifth Birthday*, eds. M.F.J Baasten and W.Th. van Peursen, 57–72. Leuven: Peeters.

Ballantyne, Tony, and Antoinette Burton. 2012. *Empires and the Reach of the Global, 1870–1945*. Cambridge, MA: Harvard University Press.

Cambon, Paul, to Sir Edward Grey, 9 May 1916, and translation, 'Arab Proposals. Amended Version', The National Archives, Kew, UK (TNA) Foreign Office records (FO) 371/2777/88317.

Clayton, Gilbert, to Foreign Office, 7 June 1918, FO 371/3396/102624.

Clayton to Sykes, 15 December 1917, F.R. Wingate Papers, Sudan Archive, Durham University Library, UK (SAD.) 147/3/92.

Clayton to the Secretary of State for Foreign Affairs, 1 July 1918, F.R. Wingate Papers, SAD. 149/1/3.

Cloarec, Vincent. 2010. *La France et la question de Syrie: 1914–1918*. Paris: CNRS.

Defrance, Albert, to Ministry of Foreign Affairs (MFA), 17 January 1918, Le Centre des Archives diplomatiques de La Courneuve (CADC), Guerre 1914–1918, Sionisme V, Vol. 1200, January–February 1918, sub-folder 2.

Eberlin, E., 'Aux Arabes: Lettre Ouverte d'un Sioniste', n.d., enclosed in Durieux to Georges Picot, 4 December 1918, CADC, Levant 1918–1929, Palestine Vol. 11, E312 4 Sionisme, October–December 1918, sub-folder 3.

El-Eini, Roza M.L. 2004. *Mandated Landscape: British Imperial Rule in Palestine, 1929–1948.* London: Routledge.

Faisal-Weizmann agreement, 3 January 1919, London Bureau Papers, Central Zionist Archives, Jerusalem, Israel (CZA) Z4/40065.

Friedman, Isaiah. 1992. *The Question of Palestine, 1914–1918: British-Jewish-Arab Relations.* New Brunswick, NJ: Transaction.

Georges Picot, François, to MFA, 27 November 1917, CADC, Guerre 1914–1918, Sionisme, Vol. 1199, folder 2, sub-folder 1.

Georges Picot to Stephen Pichon, 15 December 1918, CADC, Levant 1918–1929, Palestine Vol. 11, E312 4 Sionisme, October–December 1918, sub-folder 3.

Gerber, Haim. 2008. *Remembering and Imagining Palestine: Identity and Nationalism from the Crusades to the Present.* Basingstoke: Palgrave Macmillan.

al-Ghayati, 'Ali. Here and There. *al-Mustaqbal*, no. 90, 30 December 1917.

Grainger, John D. 2006. *The Battle for Palestine 1917.* Woodbridge: Boydell Press.

Gribetz, Jonathan Marc. 2014. *Defining Neighbors: Religion, Race, and the Early Zionist-Arab Encounter.* Princeton, NJ: Princeton University Press.

Hajjat, Abdellali, and Marwan Mohammed. 2013. *Islamophobie: Comment les élites françaises fabriquent le "problème musulman".* Paris: La Découverte.

Heschel, Susannah. 2008. *The Aryan Jesus: Christian Theologians and the Bible in Nazi Germany.* Princeton, NJ: Princeton University Press.

Hess, Jonathan M. 2000. Johann David Michaelis and the Colonial Imaginary: Orientalism and the Emergence of Racial Antisemitism in Eighteenth-Century Germany. *Jewish Social Studies* 6(2) (Winter): 56–101.

Jewish National Movement. The Commission in Palestine. Jews, Arabs, and Armenians. Statements by Dr. Weizmann. *Jewish Chronicle*, 10 May 1918, p. 11.

Kalmar, Ivan Davidson. 2009. Anti-Semitism and Islamophobia: The Formation of a Secret. *Human Architecture: Journal of the Sociology of Self-Knowledge* 7(2) (Spring): 135–144.

———. 2012. *Early Orientalism: Imagined Islam and the Notion of Sublime Power.* London: Routledge.

———. 2013. Arabizing the Bible: Racial Supersessionism in Nineteenth Century Christian Art and Biblical Scholarship. In *Orientalism Revisited: Art, Land and Voyage*, ed. Ian Netton. London: Routledge.

Khalidi, Rashid. 1997. *Palestinian Identity: The Construction of Modern National Consciousness.* New York: Columbia University Press.

———. 2007. *The Iron Cage: The Story of the Palestinian Struggle for Statehood.* Oxford: Oneworld.

Laurens, Henry. 1999. *La Question de Palestine. Tome premier, 1799–1922, L'invention de la Terre sainte.* Paris: Fayard.

Lawrence, T.E., to his mother, 2 August 1909. In *Selected Letters of T.E. Lawrence*, 2nd edn, ed. David Garnett. 1941. London.
Lawrence, T.E. 1997. *Seven Pillars of Wisdom*. London: Wordsworth.
Marchand, Suzanne L. 2009. *German Orientalism in the Age of Empire: Religion, Race, and Scholarship*. Cambridge: Cambridge University Press.
Massignon, Louis, to G. Maugras: Report on the state of French propaganda among the Arab Legion, 26 October 1917, le Centre des Archives diplomatiques de Nantes, France, 294PO/B/149 Jerusalem Consulat General, Beyrouth: 2367- Correspondence of Georges Picot, 1917–1919, sub-folder 10.
Mosse, George L. 1978. *Toward the Final Solution: A History of European Racism*. New York: H. Fertig.
Motadel, David. 2014. *Islam and Nazi Germany's War*. Cambridge, MA: Harvard University Press.
Nirenberg, David. 2013. *Anti-Judaism: The Western Tradition*. New York: W.W. Norton.
Olender, Maurice. 1989. *Les Langues du Paradis—Aryens et sémites: un couple providential, Préface de Jean-Pierre Vernant*. Paris: Gallimard/Le Seuil.
Ormsby-Gore, William, report, 22 August 1918. TNA FO 371/3395/152266.
Palestine Censorship. A Summary of Political and Economic Information obtained from the Censorship of Civil Correspondence in Palestine during the period April 1918, to January 1919, TNA FO 371/4229/83514.
Palestine and the Jews—by an Arab Son of Shem. *al- Kawkab*, no. 84, 5 March 1918. TNA FO 371/3398/57971.
Palestine Royal Commission Report. 1937. Cmd. 5479. London: H.M Stationery Office.
Pasto, James. 1998. Islam's "Strange Secret Sharer": Orientalism, Judaism, and the Jewish Question. *Comparative Studies in Society and History* 40(3) (July): 437–474
Priest, Robert D. 2015. Ernest Renan's Race Problem. *The Historical Journal* 58(1): 309–330.
Psichari, Henriette (ed). 1947. *Oeuvres complètes de Ernest Renan*. Paris: Calmann-Lévy.
Renan, Ernest. 1855. *Histoire générale et système comparé des langues sémitiques*. Paris: Imprimerie impériale.
———. 1887. *Histoire du peuple d'Israël. Tome premier*. Paris: Calmann-Lévy.
———. 1889. *Histoire du peuple d'Israël. Tome deuxième*. Paris: Calmann-Lévy.
———. 1893. *Histoire du peuple d'Israël. Tome cinquième*. Paris: Calmann-Lévy.
Renton, James. 2007a. *The Zionist Masquerade: The Birth of the Anglo-Zionist Alliance, 1914–1918*. Basingstoke: Palgrave Macmillan.
———. 2007b. Changing Languages of Empire and the Orient: Britain and the Invention of the Middle East, 1917–1918. *The Historical Journal* 50(3): 645–667.

———. 2010. Flawed Foundations: The Balfour Declaration and the Palestine Mandate. In *Britain, Palestine and Empire: The Mandate Years*, ed. Rory Miller. Farnham: Ashgate.

———. 2013. The Age of Nationality and the Origins of the Zionist-Palestinian Conflict. *The International History Review* 35(3): 576–599.

Resolutions are Measured Against Those Who Make Them. *al-Qibla*, Issue 151, 18 Shawwal 1336/31 January 1918.

Rubin, Andrew N. 2015. Orientalism and the History of Western Anti-Semitism: The Coming End of an American Taboo. *History of the Present* 5(1) (Spring): 95–108.

Said, Edward W. 1995. *Orientalism: Western Conceptions of the Orient- Reprinted with a New Afterword*. London: Penguin.

Samuel, Sir Herbert, note, 9 November 1914. Herbert Samuel papers, copies. Parliamentary Archives, UK.

Satia, Priya. 2008. *Spies in Arabia: The Great War and the Cultural Foundations of Britain's Covert Empire in the Middle East*. Oxford: Oxford University Press.

Schlözer, August Ludwig. 1781. Von den Chaldäern. *Repertorium für biblische und morgenländische Litteratur* 8: 113–176.

Sidebotham, Herbert, memorandum, 9 March 1917. Weizmann Archives, Rehovot, Israel (WA).

Sieff, Israel, to Weizmann, 19 February 1917. WA

Smith, George Adam. 1910. *The Historical Geography of the Holy Land*, 16th edn. London: Hodder and Stoughton.

———. 1915. *Atlas of the Historical Geography of the Holy Land*. London: Hodder and Stoughton.

Sokolow, Nahum, and Chaim Weizmann to Jacques Mosseri, 11 December 1917, no. 22. In *The Letters and Papers of Chaim Weizmann*, Vol. VIII, Series A, ed. Dvorah Barzilay and Barnet Litvinoff. New Brunswick, NJ: Transaction.

Sokolow to Sykes, 28 March 1917. Sledmere Papers, copy, WA.

Storrs, Ronald. 1937. *Orientations*. London: I. Nicholson & Watson.

Sykes, Sir Mark, meeting with Zionists, memorandum, 7 February 1917. p. 15. CZA A226/30/1.

———. Memorandum on the Asia-Minor Agreement, 14 August 1917. Sykes Collection, Middle East Centre Archive, Oxford (MECA).

———. minute, c. 13 June 1918, TNA FO 371/3398/105824.

Sykes, on behalf of the Joint Committee, to the Syria Welfare Committee, Cairo, 15 February 1918. TNA FO 371/3398/27647.

Sykes to Gilbert Clayton, 14 November 1917, Sykes Collection, MECA.

Sykes to Clayton, draft, 16 November 1917, Sykes Collection, MECA.

Sykes to Sir Percy Cox, 23 May 1917. Sykes Collection, MECA.

Thomas, Martin. 2007. *Empires of Intelligence: Security Services and Colonial Disorder after 1914*. Berkeley: University of California Press.

Thornton, Lynne. 1993. *Les Orientalistes: peintres voyageurs, 1828–1908*. Paris: ACR edition/Pochecouleur.
Tibawi, A.L. 1978. *Anglo-Arab Relations and the Question of Palestine, 1914–1921*. London: Luzac.
Tromans, Nicholas (ed). 2008. *The Lure of the East: British Orientalist Painting*. London: Tate.
Tolkowsky, Shmuel, diary, 7 August 1917. CZA A248/2.
Vital, David. 1987. *Zionism: The Crucial Phase*. Oxford: Clarendon.
Weizmann, Chaim, to Felix Frankfurter, 17 August 1917, no. 479. In *The Letters and Papers of Chaim Weizmann*, Vol. VII, Series A, ed. Leonard Stein. 1975. Jerusalem: Israel Universities Press.
Weizmann, Chaim, to Aaron Aaronsohn, 16 January 1918, no. 66. In *The Letters and Papers of Chaim Weizmann*, Vol. VIII, Series A, ed. Dvorah Barzilay and Barnet Litvinoff. 1977. New Brunswick, NJ: Transaction.
Weizmann and Sokolow to Aaron Aaronsohn, 16 November 1917, no. 8. In *The Letters and Papers of Chaim Weizmann*, Vol. VIII, Series A, ed. Dvorah Barzilay and Barnet Litvinoff. 1977. New Brunswick, NJ: Transaction.
Weizmann and Sokolow to Jacques Mosseri, 11 December 1917, no. 22. In *The Letters and Papers of Chaim Weizmann*, Vol. VIII, Series A, ed. Dvorah Barzilay and Barnet Litvinoff. 1977. New Brunswick, NJ: Transaction.
Weizmann to Arthur J. Balfour, 30 May 1918, no. 208. In *The Letters and Papers of Chaim Weizmann*, Vol. VIII, Series A, ed. Dvorah Barzilay and Barnet Litvinoff. 1977. New Brunswick, NJ: Transaction.
Weizmann to Vera Weizmann, 20 May 1918, no. 200. In *The Letters and Papers of Chaim Weizmann*, Vol. VIII, Series A, ed. Dvorah Barzilay and Barnet Litvinoff. 1977. New Brunswick, NJ: Transaction.
———, 30 April 1918, no. 181. In *The Letters and Papers of Chaim Weizmann*, ed. Dvorah Barzilay and Barnet Litvinoff, Vol. VIII, Series A. 1977. New Brunswick, NJ: Transaction.
———, 17 June 1918, no. 213. In *The Letters and Papers of Chaim Weizmann*, ed. Dvorah Barzilay and Barnet Litvinoff, Vol. VIII, Series A. New Brunswick, NJ: Transaction.
Wisdom is the Legacy of the Believer, He Has a Right to It Wherever He Finds It. *al-Qibla*, 13 Sha'ban 1336/23 May 1918.
Zionist Organisation. 1918. *Great Britain, Palestine and the Jews: Jewry's Celebration of its National Charter*. London.
Zionist propaganda committee (London). 1918. Report of Meeting of Propaganda Committee, 14 December–30 June 1918, CZA Z4/243.

PART III

Divergence

CHAPTER 6

The Case of Circumcision: Diaspora Judaism as a Model for Islam?

Sander L. Gilman

Two moments in recent history: a religious community in France is banned from wearing distinctive clothing in public schools, as that is seen as an egregious violation of secular society; a religious community in Switzerland is forbidden from ritually slaughtering unstunned animals, as such slaughter is seen as a cruel and unnatural act. These acts take place more than a hundred years apart: the former recently in France, the latter more than a century ago in 1893 in Switzerland (where the prohibition against ritual slaughter still stands and has now been adopted widely, from New Zealand in 2010 to Sweden, where even fish must be stunned). But who are these religious communities? In France (among other countries), the order banning ostentatious religious clothing and ornaments in schools and other public institutions has as much impact on religious Jewish men who cover their heads (and perhaps even religious Jewish married women who cover their hair) as it does the evident target group,

S.L. Gilman (✉)
Emory University, Atlanta, GA, USA

J. Renton, B. Gidley (eds.), *Antisemitism and Islamophobia in Europe*, DOI 10.1057/978-1-137-41302-4_6

Muslim women. (The law is written in such a politically correct [PC] way as also to ban the ostentatious wearing of a cross: 'Pierre, you can't come into school carrying that six-foot-high cross on your back. You will have to simply leave it in the hall.') Ironically, in 2013 such a law was proposed for civil servants in the Province of Quebec for much the same reasons, but with a twist: 'The government went as far … as publishing sartorial dos-and-don'ts, with pictograms of Sikh turbans and Muslim face veils in the verboten category, and discreet cross pendants or Star of David rings on the acceptable list.'[1] Both Jewish and Muslim groups objected to the state's definition of what is or is not a religious ritual object.

In Switzerland, even today the prohibition against kosher Jewish slaughter (*shechita*) also covers the slaughter of meat by Muslims who follow the ritual practice (*dhabiha*) that results in halal meat. The Swiss are now even contemplating banning the importation of such meat. In Great Britain these debates raged in the nineteenth century and in Poland in the twenty-first. The Jewish practice was banned by the Nazis in Germany with the Gesetz über das Schlachten von Tieren (Law on the Slaughtering of Animals) of 21 April 1933; it was sporadically permitted after 1945 through exceptions; only in 1997 were these exceptions made part of the legal code. The Islamic practice was outlawed in Germany until 1979 and even today is tolerated but not sanctioned.[2] These prohibitions affect Jews and Muslims in oddly similar ways when Western responses to 'slaughter' are measured. What is very different is how the meat is used, whether in 'traditional' dishes or in a Big Mac. The question is: How did and will these two groups respond to such confrontation with the secular, 'modern' world?[3]

Why should the focus of concern in secular Europe from the Enlightenment to today be on the practices and beliefs of Jews and Muslims? Indeed, when Sikhs in France raised the question of whether their turbans were 'cultural' or 'religious' symbols under the terms of the new regulations, the official French spokesperson asked, in effect: Are there Sikhs in France? Indeed there are.

Yet in September 2004, two French journalists were seized in Iraq and threatened with death unless the law limiting headscarves was not instituted the following week, when school was to begin in France. The reaction was not a sense of support for the struggle for an Islamic identity in France. Indeed, virtually all of the French Muslim institutions, from the official French Council of the Muslim Faith to the radical Union of Islamic Organizations in France (UOIF), spoke out against the outside pressure, even though it came from the 'Islamic' world. As Olivier Roy, a leading French scholar of Islam, noted, 'They may disagree on the law of the veil, but they are saying, "This is our fight and don't interfere." This is a pivotal moment.'[4] Indeed, Lhaj Thami Breze, the head of UOIF, who had been opposed to the law, proposed a compromise in which a moderate interpretation of the law would permit 'modest head covering'.[5] It was indeed a public change of attitude, as the unity of the Islamic community in France in opposition to 'foreign' interference concerning the 'law of the veil' suddenly was seen as a sign of the development of a secular consciousness in this religious community. What was striking is that the majority of Muslim schoolgirls did not wear or quickly removed their head coverings the day school began. Only about two hundred to two hundred and fifty girls, mainly in Alsace, wore their scarves to school, and all but about one hundred took them off before entering the buildings. The girls who kept them on were removed from the classroom and provided with 'counselling' in school. For them the *hijab*, which had been seen as 'a way to reconcile modernity, self-affirmation and authenticity', was a sign of the Western rights that they demanded as Muslims.[6] These were less central than the rule of law. Three male Sikh students in Bobigny, a Paris suburb, were sent home on the first day of class for wearing their traditional head covering. The irony was that a law aimed at Muslim students initially impacted non-Muslims. This was true whether they saw the headscarf as a political, ethnic or religious symbol. The demand that one see oneself as a citizen with the rights of the citizen to contest the claims of the secular state overrode any sense of the primary identification as a member of the Ummah, the Islamic religious community. Jacqueline Costa-Lascoux, research director at the Political Science Center of the National Center for Research in Paris (CEVIPOF), noted that 'the hostage taking has helped the Muslim community in France, mainly the young people, to understand that they can live in a democratic society and still be Muslims'.[7] The operative terms here are 'democratic society' and 'Muslims'. In the recent past this has been the argument for the ending of the ban on headscarves by the Islamic government in Turkey. It is the constitution of the modern secular state and the need for religions such as Islam and Judaism to adapt to it that are at the heart of the matter.

What does it mean to be a Muslim in this secular world of modern France or Francophone Canada? Scratch secular Europe today, and you find all of the presuppositions and attitudes of Christianity concerning Jews and Muslims present in subliminal or overt forms. Secular society in Europe has absorbed Christianity into its very definition of the secular.[8]

Indeed, one can make an argument that 'secular' society as we now see it in Europe is the result of the adaptation of Christianity to the model of secularism that arose as a compromise formation out of the wars of religion following the Reformation. The integration of the Jews into Enlightenment Europe, as Adam Sutcliffe has shown in his *Judaism and Enlightenment*, was integration into Christian Europe (with Christianity having different textures in England than in Holland than in Bavaria, and so on).[9] Whether one thinks that this provided an ideal model for all modern states, as does the philosopher Charles Taylor when he claims that secularisation provides 'people of different faiths, or different fundamental commitments', with the ability to co-exist; or whether one is leery of such claims, as is Talal Asad, who sees this merely as a 'political strategy', the 'Jewish template' may well provide a clue to the potentials for the processes that religious communities with specific ritual beliefs and practices confront.[10]

The veneer was that of a secular state, a veneer that did alter the nature of Christianity itself. Even if today it is true, as Richard Bulliet claims, that 'Christianity and Judaism pass by definition the civilizational litmus tests proposed for Islam even though some of their practitioners dictate women's dress codes, prohibit alcoholic beverages, demand prayer in public schools, persecute gays and lesbians, and damn members of other faiths to hell',[11] this was certainly not the case for Jews in the secularising Christian world of the European nations and their colonies following the Reformation. Indeed, Jews were regularly seen as being inherently unable to pass 'civilizational litmus tests' in the Western Diaspora in virtually all areas.

Yet even today there are odd and arcane echoes of older views about the meaning of Jewish ritual. In the mid-1990s, there was a general acknowledgement in the Catholic Church that the *Bible for Christian Communities* (*La Bible Latino-américaine*), written by Bernard Hurault, a Catholic missionary based in Chile, to combat the rising tide of Evangelical Christianity, was blatantly anti-Jewish. Eighteen million copies in English and Spanish were distributed in South America, and hundreds of thousands were sold in France and Belgium following its publication in May 1994. According to the text, the Jewish people killed Jesus Christ because they 'were not able to control their fanaticism' and thus showed a true lack of decorum. It was also clear that Judaism was represented as a religion of meaningless rituals, mere 'folkloric duties involving circumcision and hats'.[12] (After a legal challenge from the French Jewish community, the text was officially withdrawn; it still circulates in South America.)

How this is contested, sometimes successfully, sometimes unsuccessfully, provides an interpretative framework through which to understand the debates about the meaning and function of Islam in the West that have taken place since 9/11.

Little has altered concerning the deep cultural legacy of Europe over the past two hundred years. German, Italian, Polish and Slovakian delegates demanded that the 'Christian heritage' of the new Europe be writ large in the (failed) European constitution of 2005. It was only the post–11 September anxiety of most states that enabled Valéry Giscard d'Estaing, as president of the convention writing the constitution, to persuade the group that such a reference would be 'inappropriate'. The demand was transformed into a reference in the preamble to the 'cultural, religious, and humanist inheritance of Europe'. No one missed what was meant. Certainly one of the things that the French and Dutch referenda about the constitution in 2005 tested was the likelihood of admitting Turkey, a majority Muslim state, into the European Union. Judaism and Islam have an all-too-close relationship to Christianity and raise questions that remain troubling in Europe.

It is important not to reduce the relationship between Judaism and Islam to the role that Jewish ideas, concepts and practices did or did not have in shaping the earliest forms of Islamic belief. It is clear that nineteenth-century Jewish scholars in Europe had a central role in examining the 'Jewish roots' of historical Islam. Scholars from Abraham Geiger in the 1830s to Ignaz Goldziher at the end of the century stressed the judaising nature of early Islam. These roots, true or not, are not sufficient to explain the intense focus on the nature of Islam in Europe today. Islam is not simply a surrogate for speaking about the Jews in today's Europe because of superficial similarities to Judaism. Among Jewish scholars in the nineteenth century, the search for the Jewish roots of Islam was certainly more than simply a surrogate for speaking about the relationship between Judaism and Christianity in the nineteenth century, as Susannah Heschel so elegantly shows in her study *Abraham Geiger and the Jewish Jesus.*[13] At one moment, the examination or construction of Islam provided one major Jewish scholar with a model for the potential reform of contemporary Judaism. One can quote Goldziher's diaries:

> I truly entered into the spirit of Islam to such an extent that ultimately I became inwardly convinced that I myself was a Muslim, and judiciously discovered that this was the only religion which, even in its doctrinal and

> official formulation, can satisfy philosophic minds. My ideal was to elevate Judaism to a similar rational level. Islam, so taught me my experience, is the only religion in which superstitious and heathen ingredients are not frowned upon by the rationalism, but by the orthodox teachings.[14]

For him, the Islam he discovered becomes the model for a new spirit of Judaism at the close of the nineteenth century.

It is the seeming closeness of these 'Abrahamic' religions and their joint history that draw attention to the real or imagined differences from the majority religion and its new form: secular society. 'The "Abrahamic" religions' is the newest PC phrase: the 'Judeo-Christian tradition' was the catchword for common aspects shared between Judaism and Christianity after the Holocaust made this an acceptable notion, whereas 'the Abrahamic religions' is the new buzzword including Islam in the Judeo-Christian fold that has become current only after 9/11. Both phrases attempt to defuse the clearly Christian aspects of modern Western secular society by expanding it, but, of course, only re-emphasise it. Here Jonathan Sacks' notion of difference is helpful: in creating categories that elide difference and stress superficial similarities, one believes that one is bridging 'differences'.[15] Actually, one is submerging them.

The closeness of Christianity to Judaism and Islam results in what Sigmund Freud called the 'narcissism of minor differences'. Those differences are heightened in this secular society, which is rooted in the mindset and often the attitudes, beliefs, social mores and civic practices of the religious community in Western Europe—Christianity. Thus, in Western Europe there is a radical secularisation of religious institutions in the course of the nineteenth century. Marriage is shifted from being solely in the control of the Church to being in the domain of the state; but this form of secularisation still maintains the quasi-religious aura about marriage, something we see in the debates in France about gay marriage. No secularising European state simply abandons marriage as a religious institution that has outlived its time, as nineteenth-century anarchists and some early twentieth-century radical Zionists claimed.[16] The new minority is promised a wide range of civil rights—including those of freedom of religion—if only they adhere to the standards of civilised behaviour as defined by the secular society. This is rooted in the desire to make sure that that society, with its masked religious assumptions, re-defines the minorities' religious practice or 'secularises' a religious minority into an 'ethnic' one.

Equally, it is vital not to confuse the experiences of contemporary Islam with the rhetoric of victimisation often heard within Muslim communities in countries such as Germany. There the evocation of the Holocaust becomes a means of identifying with the iconic victims of German history, the Jews. Y. Michal Bodemann and Gökce Yurdakul have noted quite correctly how the competition for the space of the victim or of the essential Other has allowed Turkish writers, such as Yadé Kara, to call on the Jews as the model, for good or for ill, for Turkish acculturation.[17] The Turkish community regularly evokes the Holocaust when it imagines itself. Thus, at the public events in Berlin on 23 November 2002 commemorating the horrendous murder of Turkish immigrants in Mölln in 1992, one heard the Turkish spokesman, Safter Çınar, evoke the experience of the Jews in the Holocaust as the model by which the contemporary experience of Turks could be measured. The power of this analogy is clear. Yet this self-conscious evocation of the experience of the Jews is only one aspect of contemporary parallels of Jews and Muslims.

Can we now look at the experiences within the various strands of Jewish religious (and therefore social) ritual practice from the late eighteenth century (which marked the beginning of civil emancipation) that parallel those now confronting Diaspora Islam in 'secular' Western Europe?[18] The similarities are striking: a religious minority enters into a self-described secular (or secularising) society that is Christian in its rhetoric and presuppositions and that perceives a 'special relationship' with this minority. The co-territorial society sees this as an act of aggression. This minority speaks a different secular language, but also has yet a different religious language. This is odd in countries that have a national language and (in some) a religious language, but not a secular language spoken by a religious minority as well as a ritual. Religious schools that teach in the languages associated with a religious group are seen as sources of corruption and illness. Religious rites are practised that seem an abomination to the majority 'host' culture: unlike the secular majority, these religious communities follow practices such as the suppression of the rights of women (lack of women's traditional education, a secondary role in religious practice, arranged marriages and honour killings); barbaric torture of animals (the cutting of the throats of unstunned animals, allowing them to bleed to death); prohibiting the creation of 'graven images' of all types, including representations of Muhammad or God; disrespect for the dead through too rapid burial; ritual excess (in the case of the Jews, drunkenness at Purim; feasting during Ramadan in the case of the Muslims); ostentatious clothing that signals religious affiliation and has ritual significance (from women's hair covering such as the Muslim *hijab* to Jewish *sheitels* to men's hats such as the Jewish *stremil* or the Muslim *taqiyah*); and, centrally relating all of these practices, a belief in the divine 'chosenness' of the group in contrast to all others. The demonisation of aspects of religious practice has its roots in what civil society will tolerate and what it will not, what it considers to be decorous and what is unacceptable as a social practice. Why it will not tolerate something is, of course, central to the story. Thus, Alan Dundes argued a decade ago that the anxiety about meanings associated with the consumption of the body and blood of Christ in the Christian Mass shaped the fantasy of the Jews as slaughtering Christian children for their blood.[19] Yet it is equally present in the anger in secular Europe more recently directed towards Jewish ritual practices such as the mutilation of children's bodies (infant male circumcision and, for some Muslims, infant female genital cutting).

One of the most noteworthy similarities of the process of integration into Western secular society is the gradual elision of the striking national differences among the various groups. Muslims in Western Europe represent multiple national traditions (South Asian in the UK, North African in France and Spain, and Turkish in Germany). But so did the Jews in Western Europe who came out of ghettos in France and the Rhineland, from the rural reaches of Bavaria and Hungary, who moved from those parts of 'Eastern Europe'—Poland, the eastern marches of the Austro-Hungarian empire—that became part of the West and from the fringes of empire to the centre. To this one can add the Sephardic Jews from the Iberian Peninsula who settled in areas from Britain (introducing fish and chips) to the fringes of the Austrian empire. The standard image of the Jews in eighteenth-century British caricature was the Maltese Jew in his oriental turban. By the nineteenth century, it was that of Lord Rothschild in formal wear receiving the prince of Wales at his daughter's wedding in a London synagogue. Religious identity (as the Jew or the Muslim) replaced national identity—by then, few (except the antisemites) remembered that the Rothschilds were a Frankfurt family who escaped the Yiddish-speaking ghetto. 'Jews' are everywhere and all alike; Muslims seem to be everywhere and are becoming 'all alike'. Even ritual differences and theological antagonism seem to be diminished in the Diaspora, where the notion of a Muslim Ummah (or community) seems to be realised. It is the ideal state, to quote Talal Asad, of 'being able to live as autonomous individuals in a collective life that exists beyond national borders'.[20] Yet this too has its pitfalls, as the 'Jewish template' shows.

Now for Jews in those lands that will become Germany, in the Austro-Hungarian empire, in France and in those lands that will become Great Britain, the stories are all different: different forms of Christianity, different expectations as to the meaning of citizenship. Different notions of secularisation all present slightly different variations on the theme of 'What do you have to give up to become a true citizen?' Do you merely have to give up your secular language (western and eastern Yiddish, Ladino, Turkish, Urdu, colloquial Arabic)? Today there has been a strong suggestion in Germany and the UK that preaching within the mosques be done only in English—for security reasons. Do you have to abandon the most evident and egregious practices; or, as the German philosopher Johann Gottlieb Fichte (1762–1814) states (echoing debates about Jewish emancipation during the French Revolution), do you have to 'cut off their Jewish heads and replace them with German ones'?[21] And that was not meant as a metaphor, but as a statement of the impossibility of Jewish transformation into Germans.

My case in point about the function of ritual in defining European (and by extension European colonial) difference can be seen in the recent debate around banning infant male circumcision. Christianity following Paul in great measure abandoned this practice. The view that the Jews who continued with this ritual were intransigent about circumcision was espoused by the Church Fathers, Eusebius and Origen, and continued through the Renaissance (Erasmus was opposed to the practice) and through the Reformation (as was Luther). One has the body one is born with, said the Church Fathers at the Council of Jerusalem in 50 CE, waiving Jewish practices such as circumcision. Nevertheless, they did not forbid it and even today Coptic Christians (living in a Muslim culture that does require circumcision) practise this rite. Yet most Christians agreed (and agree) with Immanuel Kant in his *Religion within the Boundaries of Mere Reason* (1793):

> The subsequent discarding of the corporeal sign which served wholly to separate these people from others is itself warrant for the judgment that the new faith [Christianity], not bound to the statutes of the old, nor, indeed, to any statute at all, was to contain a religion valid for the world and not for one single people.[22]

By abandoning such practices one becomes universal and thus truly and authentically human. This view of a Jewish particularism defined by circumcision is held even by liberals such as the Italian physician Paolo Mantegazza, clearly, if mockingly, an advocate of hygiene:

> Circumcision is a shame and an infamy; and I, who am not in the least anti-Semitic, who indeed have much esteem for the Israelites, I who demand of no living soul a profession of religious faith, insisting only upon the brotherhood of soap and water and of honesty, I shout and shall continue to shout at the Hebrews, until my last breath: Cease mutilating yourselves: cease imprinting upon your flesh an odious brand to distinguish you from other men; until you do this, you cannot pretend to be our equal. As it is, you, of your own accord, with the branding iron, from the first days of your lives, proceed to proclaim yourselves a race apart, one that cannot, and does not care to, mix with ours.[23]

While circumcision is seen as creating a communitarian identity, this is considered negatively, separating the Jews from their peers.

By the nineteenth century, circumcision in Europe was viewed, for better or for worse, as a Jewish practice. This debate about the special relationship between Jews and ritual circumcision and the health exception is reflected in the Verein der Reformfreunde (Society for the Friends of Reform) in Frankfurt in 1843, which said that ritual infant male circumcision was neither a religious obligation nor a symbolic act.[24] This was in response to the 8 February 1843 finding of the Frankfurt public health authority that circumcision had to be carried out under medical supervision. However, the end result of concerns over hygiene and deformation was that ritual circumcision was less and less undertaken by acculturated Jews in Central and Western Europe.

The idea that the circumcised were inherently different because they were marked as Jewish dominates the nineteenth-century discussion. Through the Enlightenment, discussions of such religious practices seemed to be colored by debates about their medical efficacy, but these are never neutral, as specific attitudes towards the Jews seem always to define this debate. As an anonymous author stated in the leading German paediatric journal, *Journal für Kinderkrankheiten*, in 1872: 'The circumcision of Jewish children has been widely discussed in the medical press as is warranted with topics of such importance. But it is usually discussed without the necessary attention to details and the neutrality that it deserves. Indeed, it has not been free of fanatic anti-Semitism.'[25]

By the twenty-first century, the group that defines difference and the rights of the child is quite another one. When in 1999 the issue of the parents' right to circumcise comes before British courts, it is the Muslim practice of infant male circumcision that is seen at putting the child at risk:

> Re 'J' (child's religious upbringing and circumcision) said that circumcision in Britain required the consent of all those with parental responsibility, or the permission of the court, acting for the best interests of the child, and issued an order prohibiting the circumcision of a male child of a non-practicing Muslim father and non-practicing Christian mother with custody. The reasoning included evidence that circumcision carried some medical risk; that the operation would be likely to weaken the relationship of the child with his mother, who strongly objected to circumcision without medical necessity; that the child may be subject to ridicule by his peers as the odd one out and that the operation might irreversibly reduce sexual pleasure, by permanently removing some sensory nerves, even though cosmetic foreskin restoration might be possible. The court did not rule out circumcision against the consent of one parent.[26]

In Britain as well as on the European continent, the presence of a large and new Muslim community means that Muslims rather than Jews have become the litmus test for bad communitarian practices. The liberal consensus against 'Islamic rituals' such as female genital cutting also easily extends itself to rituals such as infant male circumcision as a violation of human rights, as in the anti-Islam campaigner Hirsi Ali's short-lived opposition in 2004 to that practice.[27]

According to a decision handed down by the Cologne regional court on 26 June 2012, circumcision of young boys is a criminal act, prohibited by law, even if parents have consented to the procedure.[28] The case brought before the court involved the circumcision of a four-year-old Muslim boy that was performed by a doctor at the parents' request. Complications occurred with the operation that resulted in the Cologne public prosecutor charging the doctor with the grievous bodily harm of the infant. The district court, hearing the case in the first instance, acquitted the doctor on the grounds that there was parental consent and that he had performed the procedure as a ritual act based on Islam.[29] The first court's position supports a view of circumcision as a procedure to be encouraged as a prophylaxis; the later ruling sees it as a non-medical procedure that is harmful and violates the infant's human rights.

The decision was grounded on the reasoning that such circumcisions cause 'illegal bodily harm' to the child, and that the child's right to physical integrity supersedes parents' rights and the freedom of religion. The Jewish response was clear and immediate: 'Circumcision is absolutely elementary for every Jew', the organization's president, Dieter Graumann, said in an interview with the *Rheinische Post.* He warned that if the Cologne ruling were to become the legal basis for determining the legality of circumcision, 'Jewish life in Germany might ultimately no longer be possible.'[30] The Conference of European Rabbis called an emergency three-day meeting in Berlin to discuss what to do. Its president, Rabbi Pinchos Goldschmidt, the chief rabbi of Moscow, called the Cologne ruling the 'worst attack on Jewish life since the Holocaust'. Citing France's ban on Muslim veils and Switzerland's ban on the construction of new minarets for mosques, Goldschmidt suggested that the Cologne decision is part of a wider trend of intolerance against religious traditions in Europe (*Der Spiegel* 2012).

The Cologne regional court upheld the lower court's ruling, but on different grounds: that the doctor believed he was acting lawfully in the context of an unclear legal situation surrounding the practice. While the court held that religious circumcisions were illegal because they violate the child's right to physical integrity and self-determination, it differentiated such acts from instances when a circumcision is medically necessary. On 12 December 2012, in an overwhelming vote (434 to 100), the German parliament voted to keep ritual circumcision legal in Germany. There were four stipulations to the law:

1. Adequate training of the practitioners for non-medical circumcision during the first six months.
2. No specific requirement for the use of an 'effective painkiller', as 'according to the standards of medical practice' also covers the 'necessary and effective treatment of pain in individual cases'. (If there is no anaesthesia the act remains bodily harm [*Körperverletzung*], but is not illegal [*rechtswidrig*].)
3. Parents are to be made aware of potential consequences.
4. The practice is not to be carried out on children who could be at risk of complications from non-medical circumcisers.

However, 'the ministry specifically avoids making any special provisions for circumcision for religious reasons, choosing instead to anchor it in legislation governing the rights of children to avoid requiring steps to determine the motivation of a parents' decision to have their children circumcised'.[31] It is clear that the six-month window established by the law allows for most Jewish and Muslim practices to occur legally (even though Muslims can and do circumcise up to the fourteenth year of life.)

In a sense, this is the beginning of a more systematic attack on religious practices that are now defined as Islamic and are seen as inherently violating human rights: the rights of the child to bodily integrity, not the rights of the parents to religious freedom. In November 2013, Anne Lindboe, Norway's Children's Ombudsmen, announced that the Norwegian government was going to introduce a law banning ritual male circumcision of pre-teen boys: 'With good information about risk, pain and lack of health benefits of the intervention, I think parents from minorities would voluntarily abstain from circumcising children.' She had earlier suggested that Jews and Muslims replace circumcision with 'a symbolic ritual'. She stated that infant male circumcision was a form of violence against children and should be punished as such.[32]

This debate certainly has to do with a contemporaneous surfacing of fierce public debates about Islam, Sharia law and civic society in the public sphere, such as the views of the then Archbishop of Canterbury, Rowan Williams, on the acceptability of Sharia law in the UK during 2008. While in the past it was the ritual practice of circumcision among Jews that was the trigger for such debates, it is now Islam. Advocates evoke the infant's health and public health authorities are placed in a position where they must either support or contradict such claims; opponents stress the violation of human rights and note the health risks.

Jews thought it possible to change Jewish religious practice such as circumcision and belief in the eighteenth and nineteenth centuries. Even where these traditions were altered, what was gained and what was lost in such debates was not always clear. The unquestioned ability of living religions to transform themselves and the understanding that all such transformations are often answered by claims of the immutability of religious practice, make the actual changes often invisible to the practitioners. The old saying, the more things change, the more they remain the same, seems here to be valid. All of these changes deal in general with the question of Jewish religious 'identity', but in a complex and often contradictory manner. The list of 'abominations' that secular Europe saw in Jewish ritual practices became the earmark for the question of what Jews were willing to change in order to better fit the various national assumptions about citizenship, as Kant observed. These were as different in the nineteenth century as the twenty-first-century debates in France about Islamic head covering, which is opposed because it violates the idea of a secular state; in

Germany, which is supportive under the very different meanings of multiculturalism; and in the UK, where in March 2005 the courts allowed full traditional South Asian clothing (the *jilbab*) as an exception to the 'school uniform' rule in a predominantly Muslim school where the dress code had been worked out with the parents.[33]

Now I know that there are also vast differences between Jews in the eighteenth and nineteenth centuries and Muslims today. There are simply many more Muslims today in Western Europe than there were Jews in the earlier period. Jews historically never formed more than 1% of the population of any Western European nation. Muslim populations form a considerable minority today. While there is no Western European city with a Muslim majority, many recent news stories predict that Marseilles or Rotterdam will be the first European city that will have one. In France today, there are six hundred thousand Jews while there are between 5 and 6 million Muslims, who make up about 10% of the population. In Germany, with a tiny Jewish population of slightly over one hundred thousand, almost 4% of the population is Muslim (totaling more than 3 million people). In Britain, about 2.5% of the total population (1.48 million people) is Muslim.[34] Demographics (and birthrate) aside, there are salient differences in the experiences of Jews and Muslims in the past and today. Jews had no national 'homeland'—indeed, they were so defined as nomads or a pariah people (*pace* Max Weber and Hannah Arendt). They lived only in the *Goles*, the Diaspora, and seemed thus inherently different from any other people in Western Europe (except perhaps the Roma). Most Muslims in the West come out of a national tradition often formed by colonialism in which their homelands had long histories disturbed but not destroyed by colonial rule. And last but not least, the Israeli–Palestinian conflict over the past century (having begun well before the creation of the state of Israel), the establishment of a Jewish homeland, as well as the Holocaust seem to place the two groups—at least in the consciousness of the West—into two antagonistic camps.[35]

Religion for the Jews of pre-Enlightenment Europe and for much of contemporary Islam, which has its immediate roots in majority Islamic states, became for many a 'heritage' in the Western, secular Diaspora. What had been lived experience in *milieux de mémoire* (environments of memory), to use Pierre Nora's often-cited phrase from 1994, becomes *lieux de mémoire* (places of memory) that re-configure meaning constantly within the Diaspora.[36] What is it that such memory of ritual and practice can or must abandon? What must it preserve to maintain its coherence for the group? The answer depends on time and place, and yet the experience of Jews in the Western European Diaspora seems to offer a model case, clearly because of the 'narcissism of minor differences' among the three Abrahamic religions. Jews maintain, in different modalities, their religious identity, even if the nature of the options explored created ruptures that produced new problems and, over time, partial resolutions and yet further conflicts and resolutions.

The central cultural crisis of the New Europe is not European integration in national terms, but the relationship between secular society and the dynamic world of European Islam. As the Syrian-born German sociologist Basam Tibi noted decades ago, it is the struggle within Islam to become a modern religion, whether within the Islamic world or in the Islamic Diaspora in the West, that is central.[37] Recently further voices, such as that of Tariq Ramadan and Feisal Abdul Rauf, have noted the need for a 'modern' Islam.[38] There are certainly moments of confrontation in which Islamic ritual and practices have changed in specific settings. One can think of the entire history of Bosnian Islam from the nineteenth century until its destruction in the past decade, and the resultant fundamentalist cast given to Bosnia since then. There is, however, a substantial difference between the contexts. Anyone interested in contemporary Europe before 11 September 2001 knew that the eight-hundred-pound gorilla confronting France, Germany and the UK—and, to a lesser extent, Spain and Italy—was the huge presence of an 'unassimilable' minority. Given Thilo Sarrazin's recent best-selling screed *Deutschland Schafft Sich Ab* ("Germany Does Away with Itself"), which decried the dilution of 'German' society by the reproductive capacity of a permanent and unassimilable underclass of 'Muslim immigrants' in the country,[39] the question of Muslims in Western Europe seemed to forecast the same set of problems. Yet, of course, exactly the same things have been said (with correction for national self-image) about Jews for two hundred years.

Notes

1. I. Peritz and L. Perreaux, 'Quebec Reveals Religious Symbols to be Banned from Public Sector', *The Globe and Mail*, 10 September 2013 (Peritz and Perreaux 2013).
2. R. Jentzsch, 'Das rituelle Schlachten von Haustieren in Deutschland ab 1933' (Ph.D. diss., Tierärztliches Hochschul, Hannover, 1998) (Jentzsch 1998).

3. For deep background, see J.M. Hess, *Germans, Jews, and the Claims of Modernity* (New Haven, CT: Yale University Press, 2002) (Hess 2002).
4. E. Sciolino, 'Ban on Head Scarves Takes Effect in a United France', *New York Times*, 3 September 2004, A9 (Sciolino 2004).
5. 'A Tragic Twist of the Scarf', *Economist*, 4 September 2004, p. 49 (A Tragic Twist of the Scarf 2004).
6. O. Roy, *Globalised Islam: The Search for the New Ummah* (London: Hurst & Co., 2004), p. 24 (Roy 2004).
7. T. Hundley, '"No Strikes, No Sit-ins" over France's Scarf Ban', *Chicago Tribune*, 8 September 2004, p. 6 (Hundley 2004).
8. Here I reflect the debates about 'secularisation' that have dominated much of the past half-century, from Carl Becker to Hannah Arendt to M.H. Abrams to Peter Berger to Hans Blumenberg's (1983) *The Legitimacy of the Modern Age* (trans. Robert M. Wallace [Cambridge, MA: MIT Press]) and beyond. Elizabeth Brient, 'Hans Blumenberg and Hannah Arendt on the "Unworldly Worldliness" of the Modern Age', *Journal of the History of Ideas*, 61 (2000), 513–30 (Blumenberg 1983; Brient 2000).
9. A. Sutcliffe, *Judaism and Enlightenment* (Cambridge: Cambridge University Press, 2003) (Sutcliffe 2003).
10. C. Taylor, 'Models of Secularism', in *Secularism and Its Critics*, ed. Rajeev Bhargava (Delhi: Oxford University Press, 1998), pp. 31–53; T. Asad, *Formations of the Secular: Christianity, Islam, Modernity* (Stanford, CA: Stanford University Press, 2003), p. 10 (Taylor 1998; Asad 2003).
11. R.W. Bulliet, *The Case for Islamo-Christian Civilization* (New York: Columbia University Press, 2004), p. 12 (Bulliet 2004).
12. Associated Press, 'French Bishop Orders Recall of Anti-Semitic Bible', 9 March 1995, AM Cycle (Associated Press 1995).
13. S. Heschel, *Abraham Geiger and the Jewish Jesus* (Chicago: University of Chicago Press, 1998) (Heschel 1998).
14. I. Goldziher, *Tagebuch*, ed. Alexander Scheiber (Leiden: Brill, 1978), p. 59 (Goldziher 1978).
15. J. Sacks, *The Dignity of Difference: How to Avoid the Clash of Civilizations* (London: Continuum, 2002). See also R. Harries, '[On] Jonathan Sacks, The Dignity of Difference; How to Avoid the Clash of Civilisations (2002)', *Scottish Journal of Theology*, 57 (2004), 109–15 (Sacks 2002; Harries 2004).

16. D. Biale, *Eros and the Jews* (Berkeley: University of California Press, 1997) (Biale 1997).
17. Y. Michal Bodemann and G. Yurdaku, 'Diaspora lernen: Wie sich türkische Einwanderer an den Juden in Deutschland orientieren', *Süddeutsche Zeitung* (2 November 2005) and Y. Michal Bodemann and G. Yurdaku, 'Geborgte Narrative: Wie sich türkische Einwanderer an den Juden in Deutschland orientieren', *Soziale Welt*, 56 (2005), 11–33 (Michal Bodemann and Yurdaku 2005a, b).
18. For an excellent case study of the adaptation of Judaism as religious practice in the American Diaspora, see J. Joselit, *The Wonders of America: Reinventing American Jewish Culture 1880–1950* (New York: Hill and Wang, 1994) (Joselit 1994).
19. A. Dundes, *The Blood Libel Legend* (Madison: University of Wisconsin Press, 1991) (Dundes 1991).
20. Asad, *Formations of the Secular*, p. 180.
21. See M. Mack, *German Idealism and the Jew* (Chicago: University of Chicago Press, 2003) (Mack 2003).
22. I. Kant, 'Religion within the Boundaries of Mere Reason', in *Religion and Rational Theology*, trans. Allan Wood (Cambridge: Cambridge University Press, 1996), pp. 155–6 (Kant 1996).
23. P. Mantegazza, *The Sexual Relations of Mankind*, trans. Victor Robinson (New York: Eugenics Pub. Co., 1935), p. 99 (Mantegazza 1935).
24. L. Zunz, *Gutachten über die Beschneidung* (Frankfurt am Main: J. F. Bach, 1844); L.A Hoffman, *Covenant of Blood: Circumcision and Gender in Rabbinic Judaism* (Chicago: University of Chicago Press, 1996); J. Katz, 'The Controversy over the *Mezizah*: The Unrestricted Execution of the Rite of Circumcision' and 'The Struggle over Preserving the Rite of Circumcision in the First Part of the Nineteenth Century', in *Divine Law in Human Hands: Case Studies in Halakhic Flexibility* (Jerusalem: Magnes Press, 1998), pp. 320–402 (Zunz 1844; Hoffman 1996; Katz 1998a, b).
25. Anonymous, 'Die rituelle Beschneidung bei den Juden und ihre Gefahren', *Journal für Kinderkrankheiten*, 59 (1872), 367–72 (Anonymous 1872).
26. See J. Schiratzki, *Best Interests of the Child*, Oxford Bibliographies in Childhood Studies, ed. Heather Montgomery (New York: Oxford University Press, 2013) (Schiratzki 2013).

27. *Rotterdams Dagblad*, 'Hirsi Ali wil verbod besnijdenis jongens', *Rotterdams Dagblad*, 4 October 2004 (*Rotterdams Dagblad* 2004).
28. Press Release, Landesgericht Köln, Urteile des Amtsgerichts und des Landgerichts Köln zur Strafbarkeit von Beschneidungen nicht einwilligungsfähiger Jungen aus rein religiösen Gründe [Judgments of the Local District Court and of the Cologne Regional Court on the Criminalization of Circumcision of Non-Consensual Boys for Purely Religious Reasons], JUSTIZ-ONLINE [Landgericht Köln website] (26 June 2012); Decision of 7 May 2012 [in German], Docket No. Az. 151 Ns 169/11, Landgericht Köln Cologne). Accessed at http://www.loc.gov/lawweb/servlet/lloc_news?disp3_l205403226_text on 14 September 2012.
29. Deutsche Akademie für Kinder-und Jugendmedizin. 2012. 'Stellungnahme zur Beschneidung von minderjährigen Jungen Kommission für ethische Fragen der DAKJ'. Accessed at http://dakj.de/media/stellungnahmen/ethische-fragen/2012_Stellungnahme_Beschneidung.pdf on 12 September 2013 (Deutsche Akademie für Kinder-und Jugendmedizin 2012).
30. *Der Spiegel*, 'The World from Berlin: Circumcision Ruling Is "a Shameful Farce for Germany"', *Der Spiegel Online*, 12 July 2012. Accessed at http://www.spiegel.de/international/germany/german-press-review-on-outlash-against-court-s-circumcision-ruling-a-844271.html on 15 September 2013 (*Der Spiegel* 2012).
31. M. Eddy, 'Proposal Sets Circumcision Regulations in Germany', *The New York Times Online*, 26 September 2012. Accessed at http://www.nytimes.com/2012/09/27/world/europe/27iht-circumcision27.html on 15 September 2012 (Eddy 2012).
32. Anon. 'Norwegian Official: Jews, Muslims Circumcise out of Ignorance', *Haaretz Online*, 25 November 2013. Accessed at http://www.haaretz.com/jewish-world/jewish-world-news/1.560094 (Anon 2013).
33. The complexity of this position is highlighted in the essays collected by J.R. Jakobsen and A. Pellegrini (eds.), 'World Secularisms at the Millennium', *Social Text*, 18, no. 3 (2000). See specifically their introduction, pp. 1–27 (Jakobsen and Pellegrini 2000).
34. See R. Samantrai, 'Continuity or Rupture? An Argument for Secular Britain', *Social Text*, 18, no. 3 (2000), 105–21 (Samantrai 2000).

35. Two polemical but informative books shape their argument about contemporary Islamic identity primarily around the rhetoric of the Israeli–Palestinian conflict rather than as part of the struggle about the modernisation of Islam both within and beyond Europe: J. Goody, *Islam in Europe* (London: Polity, 2004); and G. Kepel, *The War for Muslim Minds: Islam and the West*, trans. Pascale Ghazaleh (Cambridge, MA: Belknap Press of Harvard University Press, 2004). See also E. Shohat, 'Columbus, Palestine and Arab-Jews: Toward a Relational Approach to Community Identity', in *Cultural Readings of Imperialism: Edward Said and the Gravity of History*, eds. Keith Ansell-Pearson, Benita Parry and Judith Squires (New York: St. Martin's, 1997), pp. 88–105; and Asad, *Formations of the Secular*, pp. 158–80 (Goody 2004; Kepel 2004; Shohat 1997).
36. P. Nora (ed.), *Les France: Conflits et partages*, Vol. 1 of *Les Lieux de mémoire* (Paris: Gallimard, 1993) (Nora 1993).
37. See B. Tibi, *Krieg der Zivilisationen* (Hamburg: Hoffmann & Campe, 1995). His work is available in English: B. Tibi, *The Challenge of Fundamentalism* (Berkeley: University of California Press, 2002); and B. Tibi, *Islam between Culture and Politics* (New York: Palgrave Macmillan, 2002) (Tibi 1995, 2002a, b).
38. T. Ramadan, *Western Muslims and the Future of Islam* (New York: Oxford University Press, 2003); F.A. Rauf, *What's Right with Islam: A New Vision for Muslims and the West* (San Francisco: HarperSanFrancisco, 2004) (Ramadan 2003; Rauf 2004).
39. T. Sarrazin, *Deutschland schafft sich ab: Wie wir unser Land aufs Spiel setzen* (München: Deutsche Verlags-Anstalt, 2010), p. 95 (Sarrazin 2010).

References

A Tragic Twist of the Scarf. *Economist*, 4 September 2004.

Anon. 2013. Norwegian Official: Jews, Muslims Circumcise out of Ignorance. *Haaretz Online*, 25 November. http://www.haaretz.com/jewish-world/jewish-world-news/1.560094

Anonymous. 1872. Die rituelle Beschneidung bei den Juden und ihre Gefahren. *Journal für Kinderkrankheiten* 59: 367–372.

Asad, T. 2003. *Formations of the Secular: Christianity, Islam, Modernity*. Stanford, CA: Stanford University Press.

Associated Press. 1995. French Bishop Orders Recall of Anti-Semitic Bible, 9 March. AM Cycle.

Biale, D. 1997. *Eros and the Jews*. Berkeley: University of California Press.

Blumenberg, Hans. 1983. *The Legitimacy of the Modern Age*, trans. Robert M. Wallace. Cambridge, MA: MIT Press.

Brient, Elizabeth. 2000. Hans Blumenberg and Hannah Arendt on the "Unworldly Worldliness" of the Modern Age. *Journal of the History of Ideas* 61: 513–530.

Bulliet, R.W. 2004. *The Case for Islamo-Christian Civilization*. New York: Columbia University Press.

Der Spiegel. 2012. The World from Berlin: Circumcision Ruling Is "a Shameful Farce for Germany". *Der Spiegel Online*, 12 July. Accessed 15 September 2013. http://www.spiegel.de/international/germany/german-press-review-on-outlash-against-court-s-circumcision-ruling-a-844271.html

Deutsche Akademie für Kinder-und Jugendmedizin. 2012. Stellungnahme zur Beschneidung von minderjährigen Jungen Kommission für ethische Fragen der DAKJ. Accessed 12 September 2013. http://dakj.de/media/stellungnahmen/ethische-fragen/2012_Stellungnahme_Beschneidung.pdf

Dundes, A. 1991. *The Blood Libel Legend*. Madison: University of Wisconsin Press.

Eddy, M. 2012. Proposal Sets Circumcision Regulations in Germany. *The New York Times Online*, 26 September. Accessed 15 September 2012. http://www.nytimes.com/2012/09/27/world/europe/27iht-circumcision27.html

Goldziher, I. 1978. *Tagebuch*, ed. Alexander Scheiber. Leiden: Brill.

Goody, J. 2004. *Islam in Europe*. London: Polity.

Harries, R. 2004. [On] Jonathan Sacks, The Dignity of Difference; How to Avoid the Clash of Civilisations (2002). *Scottish Journal of Theology* 57: 109–115.

Heschel, S. 1998. *Abraham Geiger and the Jewish Jesus*. Chicago: University of Chicago Press.

Hess, J.M. 2002. *Germans, Jews, and the Claims of Modernity*. New Haven, CT: Yale University Press.

Hoffman, L.A. 1996. *Covenant of Blood: Circumcision and Gender in Rabbinic Judaism*. Chicago: University of Chicago Press.

Hundley, T. 2004. "No Strikes, No Sit-ins" over France's Scarf Ban. *Chicago Tribune*, 8 September.

Jakobsen, J.R., and A. Pellegrini (eds). 2000. World Secularisms at the Millennium. *Social Text* 18(3): 1–27.

Jentzsch, R. 1998. Das rituelle Schlachten von Haustieren in Deutschland ab 1933. Ph.D. diss., Tierärztliches Hochschul, Hannover.

Joselit, J. 1994. *The Wonders of America: Reinventing American Jewish Culture 1880–1950*. New York: Hill and Wang.

Kant, I. 1996. Religion within the Boundaries of Mere Reason. In *Religion and Rational Theology*, trans. Allan Wood. Cambridge: Cambridge University Press.

Katz, J. 1998a. The Controversy over the Mezizah: The Unrestricted Execution of the Rite of Circumcision. In *Divine Law in Human Hands: Case Studies in Halakhic Flexibility*. Jerusalem: Magnes Press.
———. 1998b. The Struggle over Preserving the Rite of Circumcision in the First Part of the Nineteenth Century. In *Divine Law in Human Hands: Case Studies in Halakhic Flexibility*. Jerusalem: Magnes Press.
Kepel, G. 2004. *The War for Muslim Minds: Islam and the West*, trans. Pascale Ghazaleh. Cambridge, MA: Belknap Press of Harvard University Press.
Mack, M. 2003. *German Idealism and the Jew*. Chicago: University of Chicago Press.
Mantegazza, P. 1935. *The Sexual Relations of Mankind*, trans. Victor Robinson. New York: Eugenics Pub. Co.
Michal Bodemann, Y., and G. Yurdaku. 2005a. Diaspora lernen: Wie sich türkische Einwanderer an den Juden in Deutschland orientieren. *Süddeutsche Zeitung*, 2 November.
———. 2005b. Geborgte Narrative: Wie sich türkische Einwanderer an den Juden in Deutschland orientieren. *Soziale Welt* 56: 11–33.
Nora, P. (ed). 1993. *Les France: Conflits et partages, Vol. 1 of Les Lieux de mémoire*. Paris: Gallimard.
Peritz, I., and L. Perreaux. 2013. Quebec Reveals Religious Symbols to be Banned from Public Sector. *The Globe and Mail*, 10 September.
Ramadan, T. 2003. *Western Muslims and the Future of Islam*. New York: Oxford University Press.
Rauf, F.A. 2004. *What's Right with Islam: A New Vision for Muslims and the West*. San Francisco: HarperSanFrancisco.
Rotterdams Dagblad. 2004. Hirsi Ali wil verbod besnijdenis jongens. *Rotterdams Dagblad*, 4 October.
Roy, O. 2004. *Globalised Islam: The Search for the New Ummah*. London: Hurst & Co.
Sacks, J. 2002. *The Dignity of Difference: How to Avoid the Clash of Civilizations*. London: Continuum.
Samantrai, R. 2000. Continuity or Rupture? An Argument for Secular Britain. *Social Text* 18(3): 105–121.
Sarrazin, T. 2010. *Deutschland schafft sich ab: Wie wir unser Land aufs Spiel setzen*. München: Deutsche Verlags-Anstalt.
Schiratzki, J. 2013. *Best Interests of the Child*. Oxford Bibliographies in Childhood Studies, ed. Heather Montgomery. New York: Oxford University Press.
Sciolino, E. 2004. Ban on Head Scarves Takes Effect in a United France. *New York Times*, 3 September.
Shohat, E. 1997. Columbus, Palestine and Arab-Jews: Toward a Relational Approach to Community Identity. In *Cultural Readings of Imperialism: Edward Said and the Gravity of History*, ed. Keith Ansell-Pearson, Benita Parry, and Judith Squires. New York: St. Martin's.

Sutcliffe, A. 2003. *Judaism and Enlightenment*. Cambridge: Cambridge University Press.
Taylor, C. 1998. Models of Secularism. In *Secularism and Its Critics*, ed. Rajeev Bhargava. Delhi: Oxford University Press.
Tibi, B. 1995. *Krieg der Zivilisationen*. Hamburg: Hoffmann & Campe.
———. 2002a. *The Challenge of Fundamentalism*. Berkeley: University of California Press.
———. 2002b. *Islam between Culture and Politics*. New York: Palgrave Macmillan.
Zunz, L. 1844. *Gutachten über die Beschneidung*. Frankfurt am Main: J. F. Bach.

CHAPTER 7

Islamophobia and Antisemitism in the Balkans

Marko Attila Hoare

Antisemitism and Islamophobia are two phenomena that have seemingly been growing in parallel across the globe in the period since the 9/11 attacks in 2001. As discussed in other chapters of this book, the question of whether they are essentially similar to one another or fundamentally different is highly controversial. This chapter will examine the question from the perspective of the Balkans. There, violence and chauvinism against Jews and against Muslims have frequently gone hand in hand, but have also diverged at times from one another. The relationship of state policy and nationalist ideology towards Muslims and Jews has been shaped by a common framework, but varied according to political circumstances.

As discussed in the Introduction to this book, there is some resistance to the term 'Islamophobia'. Among liberal intellectuals, it has been argued that as Islam is a religion it is, therefore, an ideology, and it is questionable whether one can be prejudiced against an ideology.[1] Yet such a distinc-

M.A. Hoare (✉)
Kingston University, Kingston upon Thames, UK

J. Renton, B. Gidley (eds.), *Antisemitism and Islamophobia in Europe*, DOI 10.1057/978-1-137-41302-4_7

tion is not satisfactory from the standpoint of a scholar of the Balkans; or, indeed, from the historical standpoint generally. To treat chauvinism against a religious community as being fundamentally different from chauvinism against an ethnic or racial group is to superimpose a modern understanding of religion onto the past. We may believe in the ideals of the separation of church and state; and of religion as a private, personal matter of conscience; but it is anachronistic to impose this liberal ideal onto past human history.

Religious and ethnic prejudice are not distinct categories, and it makes no historical sense to see them as such. This fact is recognised in the text of the United Nations Convention on the Prevention and Punishment of the Crime of Genocide, Article 2 of which states: 'In the present Convention, genocide means any of the following acts committed with intent to destroy, in whole or in part, a national, ethnical, racial or religious group, as such.'[2] Religious and racial antisemitism are distinct yet closely related phenomena; even the Nazis used religious background to determine who was Jewish.[3] In the Balkans, ethnicity and religion are historically closely related, and the model for chauvinism that antisemitism provides—in which prejudice against a religious community evolves into an ethnic or racial prejudice—is the rule rather than the exception.

Balkan Nationalisms and Non-Christian Minorities at the End of the Ottoman Empire

The Ottoman empire ruled over much of the Balkans from the late Middle Ages until the early twentieth century, and it was the Ottoman system that laid the basis for modern ethnicity and nationality in the Balkans. The Ottoman empire was organised on the basis of different legal statuses for Muslims and non-Muslims, in which Muslims were the dominant and privileged group, but Christians and Jews nevertheless enjoyed a degree of communal autonomy. As Barbara Jelavich writes:

> despite its close connections with the Ottoman government and the corruption in its operation, the Orthodox church did provide important services for the Christian people. Most significant was the fact that it kept the Christian community almost unchanged in an ideological sense until the age of the national movements. Certainly, the church preserved carefully the idea of Christian exclusiveness. It taught that the Ottoman Empire had been victorious because the sins of the Christians had called down God's

> punishment. Muslim rule was, however, ephemeral; a new age would soon arrive when the Christian people would again emerge triumphant. Although the Christian was a second-class citizen in the Muslim state, his religious leaders taught him that on a higher moral basis he was infinitely superior to his conquerors.[4]

The Ottoman empire's administrative and social division along religious lines laid the basis for the different religious communities to evolve into separate nationalities.

When the Orthodox nationalities of the Balkans rose up against their Ottoman imperial masters during the nineteenth century with the goal of establishing their independence from the empire, the process involved the expulsion or extermination of much of the non-Christian population, which was identified as an alien, non-national element. This process of ethnic or religious cleansing was directed primarily against the Muslim population that was concentrated in the towns. Yet it targeted the Jews too, who were also concentrated in the towns and who were, in the eyes of the predominantly peasant and Christian rebels, equally alien and part of the Ottoman presence.

This was seen in the violence that accompanied the uprisings themselves, with rebels massacring non-Christians, often in a pre-meditated and cold-blooded manner. The so-called First Serbian Uprising of 1804–13 involved large-scale massacres of the predominantly Muslim populations of the towns. In his account of the uprising, the famous nineteenth-century German historian Leopold von Ranke thus describes the Serbian rebels' intentions, as they prepared to occupy Belgrade in late 1806: 'Nevertheless, it is probable that even at this time all the Turks were destined to be put to death.' The rebels massacred the Ottoman commander and defenders of the Belgrade citadel, after violating their own promise of a safe evacuation. Then,

> The massacre immediately extended to Belgrade. For two days the Turks, who had endeavoured to conceal themselves, were sought out and slaughtered ... In such fearful acts of cruelty did their hatred against the Turks vent itself: hatred long suppressed, but strengthened by mutual animosities, and by the war; and at last thus fiercely bursting forth.[5]

Muslims and Jews were slaughtered, expelled or forced to convert to Christianity.[6]

Continued forced removal of non-Christians in Serbia took place more quietly in the decades that followed the establishment of Serbian autonomy in 1815–30. In the 1830s, Serbia's Prince Miloš assisted the Ottomans in suppressing a Muslim revolt in neighbouring Bosnia, in return for which the Ottomans granted him several concessions: an Ottoman *hatti serif* (edict) of May 1833 ruled, among other things, that all Muslims in Serbia would be removed within five years, except those in the Ottoman fortress towns. Through various forms of persecution and harassment, Miloš succeeded in encouraging most of the Muslim landlords to leave Serbia.[7] The Serbian capital of Belgrade had been largely Muslim before the nineteenth century, but following the establishment of the autonomous Serbian principality, the Muslim population was mostly expelled and most of its mosques were destroyed or dismantled. Similarly, although Miloš was relatively sympathetic to the Jews, his successors were less so, and the Jewish communities underwent restrictions they had not suffered in the Ottoman period, and were expelled or re-located from the towns outside Belgrade.[8]

During the Greek War of Independence in the 1820s, which established an independent Greece, over 25,000 Ottoman Muslims may have been killed by the Greek revolutionaries, producing a homogenous Greek Christian population.[9] The Greek rebels also targeted Jews: they eradicated the entire Jewish population of the town of Tripolis in November 1821, involving a death toll that some contemporary sources placed in the thousands. Jews were treated in a similar fashion throughout Greece, so that none were left in the Peloponnese by the end of the war; 'The sons of Isaac, and the sons of Ishmael, … as on every occasion during the Greek Revolution, met with a common fate', as one foreign eyewitness wrote in 1831.[10]

In subsequent decades, Greece's active pursuit of irredentism, involving the acquisition of vast new territories in the Balkan Wars and the First World War during the 1910s, resulted in renewed pressure on its Muslim and Jewish minorities. Members of these and other minorities tended to give their support to the less irredentist anti-Venizelist political camp based primarily in the lands of the old Greece, while the more aggressively nationalistic Venizelists, which derived their strongest support from the ethnic Greek population of the newly acquired lands, spearheaded policies aimed at restricting the autonomy and voting power of minorities.[11] As prime minister, Eleftherios Venizelos had signed the Treaty of Lausanne with Turkey on 24 July 1923 following its defeat of Greece in Anatolia the year before. The treaty provided for the exchange of populations between the two states, resulting in the transfer of about 400,000 Greek Muslims to Turkey, mostly from the lands newly acquired in the 1910s, in return for the transfer of at least 1.2 million Turkish Christians to Greece.[12] The assimilation of these largely Turkish-speaking refugees consolidated Greece as an ethnically Greek Orthodox state, in which the remaining minorities were the objects of further assimilation and intolerance. Reacting to perceived Jewish support for his opponents, Venizelos complained in July 1933:

> The attitude of the Jewish element, which voted for the government ticket as a group, and by order of its communal leadership and of the rabbinate, constitutes an act of hostility against half of Greece ... This has fatally created an intolerable situation, which forces the Opposition to consider the matter more radically, in time. The Jews should have been grateful until now to the old republican parties, which, although they governed the country continually, forgot even the votes of the Jews of 1 November 1920, which, however, contributed to the overthrow of the Liberals and to the destruction of Great Greece.[13]

The Bulgarians achieved autonomous statehood in the Balkan conflict of the 1870s, when Russia waged war against the Ottoman empire to enable it to carve out a state of its own. In this conflict, according to one calculation, about 260,000 Bulgarian Muslims were killed or died of disease, starvation and cold; by 1879, 17% of the Muslims of Bulgaria had perished. As many as half a million may have been permanently expelled from the country. According to Edmund Calvert, the UK's acting consul writing in September 1878, the 'Russian government allows the Christians to take the law into their own hands and to visit the Turkish Community at large with present and indiscriminate bloodshed, rapine and pillage', while the Bulgarians in the Kyzanlik district had engaged in the 'deliberate and partly successful attempts to exterminate the adult male Turkish population of that district by wholesale and cold-blooded executions'. The Bulgarian rebels massacred, pillaged and expelled the Jews along with the Muslims. According to one contemporary account, 'in some instances the Jews suffered even more than the Mahomedans from the savagery of the Bulgarians'.[14]

Following the establishment of autonomous Bulgarian statehood in 1878, 600,000 hectares of Muslim land were bought by Christians by 1900, and 175 Muslim villages were abandoned, of which 118 between 1878 and 1885. The Turkish population, which had comprised 26%

of the Bulgarian total in 1878, shrank by 1900 to 14% and by 1910 to 11.63%.[15] According to the pre-eminent study of Bulgaria's relationship to its Muslim minorities, 'There can be little doubt that intensification of Bulgaro-Muslim antipathy was one of the unfortunate results of the Bulgarian appropriation of European modernity.'[16] The growth of the state's power and intrusiveness vis-à-vis its citizens' lives following the establishment of a Communist dictatorship after the Second World War resulted in heightened persecution of Bulgarian Muslims—Pomaks (Bulgarian-speaking Muslims) and ethnic Turks. Around 140,000 Turks were expelled from Bulgaria in 1950–51. Around 350,000 Turks—half the Turkish population of Bulgaria—were expelled in 1989.[17]

Of course, the extent to which Muslims or Jews were massacred, expelled or persecuted varied according to country and period. This was not a matter of Nazi-style total extermination. Persecution and expulsion alternated and overlapped with efforts at co-option, assimilation and toleration. However, the model of nationhood remained very much one that was based on the Orthodox Christian population, in which the non-Orthodox were, at best, viewed as less national than the Orthodox.

Non-Muslim Minorities in the Muslim Balkans

This model of religiously determined nationhood was adopted not only by Orthodox Christians, but also by the Muslim Turks. The establishment of a Turkish nation-state in the 1910s and 1920s involved the extermination or expulsion of literally millions of Christians. A million Armenians were murdered in the Armenian Genocide of 1915, amounting to 50% of the pre-war Armenian population of the Ottoman Empire, while another half million were deported but survived, according to the estimate of Donald Bloxham.[18] The remaining Christian population of Anatolia, above all Greek, was mostly exterminated or expelled during the 1920s Turkish War of Independence and the subsequent population exchanges. When Turkish troops re-captured Smyrna on the western Anatolian coast from the Greeks in September 1922, the Greek archbishop was lynched at the instigation of the Turkish commandant in the town, Nurettin Pasha, after which the Christian quarters of the town were burned down, and perhaps 213,000 Christians evacuated by the Allies. The future Turkish president Mustafa Kemal subsequently criticised Nurettin for trying to claim sole credit for 'the patriotic effort of all members of the army to expel non-Muslims from western Anatolia'.[19] The destruction of Christian

Smyrna claimed tens if not hundreds of thousands of Christian victims either killed or deported into the Anatolian interior, from where most were never heard of again.[20] Formally, the Anatolian Christian victims of Turkish nationalism were Greeks or Armenians. But this included Turkish-speaking Christians who were excluded from the Turkish nation solely because of their religion. Turkish nationhood, therefore, was based on the Muslim religion: it was inclusive of Kurds and other non-Turkish-speaking Muslims who inhabited Anatolia, but it was exclusive of Turkish-speaking Christians.

After establishing their nation-state, the Turks initially had a better record of treating their Jewish citizens than did the Balkan Christians. This was a legacy of the fact that the Muslims, as the elite group in the Ottoman empire, had not viewed the Jews as outsiders in the same way that the Christians had. Yet there was still some anti-Jewish activity on the part of the Turkish state that, with Nazi encouragement, reached its peak during the Second World War. In November 1942, a Turkish law was passed to force Christian and Jewish businessmen to pay a capital levy. Some 1,400 of these businessmen, mostly Jews, who were unable to pay were imprisoned that winter in a camp at Aşkale in eastern Anatolia. In this period, for the first time in Ottoman and Turkish history, state discrimination also targeted the descendants of Jewish converts to Islam, with pro-government papers publishing Nazi-style antisemitic cartoons. Nearly half of Turkey's Jewish community, numbering around 80,000 in 1945, emigrated to Israel following the latter's foundation in 1948.[21] Furthermore, in the great anti-Greek pogrom in Istanbul in 1955, Jews were again targeted; according to the Istanbul police, 523 Jewish, 741 Armenian and 2,572 Greek businesses were destroyed.[22]

There were some exceptions to the general rule of religiously based nationhood in the Balkans. The Albanians are the only major example of a Balkan nation for whom religion is not the determining factor. The most likely explanation is that Albanian nationalism originated with the Catholic population among Albanian speakers. And the Catholics were not legally and economically subordinate to Muslim landlords in the way that Orthodox peasants throughout the Balkans were subordinate to Muslim landlords. So there was not the same degree of class oppression tied into the religious divide between Catholics and Muslims among Albanian speakers as there was between Orthodox and Muslims among the Slavic-, Greek- and Turkish-speaking peoples.[23]

Perhaps not coincidentally, the Albanians' record with regard to the Jews during the Holocaust was about the best in all of Nazi-occupied Europe: Albanians sheltered Jews more solidly than almost any other occupied people.[24] According to the website *Yad Vashem*:

> In the beginning of 1944 the Germans ordered the Jews to register, but Albanians, including government officials, helped the Jews to flee from Tirana. They found refuge with Albanian families and with partisans. We know only of two cases where Jews were captured and deported. Mrs. Bachar and her children were deported to Bergen Belsen, but survived. Yitzhak Arditi was deported with his wife and four children—only the father survived. All the other Jews survived the war. The assistance afforded to the Jews may have been grounded in an Albanian code of honor—'Besa'. Besa literally means 'to keep the promise'; its significance was that once a family was hosted by Albanians, they could trust them with their lives.[25]

In contrast to Serbia, in Bosnia (known from the 1870s as Bosnia-Hercegovina) the Orthodox Serbs in the nineteenth century comprised a minority. Like Albania, Bosnia-Hercegovina was inhabited by Orthodox, Muslims and Catholics. The two largest ethno-religious groups were the Orthodox Serbs and the Muslim Bosniaks. Bosnian Serb nationalism consequently manifested the conflicting desires both to define itself against the Muslims and to include them within the Serb nation. The Bosnian Serb socialist nationalist Vaso Pelagić celebrated Bosnia-Hercegovina's Islamic heritage and spoke glowingly of the Islamic religion: 'the Turkish religion is of a democratic nature'; 'In Islam there are no princes or paupers, according to the Qur'an, but only Muhammedans.' Pelagić believed that medieval Bosnia was inhabited by members of the Bogumil Christian sect and that, on account of the 'democratic nature of the Muhammedan religion and administration', members of this sect turned Turk en masse 'because due to their faith and freedom of thought they had been persecuted by both Orthodox and Catholics'; conversely, the Ottoman government had been 'towards other faiths much more tolerant and human than many Christian governments and states'.[26] In 1871, Pelagić called on Orthodox Serbs to 'unite fraternally with Serbo-Muhammedans and Serbo-Catholics' in the struggle against Ottoman rule. Pelagić was, however, the author of an antisemitic tract in which he promised to inform readers of 'the horrors that the kikes and their gospels—the Talmud, are preparing for the entire non-kike world'.[27]

Opposition to Catholic Austro-Hungarian rule over Bosnia-Hercegovina, which began in 1878, united Serbs and Muslims. At the turn of the century, Bosnian Serb political leaders such as Gligorije

Jeftanović, who headed the movement for Bosnian Serb autonomy under Austria-Hungary, wore the fez as a mark of their Serbdom, shunning the hat as a symbol of Viennese rule.[28] However, unlike the case with the Albanians, Serb nationalism in Bosnia-Hercegovina was firmly rooted in the Orthodox population, the mass of whose peasants were legally and socially subordinate to the Muslim landlord class; Serb nationalism proved unable to transcend this class dichotomy, and Orthodox and Muslims in Bosnia-Hercegovina developed into wholly separate nations.

Without a single dominant nationality into which they could assimilate, members of the Sephardic Jewish community in Bosnia-Hercegovina developed a distinct sense of nationality of their own. They saw themselves as distinct from the Ashkenazim, who were culturally different. And as they were not oppressed by a dominant nationality that treated them as outsiders, they were less receptive to Zionism than were the Jews of most Central European countries.[29] So the Bosnian Sephardim followed the general Bosnian pattern, whereby the different religious communities evolved into different nationalities. Immediately following the establishment of the Yugoslav state in 1918, the Political Committee of Jews of Bosnia-Hercegovina, which was predominantly Sephardic, issued a statement expressing their Bosnian Jewish national identity:

> We Jews of Bosnia-Hercegovina, who have always lived in brotherly communication with the people of this land and have shared with them all fates in joy and misfortune, following with best wishes the political aspirations of the Yugoslav peoples, feel it is our duty to make the following statement: As self conscious and nationalist Jews, who always value highly the great idea of self-determination of nations and democracy, we join the program of the National Council of Serbs, Croats and Slovenes contained in the proclamation of October 19, 1918, and as sons of this land we see guaranteed in this proclamation the free development of the Jews of Bosnia-Hercegovina.[30]

Croatia and Serbia

Croatia had not been ruled and oppressed by the Ottomans, but the effect of Ottoman expansion in the fifteenth and sixteenth centuries had been to reduce the territorial extent of the historical Croatian kingdom. Croatian nationalism was therefore not structurally conditioned by the experience of Islamic domination, but rather by the desire to claim or re-claim territories that were now inhabited by Muslims who spoke the same language as the Croats. Consequently, Croat nationalists were almost unique in Europe in the extent to which they were ready to embrace Muslims. Ante Starčević, the father of integral Croat nationalism, viewed the Bosnian Muslims as the racially purest Croats.[31] He said that of all religions, 'only the Turkish is worth something; all the others absolutely nothing'.[32] To Starčević, 'The Mohammedans of Bosnia-Hercegovina have nothing in common with the Turkish, Mohammedan breed; they are of Croat breed; they are the oldest and cleanest nobility that Europe has.'[33] According to the tradition he established, the Bosnian Muslims were the 'flower of the Croat nation'. This was possible for Croat nationalists because, unlike the Orthodox peoples of the Balkans, Croatia had not been ruled and oppressed by the Ottomans. The Islamophile character of integral Croat nationalism was, of course, a way for it to lay claim to Bosnia, where the Catholics were only a small minority.

There is some controversy over the extent of Ante Starčević's antisemitism. He made antisemitic statements, but chose a Jew, Josip Frank, as his successor as leader of his political party, the Pure Party of Right. In the inter-war period, the mainstream Croatian national movement embodied in the Croat Peasant Party upheld what its leaders termed 'a-Semitism', thus defined: 'Instead of anti-Semitism, we should therefore strictly carry out a-Semitism: instead of an unworthy struggle against the Jews, unremitting work without the Jews.' This was an expression of racial pan-Slavism or Yugoslavism.[34] A much more intense form of anti-Jewish ideology was represented by the extremist Croatian Ustasha movement, whose racial theory emerged both from, and as a reaction against, racial Yugoslavism. It found expression in the Ustasha genocide against the Jews during the Second World War.[35]

The Ustashas pursued a genocidal policy also against Orthodox Serbs—but not against Muslims, as the policy of the Ustashas was to treat Bosnian Muslims as Islamic Croats.[36] The Ustasha newspaper *Hrvatska Krajina*, published in the Bosnian town of Banja Luka, stated in April 1941 that the 'little Croat nation is divided between two worlds, a majority Western and Catholic and a minority Eastern and Islamic. We are the only nation in Europe that embraces two such different cultural and religious elements.' Of these two wings of the Croat nation, the Muslim part was the 'most pure-blooded, for while the Catholic part of the Croats was considerably infiltrated by the influx of foreign elements—German, Czech, Magyar, Italian, Slovene and so forth—the Bosnian-Hercegovinian Muslims intermarried exclusively among themselves'.[37] In August 1941, the Ustasha regime ordered the construction of a mosque in the Croatian capital of Zagreb, and it was finally opened in August 1944.

The Serb Chetniks in the Second World War were the counterpart of the Croat Ustashas: an extreme nationalist movement that systematically persecuted and killed the non-Orthodox population in Bosnia: Muslims, Croats and Jews. The Chetniks were engaged in a vicious war against the Yugoslav Partisans, who were a multi-national resistance movement led by the Communist Party of Yugoslavia. The Chetniks identified the Communists with the Jews, but also with the Muslims. According to a Chetnik pamphlet endorsed by Boško Todorović, the Chetnik commander for East Bosnia and Hercegovina:

> When it achieves freedom, a golden Serb freedom, then the Serb nation will—freely and without bloodshed, by means of the free elections to which we are accustomed in the Serbia of King Peter I—take its destiny into its own hands and freely say, whether it loves more its independent Great Serbia, cleansed of Turks and other non-Serbs, or some other state in which Turks and Jews will once again be ministers, commissars, officers and 'comrades'.

Another Chetnik pamphlet claimed:

> The Supreme Commander of all Communist forces in the country is some Comrade Tito, whose real name nobody knows, but we know only that he is a Zagreb Jew. His leading collaborators are Moše Pijade, a Belgrade Jew; Frano Vajner, a Hungarian Jew; Azija Kokuder, a Bosnian Turk; Safet Mujije, a Turk from Mostar; Vlado Šegrt, a former convict; and many others similar to them. Their names best testify as to whom they are and to how much they fight from their heart for our people.

One senior Chetnik even accused the Communists of having 'destroyed Serb churches and established mosques, synagogues and Catholic temples'.[38]

In the Second World War, however, it was still possible for the Chetniks to waver between massacring Muslims and attempting to co-opt them on the grounds that Bosnian Muslims were 'really' Serbs. For all that he and his movement had incited anti-Muslim hatred and carried out huge massacres of Muslims, the Chetnik leader Draža Mihailović came to appreciate the need to win over Muslim opinion in Yugoslavia, in light of the advantage that his Communist and Partisan opponents had derived from their successful recruitment of Muslims. In April 1944, Mihailović issued an appeal to members of the Muslim elite whom he considered sympathetic: 'With the aim of reaching as soon as possible a more earnest drawing together of Serbs and Muslims'. He assured the latter: 'We wish for Islam to be a recognised religion within our renewed state and to be, within the Serb federal unit, an equal state religion.' Furthermore, 'State policy should aim, in the national interest, that in the city of Sarajevo there develop a great Islamic spiritual centre for the whole of Europe, so that Islam would be represented to Europe with dignity, by our country, and so that the Islamic living space in our country would receive a definite advantage.'[39] So as late as the Second World War, both Serb and Croat nationalists could still make some pretence at treating Muslims as a religious group within their respective nations. This may be compared to the confusion among modern antisemites, until quite late in the day, as to whether Jews were a religious or a racial group.

Yugoslavia

The Communist-led Partisan movement represented the principal opposition to the nationalist extremism of the Croat Ustashas and Serb Chetniks, and succeeded in conquering power in Yugoslavia during the war years of 1941–45. It was a multi-national movement that drew support from all Yugoslavia's principal nationalities; its rank-and-file was disproportionately Serbian, while its leadership was disproportionately Croatian.[40] Although the great majority of Yugoslav Jews were murdered in the Holocaust, those who survived became one of the most staunchly pro-Partisan of the country's ethnic groups: 4,572 Jews fought as Partisans, of whom 1,318 were killed.[41] The Communists also acted as protectors of Bosnia-Hercegovina's Muslims. Their propaganda stressed the equality of Bosnia-Hercegovina's Serbs, Croats and Muslims and the right to self-rule of their common Bosnian homeland. In response, Muslims joined the Partisans in large numbers, particularly from the autumn of 1943. The Partisan victory involved the establishment of Bosnia-Hercegovina as one of the new federal Yugoslavia's six constituent republics.[42]

Eventually, in 1968, the Yugoslav Communist regime under Josip Broz Tito recognised the Muslims as a nation in their own right.[43] The regime also granted extensive autonomy to the Socialist Autonomous Province of Kosovo, formally part of Serbia but inhabited overwhelmingly by Muslim Albanians. Consequently, the Serb nationalist backlash against the Yugoslav federal order, which gathered momentum following Tito's death in 1980 and found expression in Slobodan Milošević's seizure of power in 1987, took an overtly Islamophobic form.[44] Although Serb nationalists in the 1980s and 1990s continued to pay lip service to the traditional nationalist view that Bosnian Muslims were really just Islamic

Serbs, in practice this kind of assimilationism was no longer possible or relevant. In the war in Bosnia-Hercegovina that broke out in full in 1992, there was no policy of forced conversion. Serb nationalists in Bosnia-Hercegovina killed, persecuted and expelled Muslims who spoke their language, much as the Serbian regime killed, persecuted and expelled Kosovo Albanians who spoke an entirely different language. They simultaneously viewed Muslims as a racially alien element, while portraying them in their propaganda as part of an international, global threat to Christian Europe—much as antisemites have viewed Jews. According to one author writing in 1991 in *Glas Crkve*, the Serbian Orthodox Church's official organ, 'for the last few decades we [Serbs] have also become known for being the target of sudden pressure of *jihad* from fundamentalist Islam'. According to another, 'Serbia and its peoples find themselves between two powerful religious internationals [Islam and Catholicism] ... In a state such as this, the national and ethnic survival of the Serbs is in great danger.'[45]

In contrast to the nationalism of the Orthodox peoples of the Balkans, it was only in the 1990s that the Croat nationalist mainstream became overtly anti-Islamic; this was due to the policy of the Croatian despot Franjo Tudjman, who aimed to join with the Serbs in partitioning Bosnia.[46] As Tudjman stated in November 1996:

> The Bosniaks were mostly—about 80%—Croat, but religion separated them from the Croat body. The majority of them speak like Croats of the Ikavian dialect; that is, therefore, the link between the Dalmatian and Slavonian Croats. I did not, therefore, wholly accept Starčević's idea that they are the flower of Croatdom, but I believe that the life of Croatia, in the geopolitical sense, cannot be without firm collaboration with the territory on which they live.

What made the difference for Croat nationalists by the 1990s, compared to the 1940s, was that by then the Muslims had been formally recognised within the Yugoslav constitutional system as a nation in their own right, distinct from the Serbs and Croats. As Tudjman continued:

> The decision of the C[ommunist] P[arty] on recognising the Muslims as a separate nation was not in the Croatian national interest, and not even Khomenei or Gaddafi agreed with it. If the Communists had not made that error, today we most probably would not have had such a war in Bosnia, and maybe this entire tragedy, which we lived through these past years, would have been different![47]

Thus, when Muslims could no longer be viewed as Islamic Croats and potentially assimilated, they became open to persecution by expansionist Croat nationalism, which switched from Islamophilic to Islamophobic. In December 1997, Tudjman claimed:

> neither Europe nor the United States of America accepted the birth of a purely Muslim entity which would favour Islamic expansion. For that reason, we accepted the [establishment of] the Croat-Muslim Federation, but on condition that it maintain links with Croatia. Otherwise the Croat minority would be Islamised. Today in Bosnia 174 mosques are being built, while Catholic churches are destroyed and only three are being restored. There is obviously a desire that that country be Islamised.[48]

Similarly, in a four-part essay in the pages of the Croatian daily *Slobodna Dalmacija*, Tudjman's literary mentor Ivan Aralica attempted to explain the burgeoning Muslim–Croat conflict in Bosnia-Hercegovina through reference to Bosnian president Alija Izetbegović's pan-Islamic manifesto *Islamic Declaration*, written in the late 1960s. He denounced

> Izetbegovic's idea, which he does not specifically mention but which is easily recognizable within the Declaration from the context, that within the secular political and social space, such as is the entire European space, both Western and Eastern, there be installed a wholly theological state. That would be impossible. Europe would never allow the transplanting of such a foreign body into its space.[49]

By this period—the 1990s—both Serb and Croat nationalists were more likely to identify with Israel on an anti-Muslim basis than they were to indulge in antisemitism.[50] Anti-Semitic statements in Franjo Tudjman's 1989 work 'Wastelands of Historical Truth' probably reflected his background as a general in the army of the fiercely pro-Arab Yugoslavia of the Non-Aligned Movement, and were edited when the work appeared in the 1990s in English translation.[51] Although the more extreme elements among Serb and Croat nationalists in the 1990s did sometimes express antisemitic views, they were generally astute enough to know the propaganda value of not being seen to be antisemitic, and they did try to appeal to Jewish opinion—though not very successfully.

It would be nonsensical to argue that the systematic destruction of mosques and the Islamic heritage in Bosnia by Serbian forces in the 1990s, combined with a propaganda that stressed the role of *mujahedin* and of foreign Islamic states, was not an expression of Islamophobia on the grounds that Islamophobia does not exist. Equally, it would be nonsensical to argue that this campaign was primarily motivated by hostility to Islam as an ideology: there was no pretence that Muslims were a danger because they might indoctrinate the Serbian population with subversive views. Serb nationalists in the 1980s and 1990s made much of the growing threat of Bosnian Muslims in Bosnia, and of Albanian Muslims in Serbia. However, the danger they presented was not that these groups would Islamify Serbia by converting Christian Serbs to Islam. Rather, the danger was that these groups would Islamify Serbia by increasingly outbreeding the Christian Serbs, and turning them into minorities in their own countries.[52]

Thus, the perceived Islamic threat was not equivalent to the Communist threat, as it was viewed in McCarthy's USA, or to the counterrevolutionary threat, as it was viewed in Stalin's USSR. Muslim children in Serb-occupied Bosnia were not simply deported along with their parents, as they might have been if they were viewed as the children of subversives. Still less were they subjected to ideological reprogramming. Rather, they were themselves singled out for rape, torture and murder. Muslim women were raped with the stated goal of making them give birth to Serb babies.[53] Biljana Plasvic, the Bosnian Serb vice-president, theorised about the Muslims being a genetically defective offshoot of the Serb nation.[54]

In conclusion, there are broad similarities between anti-Muslim and anti-Jewish chauvinism in the Balkans, insofar as both are directed against ethnic groups that have their origins in religious differences. Muslims, like Jews, have been widely treated and persecuted as ethnically alien, not simply as a religious community. There are, of course, differences between the two forms of chauvinism: antisemites traditionally portray the global Jewish conspiracy in terms of sneaky, intelligent puppet-masters working behind the scenes, whereas Balkan Islamophobes portray the global Islamic conspiracy in terms of mindless but fully visible—indeed, visually striking—fanaticism. Hatred of Islam and Muslims has, for all its intensity as felt by Balkan Christian nationalists, never quite achieved the intensity of being an all-consuming end in itself, as it has for some antisemites. And of course, Balkan Islamophobes do not formally treat global Islam as a race, in the way that antisemites treat global Jewry as a race. Islamophobia in the Bosnian war was an expression of hatred directed against a national group or groups. One of the paradoxes of this is that for all the hatred directed against the Balkan Muslim peoples by Balkan Christian nationalists, and indeed by the anti-Muslim bigots in the West who supported them, the Bosnian Muslims and Kosovo Albanians are among the most secularised Muslim peoples in the world. Just as Jewish atheists will always be the Christ-killers or ritual slaughterers of Christian children in the eyes of certain antisemites, so Bosnian Muslim and Albanian atheists will always be *jihadis* in the eyes of certain Islamophobes.

Notes

1. Christopher Hitchens complained of 'the stupid neologism "Islamophobia," which aims to promote criticism of Islam to the gallery of special offenses associated with racism' ('Facing the Islamist Menace', *City Journal*, Winter 2007 [Facing the Islamist Menace 2007]).
2. 'Convention on the Prevention and Punishment of the Crime of Genocide. Adopted by the General Assembly of the United Nations on 9 December 1948', United Nations Treaty Collection website, accessed at https://treaties.un.org/doc/Publication/UNTS/Volume%2078/volume-78-I-1021-English.pdf on 8 July 2015.
3. R. Hilberg, *The Destruction of the European Jews*, Vol. 1 (New York: Holmes and Meier, 1985), p. 9 (Hilberg 1985).
4. B. Jelavich, *History of the Balkans: Eighteenth and Nineteenth Centuries* (Cambridge: Cambridge University Press, 1983), p. 52 (Jelavich 1983).
5. L. von Ranke, *History of Servia and the Servian Revolution* (New York: De Capo Press, 1973), pp. 177–9 (von Ranke 1973).
6. D.J. Popović, *Beograd kroz vekove* (Belgrade: Turistička štampa, 1964), pp. 29, 316–18, 412 (Popović 1964).
7. M. Boro Petrovich, *A History of Modern Serbia, 1804–1918*, Vol. 1 (New York and London: Harcourt Brace Jovanovich, 1976), pp. 105–6 (Boro Petrovich 1976).
8. P.J. Cohen, *Serbia's Secret War: Propaganda and the Deceit of History* (College Station: Texas A&M University Press, 1996), pp. 64–9 (Cohen 1996).
9. J. McCarthy, *Death and Exile: The Ethnic Cleansing of Ottoman Muslims, 1821–1922* (Princeton, NJ: The Darwin Press, 1995), p. 12 (McCarthy 1995).
10. K.E. Fleming, *Greece: A Jewish History* (Princeton, NJ: Princeton University Press, 2008), pp. 16–17 (Fleming 2008).

11. G.Th. Mavrogordatos, *Stillborn Republic: Social Coalitions and Party Strategies in Greece, 1922–1936* (Berkeley: University of California Press, 1983), pp. 226–62 (Mavrogordatos 1983).
12. B. Clark, *Twice a Stranger: How Mass Expulsion Forged Modern Greece and Turkey* (London: Granta Books, 2006), pp. xi–xii (Clark 2006).
13. Mavrogordatos, *Stillborn Republic*, p. 260.
14. McCarthy, *Death and Exile*, pp. 86–7, 90–2.
15. R.J. Crampton, *Bulgaria* (Oxford: Oxford University Press, 2007), p. 426 (Crampton 2007).
16. M. Neuburger, *The Orient Within: Muslim Minorities and the Negotiation of Nationhood in Modern Bulgaria* (IIthaca, NY: Cornell University Press, 2004), p. 17 (Neuburger 2004).
17. Ibid., pp. 67–8, 81–2.
18. D. Bloxham, *The Great Game of Genocide: Imperialism, Nationalism and the Destruction of the Ottoman Armenians* (Oxford: Oxford University Press, 2005), p. 1 (Bloxham 2005).
19. A. Mango, *Ataturk* (London: John Murray, 2004), pp. 345–6 (Mango 2004a).
20. G. Milton, *Paradise Lost—Smyrna 1922: The Destruction of Islam's City of Tolerance* (London: Sceptre, 2009), p. 372 (Milton 2009).
21. A. Mango, *The Turks Today* (London: John Murray, 2004), pp. 33–5 (Mango 2004b).
22. A. de Zayas, 'The Istanbul Pogrom of 6–7 September 1955 in the Light of International Law', *Genocide Studies and Prevention*, 2, no. 2 (August 2007), 148 (de Zayas 2007).
23. See S. Skendi, *The Albanian National Awakening, 1878–1912* (Princeton, NJ: Princeton University Press, 1967) (Skendi 1967).
24. B.J. Fischer, *Albania at War, 1939–1945* (London: Hurst and Co., 1999), p. 187 (Fischer 1999).
25. 'Albania', Yad Vashem website, accessed at http://www.yad-vashem.org/yv/en/righteous/stories/historical_background/albania.asp on 9 July 2015.
26. V. Pelagić, *Istorija bosansko-ercegovačke bune u svezi sa srpsko- i rusko-tursko ratom* (Budapest: Štamparija Viktora Hornjanskoga, 1879), pp. 74, 39–41 (Pelagić 1879).
27. V. Pelagić, *Vjerozakonsko ućenje Talmuda ili ogledalo čivutskog poštenje* (Belgrade: T. Jovanović, 1878), p. 4 (Pelagić 1878).

28. V. Zrnić, 'Uspomene iz gimnazije', in *Milan Srškić 1880–1937*, ed. M. Popović, Odbor za izdavanje Spomenice pok (Sarajevo: M. Srškiću, 1938), p. 19 (Zrnić 1938).
29. H. Press Freidenreich, *The Jews of Yugoslavia: A Quest for Community* (Philadelphia: Jewish Publication Society of America, 1979), pp. 146–7 (Press Freidenreich 1979).
30. Ibid., p. 146.
31. N. Bartulin, *The Racial Idea in the Independent State of Croatia: Origins and Theory* (Leiden: Brill, 2014), pp. 37–8 (Bartulin 2014).
32. Ivo Banac, *The National Question in Yugoslavia: Origins, History, Politics* (Ithaca, NY: Cornell University Press, 1984), pp. 108, 363–4 (Banac 1984).
33. F. Sultaga, *Bosna i Bošnjaci u hrvatskoj nacionalnoj ideologiji* (Sarajevo: Salfu, 1999), p. 88 (Sultaga 1999).
34. Bartulin, *The Racial Idea in the Independent State of Croatia*, pp. 66–9.
35. J. Tomasevich, *War and Revolution in Yugoslavia, 1941–1945: Occupation and Collaboration* (Stanford: Stanford University Press, 2001), pp. 592–604 (Tomasevich 2001).
36. See M.A. Hoare, *Genocide and Resistance in Hitler's Bosnia: The Partisans and the Chetniks, 1941–1943* (Oxford: Oxford University Press, 2006) (Hoare 2006).
37. Ž. Rukavina, 'Bosansko-hercegovački muslimani i Hrvatstvo', *Hrvartska Krajina*, yr 1, no. 2, 23 April 1941 (Rukavina 1941).
38. Hoare, *Genocide and Resistance in Hitler's Bosnia*, pp. 157–61.
39. D.M. Mihailović, *Rat i mir đenerala: Izabrani ratni spisi* (Belgrade: Srpska reč, 1998), Vol. 2, p. 22 (Mihailović 1998).
40. See M.A. Hoare, 'Whose is the Partisan Movement? Serbs, Croats and the Legacy of a Shared Resistance', *Journal of Slavic Military Studies*, 15, no. 4 (December 2002), 24–41 (Hoare 2002).
41. Tomasevich, *Occupation and Collaboration*, p. 605.
42. See M.A. Hoare, *The Bosnian Muslims in the Second World War: A History* (London: Hurst and Company, 2013) (Hoare 2013).
43. See S.P. Ramet, *Nationalism and Federalism in Yugoslavia, 1962–1991*, 2nd edn. (Bloomington: Indiana University Press, 1992), pp. 168–77 (Ramet 1992).

44. N. Cigar, *Genocide in Bosnia: The Policy of "Ethnic Cleansing"* (College Station: Texas A&M University Press, 1995), pp. 22–37 (Cigar 1995).
45. Ibid., pp. 30–1.
46. M.A. Hoare, 'The Croatian Project to Partition Bosnia-Hercegovina, 1990–94', *East European Quarterly*, 31, no. 1 (March 1997), 121–38 (Hoare 1997).
47. *Bosna i Hercegovina i Bosnjaci u politici i praksa Dr Franje Tuđmana, Vijeće Kongresa bošnjačkih intelektualaca*, Sarajevo, 1998, p. 55 (*Bosna i Hercegovina i Bosnjaci u politici i praksa Dr Franje Tuđmana* 1998).
48. Ibid., p. 58.
49. Ivan Aralica, 'Lebdenje svijetom kao na carobnom sagu', *Slobodna Dalmacija*, 18 March 1993 (Aralica 1993).
50. Cigar, *Genocide in Bosnia*, p. 124; Cohen, *Serbia's Secret War*, pp. 116–17.
51. Franjo Tudjman, 'Bespuća povijesne zbiljnosti: Rasprava o povijesti i filozofiji zlosilja', Nakladni zavod Matice Hrvatske, Zagreb, 1989; 'Horrors of War: Historical Reality and Philosophy', M. Evans and Co., New York, 1996.
52. B. Magaš, *The Destruction of Yugoslavia: Tracking the Break-up 1980–92* (London: Verso, 1993), pp. 59–60; Cigar, *Genocide in Bosnia*, pp. 27–30 (Magaš 1993).
53. M.A. Sells, *The Bridge Betrayed: Religion and Genocide in Bosnia* (Berkeley: University of California Press, 1996), pp. 21–4 (Sells 1996).
54. S. Inić, 'Biljana Plavšić: Geneticist in the Service of a Great Crime', *Bosnia Report*, no. 19, June–August 1997 (Inić 1997).

References

Aralica, Ivan. 1993. Lebdenje svijetom kao na carobnom sagu. *Slobodna Dalmacija*, 18 March.

Banac, Ivo. 1984. *The National Question in Yugoslavia: Origins, History, Politics.* Ithaca, NY: Cornell University Press.

Bartulin, N. 2014. *The Racial Idea in the Independent State of Croatia: Origins and Theory*. Leiden: Brill.

Bloxham, D. 2005. *The Great Game of Genocide: Imperialism, Nationalism and the Destruction of the Ottoman Armenians.* Oxford: Oxford University Press.

Boro Petrovich, M. 1976. *A History of Modern Serbia, 1804–1918*. New York and London: Harcourt Brace Jovanovich.

Bosna i Hercegovina i Bosnjaci u politici i praksa Dr Franje Tuđmana, Vijeće Kongresa bošnjačkih intelektualaca, Sarajevo, 1998.

Cigar, N. 1995. *Genocide in Bosnia: The Policy of "Ethnic Cleansing"*. College Station: Texas A&M University Press.

Clark, B. 2006. *Twice a Stranger: How Mass Expulsion Forged Modern Greece and Turkey*. London: Granta Books.

Cohen, P.J. 1996. *Serbia's Secret War: Propaganda and the Deceit of History*. College Station: Texas A&M University Press.

Crampton, R.J. 2007. *Bulgaria*. Oxford: Oxford University Press.

de Zayas, A. 2007. The Istanbul Pogrom of 6–7 September 1955 in the Light of International Law. *Genocide Studies and Prevention* 2(2) (August): 148.

Facing the Islamist Menace. *City Journal*, Winter 2007.

Fischer, B.J. 1999. *Albania at War, 1939–1945*. London: Hurst and Co.

Fleming, K.E. 2008. *Greece: A Jewish History*. Princeton: Princeton University Press.

Hilberg, R. 1985. *The Destruction of the European Jews*. New York: Holmes and Meier.

Hoare, M.A. 1997. The Croatian Project to Partition Bosnia-Hercegovina, 1990–94. *East European Quarterly* 31(1) (March): 121–138.

———. 2002. Whose is the Partisan Movement? Serbs, Croats and the Legacy of a Shared Resistance. *Journal of Slavic Military Studies* 15(4) (December): 24–41.

———. 2006. *Genocide and Resistance in Hitler's Bosnia: The Partisans and the Chetniks, 1941–1943*. Oxford: Oxford University Press.

———. 2013. *The Bosnian Muslims in the Second World War: A History*. London: Hurst and Company.

Inić, S. 1997. Biljana Plavšić: Geneticist in the Service of a Great Crime. *Bosnia Report*, no. 19, June–August.

Jelavich, B. 1983. *History of the Balkans: Eighteenth and Nineteenth Centuries*. Cambridge: Cambridge University Press.

Magaš, B. 1993. *The Destruction of Yugoslavia: Tracking the Break-up 1980–92*. London: Verso.

Mango, A. 2004a. *Ataturk*. London: John Murray.

———. 2004b. *The Turks Today*. London: John Murray.

Mavrogordatos, G.Th. 1983. *Stillborn Republic: Social Coalitions and Party Strategies in Greece, 1922–1936*. Berkeley: University of California Press.

McCarthy, J. 1995. *Death and Exile: The Ethnic Cleansing of Ottoman Muslims, 1821–1922*. Princeton, NJ: The Darwin Press.

Mihailović, D.M. 1998. *Rat i mir đenerala: Izabrani ratni spisi*, vol. 2. Belgrade: Srpska reč.

Milton, G. 2009. *Paradise Lost—Smyrna 1922: The Destruction of Islam's City of Tolerance*. London: Sceptre.

Neuburger, M. 2004. *The Orient Within: Muslim Minorities and the Negotiation of Nationhood in Modern Bulgaria*. Ithaca, NY: Cornell University Press.

Pelagić, V. 1878. *Vjerozakonsko učenje Talmuda ili ogledalo čivutskog poštenje*. Belgrade: T. Jovanović.

———. 1879. *Istorija bosansko-ercegovačke bune u svezi sa srpsko- i rusko-tursko ratom*. Budapest: Štamparija Viktora Hornjanskoga.

Popović, D.J. 1964. *Beograd kroz vekove*. Belgrade: Turistička štampa.

Press Freidenreich, H. 1979. *The Jews of Yugoslavia: A Quest for Community*. Philadelphia, PA: Jewish Publication Society of America.

Ramet, S.P. 1992. *Nationalism and Federalism in Yugoslavia, 1962–1991*, 2nd edn. Bloomington: Indiana University Press.

Rukavina, Ž. 1941. Bosansko-hercegovački muslimani i Hrvatstvo. *Hrvartska Krajina*, yr 1, no. 2, 23 April.

Sells, M.A. 1996. *The Bridge Betrayed: Religion and Genocide in Bosnia*. Berkeley: University of California Press.

Skendi, S. 1967. *The Albanian National Awakening, 1878–1912*. Princeton, NJ: Princeton University Press.

Sultaga, F. 1999. *Bosna i Bošnjaci u hrvatskoj nacionalnoj ideologiji*. Sarajevo: Salfu.

Tomasevich, J. 2001. *War and Revolution in Yugoslavia, 1941–1945: Occupation and Collaboration*. Stanford, CA: Stanford University Press.

Tudjman, F. 1989. *Bespuća povijesne zbiljnosti: Rasprava o povijesti i filozofiji zlosilja*. Zagreb: Nakladni zavod Matice Hrvatske.

———. 1996. *Horrors of War: Historical Reality and Philosophy*. New York: M. Evans and Co.

von Ranke, L. 1973. *History of Servia and the Servian Revolution*. New York: De Capo Press.

Zrnić, V. 1938. Uspomene iz gimnazije. In *Milan Srškić 1880–1937*, ed. M. Popović, Odbor za izdavanje Spomenice pok. Sarajevo: M. Srškiću.

CHAPTER 8

Antisemitism and Its Critics

Gil Anidjar

I might as well admit it. I am one of those who struggle against antisemitism. I tend to think about it a lot. I read and reflect; I write about it sometimes. I take action when I can. I even formulated some ideas, a theory of sorts, playing my part, adding my bit to the growing number of accounts of it. You could say that I have been moved, nay, mobilised to criticise antisemitism, fight against it. I am no imaginary Jew, I do not think, not one of those Alain Finkielkraut used to think of at least; nor am I as empirically ignorant of antisemitism as Yoav Shamir claimed to be before his documentary *Defamation* (and perhaps after it as well).[1] I could easily count myself among those who think that it 'is right to voice concern about rising anti-Semitism, and every progressive Jew, along with every progressive person, ought to be vigorously challenging anti-Semitism wherever it occurs'.[2] I am for myself, then, 'anti-antisemite'. Most definitely, yes. Yes, I am. And I do believe that Jonathan Judaken has offered a strong justification for this admittedly awkward term when he wrote that 'anti-antisemitism clearly denotes an opposition to prejudices and stereotypes

G. Anidjar (✉)
Columbia University, New York, NY, USA

J. Renton, B. Gidley (eds.), *Antisemitism and Islamophobia in Europe*, DOI 10.1057/978-1-137-41302-4_8

related to Jews, Judaism, and Jewishness, and anti-antisemites resist the institutionalization of discrimination against Jews'.[3] It is true that I have felt inclined to empathise with Hannah Arendt, with her infamous reservations about her love of the Jewish people, but this too I must confess: I come very close to being a philosemite. Although this word too, Judaken points out, may be troubling: 'The term philosemitism implies a love of Jews and Judaism. However, its usage almost always refers to those who oppose antisemitism but who often lack an understanding of the history, culture, and religion of the Jews' (p. 20). Just the same, I trust that the notion may be capacious enough to include someone like me. But I wish to insist on keeping the two tendencies distinct. The philosemite in me is not the anti-antisemite in me.

Becoming Who One Is

Now, given my trade, I have not been particularly well trained to be self-effacing, nor am I especially adept at relinquishing personal, even individualistic (not to say narcissistic) explanations for my being or becoming what I am now, an anti-antisemite. In fact, I am quite readily prepared to embrace an account that, appropriately biographical, reaches back into my childhood, my formative experiences, the nature of my upbringing and my life trajectory; or one that attends to the way I have taken hold of my current situation, the freedom I have learned to enjoy and exercise, a comfortable social position, a peculiar sensibility and a religious or spiritual inclination among other character traits; and then there is the professional research I have done, which some might consider an achievement of sorts. I am not that special, of course, but I am confident in my individuality. I try to take full responsibility for acts I engage in, and for my motivations too. That is why today I should wish in my own case to dig further and 'explore the relationship between the immanent form a normative act takes, the model of subjectivity it presupposes (specific articulations of volition, emotion, reason, and bodily expression), and the kinds of authority upon which such an act relies'.[4] Yet the question I must also ask concerns my earlier assertion that 'I myself have been moved and even mobilised' against antisemitism. For with all due respect to 'the deeply individualistic language which speaks of atomistic individuals who enter into relations with each other on the basis of a purely rational calculation', I am forced to acknowledge that I am not alone.[5] I could not claim, could I, that I have moved or mobilised myself after all, autonomously and on my own? Insofar as I struggle against antisemitism, I vaguely sense that I am participating in something greater than myself. I am, as it were, 'framed'.[6] I follow certain rules. I play, indeed, I want to play a part in a larger dynamic that includes but also exceeds social confines. I join—at times I seem caught—in broader trends and movements. One could indeed say, and with good reason, that there is a spirituality to it all, and though I know too little of it, I would like to think that I have inherited a certain history, a tradition even. Like many others, individual and social actors, I respond to antisemitism and, perhaps to a comparable extent, to the forces that, however tenuous or fragmented, are gathered against it. I experience these things.[7] I feel, at any rate, interpellated and in some measure supported.

Do not misunderstand me. I entertain few doubts in the matter, nor am I here to raise disturbing questions. It is to my mind indisputably the case that antisemitism must be confronted and fought. But I would be remiss in considering this comportment necessary or unavoidable, if I did not examine some of the general conditions under which this struggle became mine.[8] I want to understand the essence of my choices and decisions, the opportunities offered me, as opposed to the struggles or paths I could have pursued but did not. Did I identify and decide on the right cause? Am I close to the root of the problem? Should I endorse, for instance, the still current distinction between racism and antisemitism (a distinction that seems to operate, at least rhetorically, even for those who might otherwise oppose it)? Should I really single out antisemitism? I would certainly wish to learn whether there is a larger struggle (or struggles) of which I am in fact a part, or another from which I perhaps broke away unintentionally; and if so, what its nature is, and whether my joining in this struggle, my being interpellated (without fully knowing by what or by whom), testifies to a desire to be normative or contentious. Who am I, finally? Am I a good or a bad subject? Is the impulse I feel generated by affirmation and consensus or by opposition and resistance? In fighting antisemitism, in other words, am I attaching myself to or detaching myself from the collective and constructive will of society? Am I joining a marginal, besieged and contesting minority? Is the struggle against antisemitism a matter of consensus, or in opposition to it? Is it local or global, a matter of concern for civil society, for the state or the international community, for national and international law? How does it relate to other struggles? Do I have my priorities right? How precisely am I responding, responsibly responding, to the complex and contradictory messages and possibilities that surround me, to the numerous emergencies that press themselves upon me? Perhaps I should simply affirm my active participation as a self-evident comportment. I cannot be everywhere at once, but I must be where I can. Still, can I be so certain that I do in fact participate in a collective endeavour? Is there at all an organised movement against antisemitism? Should there be one? And if so, what kind of movement would it be? A religious, a social or a political movement?[9]

But since I have alluded to my anti-antisemitism as a kind of spiritual practice, I should like to linger with that a little more. It seems clear that there is indeed a higher calling at work here, a measure of magnitude and of gravity that is undeniable. All the more reason, I think, to endorse 'the view that spirituality, whatever it is and however it is defined, is entangled in social life, in history, and in our academic and nonacademic imaginations'.[10] It would be misleading, in other words, even mistaken, simply to extract my experience of antisemitism, my commitment and struggle against it, 'from the institutions where it is lived out' and to do so at the outset. It moreover runs the risk of distorting and mischaracterising the phenomenon, drawing 'attention away from the conundrum it poses and the possibilities it allows' (ibid.).

Whatever account I myself could give for my actions, whatever motivations or thoughts I entertain, it will not do to treat it all as a mere individual matter, much less as some consensual phenomenon. If fighting antisemitism is indeed a kind of spiritual practice, grounded as it may be in a personal experience, it still demands that we also think 'about the location of contemporary spirituality not just in organizational terms but also in geographical and historical terms' (p. 3). At the same time, and as with other forms of spirituality, it seems necessary to address my struggle against antisemitism, whatever its precise nature might be, and to make sense of it 'historically, institutionally, and imaginatively without pulling it completely together into a single thing' (p. 6). I might have to establish, at any rate, 'that what we think of as the spiritual is actively produced' together with a network of larger trends and institutions. However isolated or inchoate my own spiritual practice, or set of practices, might appear to be, I should at least entertain the possibility that 'it is not unorganized, but rather organized in different ways, within and adjacent to a variety of religious and secular institutional fields'. Furthermore, and 'against the view that people learn to be spiritual practitioners on their own, or purely through the mediations of books and literature or as shoppers in some kind of undifferentiated market', I might have to bring myself to acknowledge, and reflect on the fact, that I have been enabled and supported, taught and guided, that I am definitely not alone and that, like many others, I have grown into my own, into the struggle against antisemitism, in and by way of a variety of institutional settings and cultural vectors (p. 23). If, then, this struggle of mine against antisemitism, along with the motivations that drive me towards and through it, is part of a larger social phenomenon—a movement—this movement might have to be understood as either religious, social or political. Yet can there be a movement, a collective or social movement, that does not know itself? Does a movement have to be concerted? Could I be part of it without my own knowledge? Could I be framed to such an extraordinary extent? But is it at all, I ask again, a movement? Is there a war against antisemitism? And if there is, what precisely am I in this fight? What kind of an actor does it make me? Am I a rebel or a conformist? Am I an intellectual, an activist or a foot-soldier? What exactly is it that I have joined?

Not to Change the Subject

I hope to be forgiven if I cater some more to my sense of self-importance (I hope it is not totally overblown). I wish to linger, in other words, with the subject. Let me make clear once again that in spite of 'the overwhelming tendency ... to conceptualize agency in terms of subversion or resignification of social norms, to locate agency within those operations that resist the dominating and subjectivating modes of power',[11] I am not particularly invested in portraying myself as a agent of protest, a subversive person, nor as a freedom fighter or as a lone ranger engaged in the transvaluation of all values. Surely, my general goal is a liberatory one, but the agency that I wish to affirm and that I see myself deploying is not necessarily 'conceptualized on the binary model of subordination and subversion'. I would certainly not want a priori to dismiss, nor even elide, those 'dimensions of human action whose ethical and political status does not map onto the logic of repression and resistance'.[12] I gladly concede that with regard to the most striking portraits that have been drawn of pertinent figures of modern subjectivity, and particularly with regard to situations where hatred is involved, the general inclination has been to focus on oppressor ('portrait of the coloniser', 'portrait of the antisemite') and victim ('portrait of the colonised', 'portrait of the Jew').[13] Indeed, I would not presume to know whether we have transcended the master–slave dialectic, nor am I entirely confident that we should. Yet this might be where the problem lies: in my attempt to understand myself, I am confronted with a puzzling absence, an emptiness of sorts. It is after all an empirical fact that in my struggle against antisemitism I have been handed no portrait to emulate, no model for guidance, no prior narrative or articulated understanding of the kind of subject that I am, or that I have become. The development, as it were, of my anti-antisemitic engagements does not seem to fall squarely into the familiar terms and oppositions otherwise handed down to me. In my fight against antisemitism, ultimately (and suspending for now the feelings I expressed earlier), I may or may not conceive of myself as personally interpellated (though, were I to go in this direction, a great number of resources would offer themselves to me). Nor would I want to locate myself unhesitatingly in a commanding position, claim sovereignty of any kind for myself, much less power or potency (were I to do so, however, countless reflections on 'the subject of power', with all the nuances of the double genitive, and mimicry to boot, would certainly be of tremendous assistance). To repeat: as I comport myself in the struggle against antisemitism and plausibly partake of it, it seems safer to suspend the determination of whether I am a normative or a subverting, resisting subject. Pointers are overwhelmingly lacking and I find myself in an unexpected conundrum. Do I study up or down?[14] And if I am grounded in a personal, spiritual experience, if that experience is in fact individual, if it is not only independent of social grounds and of tradition but is in fact the sole and unprecedented engine of my actions, then it makes sense 'to investigate experience not with sociological tools but rather with psychological or psychical (or scientific) tools and methods'.[15] Besides, the sheer rarity of a general, reflexive dimension among all those who struggle, as I do, against antisemitism has all but naturalised what is, after all, the consequence of historical, and highly historicised, events. Except that I find myself there as well bereft of assistance. If there is a collective struggle, a 'we' to which I belong, I know nothing of it. I know nothing of this movement nor of the constituency it gathers or constitutes—and even less of its history. Consider that scholars (and not only scholars) have elsewhere 'captured the indefinable moment when a group of separate individuals became a collective actor'.[16] They have further identified many reasons why people join social movements: 'because it is fun, because their sense of solidarity with people they know who are already in the movement demands it; because if they don't do it, no one else will; because they are morally shocked and compelled by an injustice, because it is who they are'.[17] I may find inspiration or solace in these, but we, we anti-antisemites, have been handed no such list of reasons, no 'sociology of accounts' nor 'scholarly tales', no explanation of whence the learning or support we have or have not received.[18] We have no knowledge of the conditions of our being, no portrait of ourselves, neither individual nor collective, not by philosophers or historians, nor by psychologists or psychoanalysts. We do not know how it is that we became what we are.[19]

But perhaps I have taken the wrong path. After all, it is not only the subversive nature of agency that has dominated the scene of reflection, but also a certain proximity between agency and identity, or victimhood.[20] Perhaps I should not so much think of myself as a subject, here, not even primarily as an agent or actor. I have an identity, to be sure, a complex set of identities even. I remain the master of my domain, in other words, whatever domain that is, but being anti-antisemitic may not have much to do with that. It probably should not be considered, nor accounted for, by way of my identity. It is not really who I am, not quite an aspect of that identity (or identities). Is it not rather, and primarily, something that I do? Whether as a professional or as a volunteer, engaged in occasional and ephemeral acts or committed to the endurance of strong lines of action, I do not really have to be anything to be anti-antisemitic. No prior identity is required, and none is thereby constituted. There is no particular subject here and anyone can therefore become part of the struggle. Besides, notwithstanding the occasional biography, do we have a fuller account of other activisms? Do we know more about the antiracist or the environmentalist, even the peace or community activist?

We may know 'a great deal about the conditions in which people mobilize on behalf of long-standing interests, but we still know relatively little about why certain areas of social life—race relations, say, or nuclear policy, or university curricula—suddenly generate new or newly conflicting interests'. We also 'know very little about where activists' beliefs about what is instrumental come from and how those beliefs eliminate options as well as opening them up'.[21] Indeed, we have little insight into 'what persuades people to participate in collective action before movement organizations with strategic recruiting pitches have been established'.[22] Like countless others, no doubt, I sometimes engage in collective action, volunteering my time and strengthening my commitments, pursuing particular vectors that may be explored by way of statistics and other instruments of measure. One would have to expect that some shared elements of background (race, class, gender and religion, education and profession too) are likely to be found in common among my fellow anti-antisemites, but these would not seriously account, not in a satisfactory and reflective manner, for the nature of the struggle I have joined. Still, over against racism (or abolitionism), and in contradistinction with feminism and queer activism, environmentalism, or community and class mobilisation, the struggle against antisemitism seems to have been generally immune to such scrutiny. It has been criticised, of course, at times assimilated to other, less commendable and usually contemporary concerns (much like antisemitism has often been reduced to 'external' factors, to social or economic reasons). Yet the struggle against antisemitism seems peculiarly devoid of history. The literature on antisemitism is vast and still growing, but it has predictably focused on understanding antisemitism rather than on proposing a reflection on its own genealogy, the conditions of its possibility, its meaning or efficacy. This situation is obviously not unique to the case at hand. Some, like Sven Lindqvist, have tried to attend to the tradition of antiracist activism over the centuries, underscoring that he wanted to remember those 'who today are often forgotten, and as far as I know have never been discussed together'. Lindqvist had hoped 'to show those who are today fighting against racism something of the long and proud tradition to which they belong', a tradition of which they are for the most part ignorant, just as I am.[23] This does not quite amount to a portrait, but it provides the rudiments of a history of antiracist struggle in its different facets. The abolitionist movement is another example, which itself elaborated its own narrative of complex ties to the biblical account of Exodus and to Christian doctrine; and the earlier arguments, however ineffective, that were inaugurated by Bartolomeo de Las Casas or Thomas Morton against the persecution of Amerindians are themselves rich with reflexive memory, and still alive and remembered.[24] Consider as well the force of anticolonialism (the appeal to universalism of Toussaint l'Ouverture) and the consistent, if all too rare and ambivalent, opposition to 'the scandal of Empire'.[25] Recall the nineteenth-century struggle against sexual inequality and later against misogyny and sexual discrimination. There is after all a reflexive history of feminist consciousness (much more than a mere oppositional force, obviously), accounts of the emergence of the queer subject or indeed agent.[26] But what about anti-antisemites?

In this fragmented and seemingly tenuous struggle, no tradition, no antecedents, appear to have been claimed, none seem to be found—not even an 'invented' or an 'imagined' tradition. There are the survivors, of course (I will have to return to that), but the history of the Holocaust is not the whole, only the culmination, of the history of antisemitism, and it does not seem to teach us much about the struggle against it. If anything, it signalled its failure rather spectacularly, even as it is now construed as the height of its triumph: Was the Second World War actually fought in order to put an end to antisemitism? Did we not inherit, in its aftermath, the major instruments of our anti-antisemitic actions? There are neither forerunners nor prior traditions invoked in the still unsurpassed interventions of Horkheimer and Adorno, Sartre and, last but not least, Arendt. These recent figureheads of the struggle against antisemitism are at once isolated and conspicuously discreet about their precursors, if there were any.[27] Arendt does retrieve out of oblivion the idiosyncratic figure of Bernard Lazare; Horkheimer and Adorno rely, for their part, on Sigmund Freud (though not on his singular attempt to counter antisemitism in *Moses and Monotheism*); and whereas Fanon could complain about the inadequacy of his predecessors in the struggle against racism and colonialism (Maran, Mannoni), Sartre—who inspired Fanon before writing a preface for *The Wretched of the Earth*—makes no similar gesture in his famed attack on 'the antisemite'.

Located elsewhere than between being and doing, indeed even prior to any specific doing, anti-antisemitism would perhaps be lacking a portrait and a history because it has none—or because it is unrelated to becoming. If I am neither master nor slave, neither antisemite nor Jew, it is because I am not, nor did I become part of this struggle. And while there are 'claims that develop' in other spiritual narratives, which 'frequently require that people appear as quite individualistic within their own stories'; while one can understand these narratives and 'representations as staking out claims and possibilities for a certain kind of authentic and authoritative experience', nothing of the sort needs be said or invoked about the struggle of which I am part.[28] There are numerous and excellent reasons to consider that this absence of history or reflexivity corresponds simply to the very realities we confront, the conditions of the struggle against antisemitism. Then again, much as individual accounts have the potential to hide larger social forces, this peculiar situation too may conceal 'participation in a history that was carried in practice rather than in other forms of memory'.[29]

Ultimately, and more importantly, the puzzle (if it is one) that I have been trying to articulate with regard to my own struggle against antisemitism 'cannot be solved by locating it within a history it refuses', minimally by failing to articulate that history. Their importance notwithstanding, 'historical narratives disentangle the various tendrils of ... practice, ideology and experience, making the object clear yet in the process obscuring the very institutional and theoretical entanglements that give it power'.[30] And yet in this context as well, it must be re-iterated that we lack more than a portrait of the anti-antisemite, more than a comparative map in which this struggle would figure. It is the very status of this struggle as constituting a larger, organised or less organised movement (in its 'institutional and theoretical entanglements' with earlier or parallel struggles and movements as well) that remains open to question.

Episodes of Contention

It might be worth repeating, then, that an account of the individual or collective struggle against antisemitism, even at a most preliminary or rudimentary level, has yet to be initiated or conducted. Is there or is there not a war on antisemitism? To the extent that there are signs of a struggle, this question warrants an answer. None has been forthcoming. Surely, the invocation of antisemitism has, in some instances, been criticised, its legitimacy questioned in particular cases and situations.[31] But never was the need to condemn, and struggle against, antisemitism and how called into question, or interrogated—only the appropriateness of the battle site. No perspective has been offered, at any rate, critical or otherwise, and certainly none that would explain the making of an anti-antisemite, what it is that makes an individual (a collective or an organisation, a state even) into a focused opponent of antisemitism. There is, as it were, no 'grammar of a discourse', much less a portrait, of the anti-antisemite.[32] Nor is there a description or an account of the struggle against antisemitism, whether there is such, in its social and institutional, cultural and political sweep, nor of its rate of success (including a measure of its failures).[33] After all, actors and 'scholars should be able to account for the activists' choices (and even predict them) by identifying the conditions in which some strategies are more likely to be effective than others'.[34] Yet everything is as if, when it comes to anti-antisemites, 'the fuzziness of their sense of community meant that it occurred to none of them to ask how many of them there were in the world, and what, if they agreed to bend their energies into common action, they would be able to wreak upon the world to their common benefit'.[35] We might nonetheless agree that, in this urgent case as well, 'a perspective that sees a social movement as a simple historical residue or as the unmediated outcome of a policy decision is incapable of explaining it, for it necessarily ends up denying the movement any social history'.[36] But in order to determine whether there is in fact a movement, a social movement dedicated to the struggle against antisemitism, and to provide ourselves with instruments for its measure and evaluation, it seems more appropriate at this juncture to reflect on and draw from the work of social science scholars who have engaged and debated similar problems. It is from these scholars that I borrow the questions I have begun to deploy so far; from them as well that I borrow the phrase 'episodes of contention', which broadly corresponds to what I am after, while endorsing the necessity of advocating and demonstrating a much-needed mindfulness with regard to the very notion of 'movement' and the difficulties associated with it.[37]

However, first a random and non-exhaustive sample of such episodes from recent years, which should suffice to illustrate for now: from an Anti-Defamation League campaign in New York City to a range of reactions to Hollywood movies, and a number of subsequent *New York Times* editorials dedicated to the issue; from groups like 'Campus Watch' and accusations of antisemitism at Columbia University and elsewhere to the numerous pieces of legislation passed by the US Congress as well as by many European parliaments; from the scholarly institutes established and expanded to study antisemitism to international conferences on the topic conducted by scholars and policy-makers (in Europe, the USA, Australia and so forth); from Harvard President Lawrence Summer's statements (raising the spectre of speech and acts that would be 'anti-Semitism in effect if not in intent') with rejoinders by intellectuals and writers, all of whom are participating in renewed reflections on and responses to antisemitism. The list is longer (I return to it in more detail later), and though it leaves aside the organisations explicitly devoted to the struggle (the French Ligue internationale contre l'antisémitisme [LICA], for example, which was quickly renamed Ligue internationale contre le racisme et l'antisémitisme—see Gordon, this volume; or the Anti-Defamation League, mentioned earlier), it seems to provide more than sufficient ground nonetheless to take stock and consider a substantial number of public acts and gestures as 'episodes of contention'.

Note that, in their inquiry, our social scientists would 'not claim that these episodes are identical, nor that they conform to a single general model. They obviously differ in a host of consequential ways.'[38] But they might be grouped under the same heading for two reasons. 'First, the study of political contention has grown too narrow, spawning a host of distinct topical literatures—revolutions, social movements, industrial conflict, war, interest group politics, nationalism, democratization—dealing with similar phenomena by means of different vocabularies, techniques, and models.' In order to search for 'parallels across nominally different forms of contention', it seems therefore necessary to attend to 'similar causal mechanisms and processes in a wide variety of struggles'. The second reason the scholars might adduce has to do with their call to 'challenge the boundary between institutionalized and noninstitutionalized politics', the difference between 'legally prescribed, officially recognized processes' and other processes and procedures. They 'recognize this difference' and even propose a distinction between 'two broad categories of contention—contained and transgressive'. Still, as they deploy the distinction, they 'insist that the study of politics has too long reified the boundary between official, prescribed politics and politics by other means'.[39]

All this is most illuminating in accounting for state and non-state actors and actions, parameters and elements of the 'episodes' that I have briefly invoked, but I should say that I am not certain we need to confine our own imagination to political movements in particular, and most certainly not to contentious ones. Whether we are dealing (indeed, *whether* we are at all dealing) with a spiritual or religious movement, or with a more complex, plural and even disconnected phenomenon that involves spiritual, political as well as institutional and even economic dimensions, we may nevertheless profit by following the vectors traced here in order to perceive what are parts or fragments of a wide, indeed collective, mobilisation. As mobilisations go, it appears from the episodes mentioned to be a fairly successful one. It seems by now plausible, at any rate, that there is, if not a concerted effort and struggle, at least a multiple and layered deployment of diverse means and interventions, united or not, minimally tactical if not always strategic, all of which gather in pursuit of the same purpose: namely, to wage a war against antisemitism. There is, there appears to be, a war against antisemitism. That would be why, with all its apparent fragmentation and haphazardness, the struggle against antisemitism must be treated as a social and political movement.

Granted that there is a war on antisemitism, then, what is it that it does? What does it involve? A quick survey should reveal that it engages national and international law, international relations too, local as well as global institutions, non-governmental organisations, museums and memorials, schools, universities and research centres, but also literature and film, media and entertainment, world-famous personalities, educational material and more. Its geographical location and span are, simply, the entire world. It has economic dimensions as well: the sources of its means and funding, as well as the social stratification to which it testifies or which it interrogates or maintains, or simply preserves and reproduces; the kinds of individuals, specific classes and groups who engage in militant or sporadic activism on its behalf; the origin and location of those who intervene as its public representatives or intellectuals; the circles of discussion and action; the platforms gained in the global public sphere; the literary and journalistic dissemination of concerns in newspapers, magazines, professional journals, books and other publications; the sorts of audiences reached or mobilised, and so forth. All these and more would certainly testify to the range and significance of the war against antisemitism. But the war on antisemitism also seems to function as an international movement (albeit not quite as an organisation) with its own attributes and particulars, its singular 'framing'. It has a centre (or more precisely, centres) and a periphery. It operates locally, of course, fighting on distinct battlegrounds that are distributed unevenly in the social cartography, and on the surface of the globe as well. It is conducted as a programme of increased vigilance and discourse that involves heads of states, political and cultural figures, institutions and media, actors, books, journals and countless other sites. It is at work in legislative assemblies and in international courts (witness the often redundant explosion of legislative activity in France, the UK and the USA, often with an international dimension)[40]; in crowded museums and in Hollywood studios (from *Shoah* and countless other documentaries to Spielberg's feature films and the scrutinisation of Mel Gibson's 'life and works', and on to the adaptations, showings of and critical flurry around Shakespeare's *The Merchant of Venice*); and on the streets of New York City ('Anti-Semitism is Anti-Me' was the motto of that Anti-Defamation League 2004 campaign I mentioned, which was disseminated throughout the city) or the streets and walls of Paris (contested counts of 'antisemitic acts', including a number of embarrassing fabrications; intense and unified emotion and repeated public condemnations; and a 2004 sensitisation campaign by the Jewish Students Organization [UEJF] with the words 'sale juif [dirty Jew]' sprayed over a representation of Jesus Christ and other iconic figures), in London and Berlin (with the intensified pursuit of memorialisation, or the controversy over Günther Grass's past and merit) as well as in Buenos Aires and Caracas, and even in Lincoln and Lynchburg. Moreover, I read numerous reports that a comparable, if highly distinct, set of battles is being fought (or in need of being fought) in Cairo, Beirut and Teheran, Jerusalem and Baghdad, in truth from Casablanca to Kuala Lumpur via Istanbul and Mumbai. But aside from a heightened awareness

(I now know I am in fact part of something larger, a social movement of sorts), what insight does this all-too-rapid sketch provide? What knowledge can it contribute? What are its consequences?

The Minimal Self

I am again compelled to return to my sense of individuality, to the suggestion the preceding pages cannot fail to make that my commitment to the struggle against antisemitism either partakes of larger, social and institutional forces (about which I confess to being still sceptical), or that commitment alone does not suffice as an account for my actions, limited though these might be, and, more important, regardless of how justified and legitimate my actions are in the face of an objective urgency. True, I am devoid of a coherent public image. I know little of what happens overseas. And there is no portrait of the anti-antisemite. But does that make me a 'strategic dupe'?[41] Did I not testify to my diminished and modest sense of self-importance? It is one thing to recognise that I am not on the front lines (I am neither master nor slave), another to have me dissolve in depersonalising social, and even global, trends. I am perhaps what Christopher Lasch has referred to as a 'minimal self', but that does not mean that I—and I do mean I—am not objectively responding to antisemitism.[42] The significance of my portrait's absence does not constitute a licence to dismiss me from relevance, nor does it necessitate imagining a 'social movement' of whatever sort. I am adamant that the puzzle I allegedly constitute cannot be solved by locating me within a history that I, once again, refuse.[43] Besides, do we really need an account, a history of the struggle against antisemitism? It is not as if the matter were so obscure. After the Holocaust, is not the struggle against antisemitism, my struggle against antisemitism, a matter of sheer survival? And not just for the Jews, or for the survivors?

Incidentally, too few are those who have reflected on the importance of the survivor; too few have reflected critically on the survivor as a figure and symbol, as a narrative ground and political ideal. I am only aware of Elias Canetti and of Christopher Lasch, who have at least provided us with highly pertinent, as well as unprecedented, reflections on the notion of survival and its vicissitudes, reflections that do more than illuminate the singular place of the survivor in an economy that effectively transcends the opposition between master and slave, between victim and oppressor. Together, they have articulated a rich account of the disseminated force of the discourse of survival, my own and others. They have in fact drawn a portrait in which I can recognise myself. The argument, in its resonance with Walter Benjamin's 'bare life' as popularised by Giorgio Agamben, might now sound so obvious that it hardly needs repeating.[44] But as Lasch explains it, the 'culture of survivalism' as a whole constitutes a substantive response to a broad range of developments (Canetti seems to locate in the possibility of a nuclear holocaust the last and formative factor), and it has pervaded every aspect of modern life.[45] 'Even opposition movements,' Lasch pertinently adds, 'the peace movement, the environmental movement—take survival as their slogan.'[46] Survival has become the ultimate order of the day. 'It has entered so deeply into popular culture and political debate that every issue, however fleeting or unimportant, presents itself as a matter of life or death.'[47]

Yet the matter goes further than the arguments of Lasch himself on the 'culture of narcissism', further than other, more prosaic litanies about post-modern individualism and egotism. The figure of the survivor corresponds rather to a deep structure of subjectivity that has embraced a novel, innermost ideal and paradigmatic, narratological figure. It begins perhaps with the institutional and ideological force of the individual as the ground and subject of the modern state ('preservation of life being the end', as Hobbes put it, 'for which one man becomes subject to another, every man is supposed to promise obedience, to him in whose power it is to save, or destroy him')[48]; is further sedimented by way of social Darwinism (the 'survival of the fittest')[49] and accelerated modes of consumption and cultural habituation ('planned obsolescence'); and is aided by the massive infusion of psychologists and psychiatrists into everyday life along with the proliferation of 'survival literature'.[50]

Now, Lasch himself demonstrates all this, as he underscores the prominence of the survivor in the generalised discourse of victimology ('We think of ourselves both as survivors and as victims or potential victims'),[51] but the current conflation of the survivor with the witness indicates, I think, something else. The survivor is not a victim.[52] In saying this, I am however not substantially veering from Lasch and his notion of the 'minimal self', but merely suggesting that his emphasis on victimology misses the essential contribution that he himself makes (and that Canetti had made before him).

But wait. Are we still talking about me? My answer will have to be an unequivocal yes, and it can be summarised by way of a formulation that will hardly surprise or break any new ground: the struggle against antisemitism is a struggle for survival. Yet this is a more complicated statement than it might appear, for it says, indeed it narrates, a number of things at once. First, the struggle is conducted on behalf of the survivors, but it posits for itself, second, the explicit goal of embracing, affirming and protecting, survival, minimally, of testifying for it; third and finally, it is itself a strategy of survival. Neither master nor slave, neither oppressor nor victim, I do turn out to have had a portrait of my own. I, who struggle against antisemitism, am a survivor. I am not a victim, in other words, and I do think that the heightened concern over the reigning victimology, what Wendy Brown has felicitously called 'wounded attachments', has obscured an important, initial insight.[53] It has to do with an issue that is larger than social movements, which Canetti articulated and Lasch elaborated (the absence of the former in the latter notwithstanding). It is the centrality of the survivor as a founding political figure of modernity. Canetti recognised the pre-modern (or 'primitive') antecedents, of course, though he was careful to further distinguish the concept of survival from the earlier 'instinct of self-preservation'.[54] More important, and within this larger political context, the nature of the survivor as fundamentally distinct, even opposed, to the victim gives us precious information about the struggle against antisemitism. For it is true that the survivor has an essential relation to mourning and loss, yet I would venture nonetheless that he does not, nor is he defined by, fear for his own life. He is not a victim now, nor does he imagine himself such in the future. But the future is of course crucial, for the temporality of the survivor associates him less with the dead (the victims) than with the horizon to come that he shares with the perpetrators. He is, to put this hypothesis starkly, indestructible.[55] If this is correct, we will have been provided, albeit without having recognised it, with a crucial confirmation. The struggle against antisemitism must be understood primarily out of the sources it claims for itself, out of the history it mourns: the history of the Holocaust (some might say, against my earlier assertions, the history as Holocaust, but the difference is irrelevant to our specific purpose).

Yet do recall that Canetti's original contribution began not with what Lasch calls 'the survival mentality', not with the psychic life of individuals, in other words, but rather with a social fact at its most primal: the crowd. The paradoxical reason why the survivor presents us with the elements of a portrait of the anti-antisemite is, first of all, because the survivor is a narratological person that, essentially related to the crowd, also explains the latter's 'disappearance' from view (recall that no account of the war on antisemitism has been offered, no theorisation of its history or efficacy, indeed, even of its existence as a social movement). The survivor is, in a certain sense, the very figure—metaphor and metonymy—of the crowd. He represents the crowd and substitutes for it, a member of the crowd and the marker of its disappearance. He thus partakes of it and transcends it as well, functioning as its remaining guardian and memorial. He embodies it, re-members it, and immortalises it as past and gone. Consider, then, that an early example of the crowd in Canetti's reflections is the 'flight crowd', a crowd that is produced or created by a threat. Danger threatens, and 'it can threaten the inhabitants of a city, or all those who belong to a particular faith, or speak a particular language'.[56] Now, as the flight crowd forms, Canetti tells us, no one, among its individual members, 'no-one is going to assume that he, out of so many, will be the victim and, since the sole movement of the whole flight is towards salvation, each is convinced that he personally will attain it' (ibid.). Each member, in other words, sees himself as a lone survivor (as Christ, Canetti later explains). And this is precisely where the analytical distinction begins to settle between victim and survivor. For more generally, crowds certainly have a strong relation to victims, but they are also radically distinct from them, collectively and individually. In fact, another crowd that occupies Canetti's attention, another paradigmatic social movement, if you will, is the 'lamenting pack', the nature of which is precisely to mourn the passing of one or more victims. As he elaborates on the nature of this lament, and follows more or less explicitly in Freud's footsteps, Canetti confronts its more disturbing aspects, aspects that demonstrate the profound relation of, as well as the essential distinction between, victim and survivor. 'For the same people who have cause to lament are also survivors. They lament their loss, but they feel a kind of satisfaction in their own survival. They will not normally admit this, even to themselves, for they regard it as improper' (p. 263). The survivor is not a victim, in other words, because over against the victim, he has triumphed over death (making evident the narrative, and Christological, structure that he enacts and embodies). He sees himself as invulnerable, and that is precisely why, participating 'in the attempt to restore narcissistic illusions of omnipotence', he is particularly inclined to mobilise against pain and loss (think 'disaster preparedness').[57] As Lasch puts it, 'limits imply vulnerability, whereas the survivalist seeks to become invulnerable, to protect himself against pain and loss'.[58] Does this not constitute the very basis of my struggle against antisemitism? What else does 'never again' mean if not the logical horizon of my story, the end of vulnerability articulated as an enduring political ideal, as the very ideal of endurance? 'Trauma', write Didier Fassin and Richard Rechtman, 'is both the product of an experience of inhumanity and the proof of the humanity of those who have endured it.'[59]

Muslims

Neither master nor slave, the survivor is at once the paradigmatic figure of modernity and the privileged memory of the struggle against antisemitism. As a metonymy of the crowd (in its disastrous disappearance), he accounts for the inchoate nature of the social movement that has occupied me here, while signalling its collective force. The metonymy 'signals a cluster of terms', while 'the relations among the terms are never specified. The relations are assumed to be obvious', yet that very assumption discourages anyone 'from considering whether the implied relations are empirically accurate'. Ultimately, 'metonymies function as a kind of causal thread in the stories that appear in fragmented form in activists' descriptions, claims, non-narrative explanations, and references'.[60] Interrogating the figure of the survivor, engaging in the kind of reflective or critical attitude that would lead to questioning its authority and its effects, should lead to a recognition of the peculiarly fragmented, unrecognised, yet potent character of the struggle against antisemitism. To uphold the survivor as a political ideal, on the other hand, is already to partake of the war on antisemitism (and vice versa), albeit without recognising its history, its political import, nor indeed its political successes. The survivor 'has been glorified as a hero and obeyed as a ruler ... His most fantastic triumphs have taken place in our own time, among people who set great store by the idea of humanity.'[61]

Neither master nor slave, the survivor is not a victim. Yet his triumph is, as it were, shadowed by another figure, which provisionally completes my portrait of the anti-antisemite, and will bring me to my conclusion. Recall that the question we have been pursuing has involved the arduous recognition of the war against antisemitism as a movement that does not know itself, as a cluster of actors and activists that have been handed no portrait of themselves. The consequences of this state of affairs have to be measured as well. If we do not know ourselves, we anti-antisemites, we remain unaware of the role we play, of the nature of our ways, of the efficacy of our engagement. We may very well enact an older history or be devoid of one, but we also remain oblivious to still current, structural relations that, rather than remaining implicit, are simply occluded. The all-too-familiar quips about the 'Semitic' character of Arabs or Muslims gain much of their illuminating pertinence here and should remind us that the subjects of antisemitism—that is, the antisemitic subjects—have long posited themselves in opposition to both Jews and Arabs, Jews and Muslims. The inextricable history of these two enemies has been signalled on a number of occasions (if in a restricted, regional manner), and most prominently by Edward Said.[62] And though (or rather, because) it has yet to gain much acceptance, it constitutes a privileged site of occlusion towards a portrait of the anti-antisemite. The formula is after all only too simple: Orientalism is antisemitism—which is to say that there is no document of one that is not at the same time a document of the other. And just as an antisemitic document is never free of Islamophobia, so Islamophobia always taints the manner in which antisemitism is transmitted from one owner to another, from one war to another.[63] Rather than rehearse this argument once again, I want to bring to a close the reflections I have attempted so far and argue finally that I, anti-antisemite, have also been provided with a negative portrait of myself, a shadow that paradoxically completes the guidance I have been seeking, and around which I and others have long been given the ability to rally, albeit silently.[64] This makes explicit the manner in which we partake of the same collective movement, the way in which we wield the same 'spiritual weapons' to which I have been attending.[65]

This once hugely popular portrait, which must truly be read to be believed, provides us with a culmination of sorts, with all the attributes of the survivor we have seen, only marked, as I said, negatively—with a different ending. It was drawn by one of its most faithful, self-appointed ethnographers during the Cold War, and it testifies to the older roots, to the endurance, of Orientalism, aka Islamophobia, aka antisemitism. For the negative portrait of the survivor is, unsurprisingly, the Muslim. Adding yet another layer to the 'ferocious irony' identified by Giorgio Agamben, which saw Jews die (rather than survive) as Muslims, *Muselmänner*, in Auschwitz,[66] Czeslaw Milosz called his 1953 portrait 'Ketman', drawing his inspiration from Joseph Arthur, Comte de Gobineau, a major personality if ever there was one in the history of modern racism. What is significant, though, is that the figure of Ketman is either the non-survivor or the wrong survivor. Unredeemed—and beyond redemption—he is at any rate a different metonymy for the crowd ('they crowd my memory with their faceless presence', famously wrote Primo Levi), for a social movement that signals, this time, a massive but negative dénouement.[67] Painting him as an actor of sorts, Milosz deploys the force of numbers to evoke the sheer magnitude of his crowding presence: 'Acting on a comparable scale has not occurred often in the history of the human race.'[68] Appropriately raising the spectre of 'countless hordes pouring out of the Euro-Asian continent' (p. 62), Milosz locates Ketman at the centre of an elaborate typology, underscoring that 'the number of varieties of Ketman is practically unlimited, the naming of deviations cannot keep pace with the weeding of a garden so full of unexpected specimens' (p. 60). It is at any rate against him, against 'Ketman as a social institution' (p. 79), that an earlier mobilisation had long taken place, with no more recognition of itself as a social, intellectual, religious and political movement. When Said named that movement 'Orientalism', he left out the Congress for Cultural Freedom and the war against Communism (that war against the East), but he did not omit antisemitism. And while they do tend to recognise some historical antecedents, the contemporary scholars who have spoken of a new Islamophobia still have a great many difficulties identifying among long-familiar faces the well-known portrait of the antisemite. For there is perhaps the surest vector towards an understanding of the war on antisemitism: namely, that it has enlisted itself in a long and extended struggle that takes as its ideal Christ over Ketman, and the survivor over the Muslim.

Notes

1. A. Finkielkraut, *The Imaginary Jew*, trans. Kevin O'Neill and David Suchoff (Lincoln: University of Nebraska Press, 1994); Defamation, Dir. Yoav Shamir (2009) (Finkielkraut 1994).
2. J. Butler, *Precarious Life: The Power of Mourning and Violence* (London: Verso, 2004), pp. 102–3 (Butler 2004).
3. J. Judaken, *Jean-Paul Sartre and the Jewish Question: Anti-antisemitism and the Politics of the French Intellectual* (Lincoln: University of Nebraska Press, 2006), p. 20 (Judaken 2006).
4. S. Mahmood, *Politics of Piety: The Islamic Revival and the Feminist Subject* (Princeton, NJ: Princeton University Press, 2005), p. 23 (Mahmood 2005).
5. S. Kaviraj, *The Imaginary Institution of India: Politics and Ideas* (Columbia University Press, 2010), p. 199; the book is dedicated to an understanding of the dynamic emergence of national movements (Kaviraj 2010).

6. E. Goffman, *Frame Analysis: An Essay on the Organization of Experience* (Boston: Northeastern University Press, 1986), and see for a review of the ensuing scholarship, R.D. Benford and D.A. Snow, 'Framing Processes and Social Movements: An Overview and Assessment', *Annual Review of Sociology*, 26 (2000), 611–39 (Goffman 1986; Benford and Snow 2000).
7. Recall that Goffman locates the matter of 'frames' precisely at an experiential level. 'I am not addressing the structure of social life,' he writes, 'but the structure of experience individuals have at any moment of their social lives' (Goffman, *Frame Analysis*, p. 13).
8. See Benford and Snow, 'Framing Processes and Social Movements', p. 620.
9. I re-visit an argument I have made elsewhere on the 'war against antisemitism' (G. Anidjar, 'When Killers Become Victims: Anti-Semitism and Its Critics', *Cosmopolis: A Review of Cosmopolitics*, 3 (2007) [http://agora.qc.ca/cosmopolis] [Anidjar 2007]).
10. C. Bender, *The New Metaphysicals: Spirituality and the American Religious Imagination* (Chicago: University of Chicago Press, 2010), p. 5 (Bender 2010).
11. Mahmood, *Politics of Piety*, p. 14.
12. Mahmood, *Politics of Piety*, p. 14.
13. On the portrait of the coloniser, antisemite and philosemite, see Judaken, *Jean-Paul Sartre and the Jewish Question*, pp. 165–6 & 259.
14. L. Nader, 'Up the Anthropologist—Perspectives Gained from Studying Up', in *Reinventing Anthropology*, ed. Dell Hymes (New York: Vintage Books, 1974), pp. 284–311 (Nader 1974).
15. Bender, *New Metaphysicals*, p. 10.
16. F. Polletta, *It Was Like a Fever: Storytelling in Protest and Politics* (Chicago: University of Chicago Press, 2006), p. 34; hereafter abbreviated as *Fever* (Polletta 2006).
17. Polletta, *Fever*, p. 35.
18. T.L. Orbuch, 'People's Accounts Count: The Sociology of Accounts', *Annual Review of Sociology*, 23 (1997), 455–78; and see Poletta on "scholarly tales" and narrative (Polletta, *Fever*, p. 167) and Kaviraj, *The Imaginary Institution of India*, 201ff (Orbuch 1997).
19. Consider, by way of a troubling contrast, P.S. Bearman and K. Stovel, 'Becoming a Nazi: A Model for Narrative Networks',

Poetics, 27 (2000), 69–90; and K.M. Blee, 'Becoming a Racist: Women in Contemporary Ku Klux Klan and Neo-Nazi Groups', *Gender and Society*, 10, no. 6 (December 1996), 680–702 (Bearman and Stovel 2000; Blee 1996).

20. 'A plethora of studies call attention to the ways in which movements identify the "victims" of a given injustice and amplify their victimization' (Benford and Snow, 'Framing Processes and Social Movements', p. 615; and see Polletta, *Fever*, p. 4 and more generally her Chapter 5, 'Ways of Knowing and Stories Worth Telling: Why Casting Oneself as a Victim Sometimes Hurts the Cause', pp. 109–40).
21. Polletta, *Fever*, p. 6.
22. Polletta, *Fever*, p. 28.
23. S. Lindqvist, *The Skull Measurer's Mistake And Other Portraits of Men and Women Who Spoke Out Against Racism*, trans. Joan Tate (New York: New Press, 1997), p. 3 (Lindqvist 1997).
24. See e.g., P. Goodman, *Of One Blood: Abolitionism and the Origins of Racial Equality* (Berkeley: University of California Press, 1998); A. Pagden, *Lords of all the World: Ideologies of Empire in Spain, Britain and France, c.1500–c.1800* (New Haven, CT: Yale University Press, 1995), and see also A.M. Stern, *Eugenic Nation: Faults and Frontiers of Better Breeding in Modern America* (Berkeley: University of California Press, 2005), esp. pp. 185–92 ("Eugenics and Its Critics") (Goodman 1998; Pagden 1995; Stern 2005).
25. C.L.R. James, *The Black Jacobins: Toussaint L'Ouverture and the San Domingo Revolution* (New York: Vintage, 1989); N.B. Dirks, *The Scandal of Empire: India and the Creation of Imperial Britain* (Cambridge, MA: Belknap Press of Harvard University Press, 2006) (James 1989; Dirks 2006).
26. As Polletta points out, Betty Friedan provided in *The Feminine Mystique* an influential—if fictional—portrait of her rise to feminist consciousness from suburban domesticity (Poletta, *Fever*, 33). Earlier, Simone de Beauvoir had already articulated her powerful account of how one becomes a women in *The Second Sex* (S. de Beauvoir, *The Second Sex*, trans. Constance Borde and Sheila Malovany-Chevallier [New York: Knopf, 2010] [de Beauvoir 2010]); and see J.W. Scott (ed.), *Feminism and History* (Oxford: Oxford University Press, 1996), as well as J. Butler, *Gender Trouble:*

Feminism and the Subversion of Identity (New York: Routledge 2006) (Scott 1996; Butler 2006).

27. And compare with the situation Kaviraj insists on, namely that 'what we with casual mundaneness call politics today is historically an entirely new activity, unlike its namesakes in earlier times. It is hardly surprising that this activity is often referred to as "politics" in a vernacularized English term, precisely because, in a strict sense, it had no historical precursors' (Kaviraj, *The Imaginary Institution of India*, p. 6).
28. Bender, *New Metaphysicals*, p. 59; in the chapter from which I extract this quote, Bender seeks to explain 'how it is that one becomes a mystic' (p. 58; and see p. 182).
29. Bender, *New Metaphysicals*, p. 2.
30. Bender, *New Metaphysicals*, p. 184.
31. See e.g., J. Marelli, 'Usages et maléfices du theme de l'antisémitisme en France', in *La république mise à nu par son immigration*, ed. Nacira Guénif-Souilamas (Paris: La Fabrique, 2006), pp. 133–59; and see E. Balibar et al., *Antisémitisme: l'intolérable chantage* (Paris: La Découverte, 2003); A. Cockburn and J. St. Clair, *The Politics of Anti-Semitism* (Oakland, CA: AK Press and CounterPunch, 2003), and see J. Butler, *Precarious Life*, pp. 101–27 (Marelli 2006; Balibar et al. 2003; Cockburn and St. Clair 2003).
32. Talal Asad makes productive use of Wittgenstein's notion of 'grammar' in his anthropological analyses (T. Asad, *Formations of the Secular: Christianity, Islam, Modernity* [Stanford, CA: Stanford University Press, 2003], pp. 25 & 25n9 [Asad 2003]). The phrase 'the grammar of a discourse' occurs on p. 161.
33. 'Only in the nationalist imagination, and not in real social history', writes Mahmood Mamdani, 'can movements emerge full-blown as the Greek goddess Athena is supposed to have done from the head of Zeus.' Mamdani pursues this argument and explains that 'the question we need to ask when assessing the democratic content of a movement is not just one concerning its geographical sweep, but also one that underlines the social character of its demands: Do they tend toward realizing equality or crystallizing privilege? Are they generalizable to other ethnic groups or can they be realized only at the expense of others? In other words, when do they signify a struggle for rights and when a demand for privilege?'

(M. Mamdani, *Citizen and Subject: Contemporary Africa and the Legacy of Late Colonialism* [Princeton, NJ: Princeton University Press, 1996], p. 203 [Mamdani 1996]).

34. Polletta, *Fever*, p. 53.
35. Kaviraj, *The Imaginary Institution of India*, p. 14.
36. Mamdani, *Citizen and Subject*, p. 188.
37. The phrase 'episodes of contention' is found in D. McAdam, S. Tarrow, and C. Tilly, *Dynamics of Contention* (Cambridge: Cambridge University Press, 2004), p. 6. Needless to say, the literature on social and political movements is vast and growing, with Tarrow and Tilly being at once foundational and prolifically contributing figures. Interestingly, according to Irfan Ahmad, 'social movements studies' have tended to neglect 'movements outside the West', a phenomenon he attributes to the lingering Orientalism of social scientists (I. Ahmad, *Islamism and Democracy in India: The Transformation of Jamaat-e-Islami* [Princeton, NJ: Princeton University Press, 2009], p. 26 [Ahmad 2009]) (McAdam et al. 2004).
38. McAdam, Tarrow, and Tilly, *Dynamics of Contention*, p. 6.
39. ibid.
40. See U.S. Congress, 'Global Anti-Semitism Review Act', 29 September 2004; and see www.csce.gov/index.cfm?Fuseaction=ContentRecords.ViewDetail&ContentRecord_id=279&Region_id=0&Issue_id=0&ContentType=H (U.S. Congress 2004).
41. It might help, Polletta suggests, 'to identify the mechanisms by which culture sets the terms of strategic action, but without treating actors as strategic dupes' (Polletta, *Fever*, p. 28).
42. C. Lasch, *The Minimal Self: Psychic Survival in Troubled Times* (New York: W.W. Norton & Co., 1984) (Lasch 1984).
43. Bender, *New Metaphysicals*, p. 184.
44. G. Agamben, *Homo Sacer: Sovereign Power and Bare Life*, trans. Daniel Heller-Roazen (Stanford, CA: Stanford University Press, 1998) (Agamben 1998).
45. Lasch, *The Minimal Self*, p. 57; E. Canetti, *Crowds and Power*, trans. Carol Stewart (New York: Farrar, Straus and Giroux, 1984); and see G. Anidjar, 'Survival', *Political Concepts: A Critical Lexicon*, 2 (Winter 2012), online at http://www.politicalconcepts.org/survival-gil-anidjar/ (Canetti 1984; Anidjar 2012).
46. Lasch, *The Minimal Self*, p. 16.

47. Lasch, *Minimal Self*, p. 60; and see W.A. Davis' parallel, and striking, reflections in his *Death's Dream Kingdom: The American Psyche since 9-11* (London: Pluto Press, 2006) (Davis 2006).
48. T. Hobbes, *Leviathan*, ed. Richard Tuck (Cambridge: Cambridge University Press, 1996), XX, p. 140; I quote from a less obvious moment in Hobbes, whose otherwise famous definition of life ('solitary, poore, nasty, brutish, and short', XIII, p. 89) does not quite highlight the paradoxical ground that self-preservation, indeed survival, provides: the submission to its potential, and legitimate, destroyer. Adding Rousseau to the list of those who embraced the 'dream of the sole survivor', one of the rare readers of Canetti, Roberto Esposito, explains that individuals, for Hobbes, 'are paradoxically sacrificed to their own survival' (R. Esposito, *Communitas*, trans. Timothy Campbell [Stanford, CA: Stanford University Press, 2010], pp. 43, 14 [Esposito 2010]) (Hobbes 1996).
49. Spencer's phrase, of course, not Darwin's, but its immense popularity is not in doubt; see e.g., R. Hofstadter, *Social Darwinism in American Thought* (Boston: Beacon Press, 1955) (Hofstadter 1955).
50. D. Haraway, *Primate Visions: Gender, Race, and Nature in the World of Modern Science* (New York: Routledge, 1989), p. 369 (Haraway 1989).
51. C. Lasch, *The Minimal Self*, p. 66.
52. On the term 'survivor', see Polletta, *Fever*, p. 116.
53. W. Brown, *States of Injury: Power and Freedom in Late Modernity* (Princeton, NJ: Princeton University Press, 1995), pp. 52–76 (Brown 1995).
54. Canetti, *Crowds and Power*, p. 250; to repeat: for Canetti, as for Hobbes, survival is a political choice, not an 'instinct'.
55. D. Fassin and R. Rechtman, *The Empire of Trauma: An Inquiry into the Condition of Victimhood*, trans. R. Gomme (Princeton, NJ: Princeton University Press, 2009) (Fassin and Rechtman 2009).
56. Canetti, *Crowds and Power*, p. 53.
57. Lasch, *The Minimal Self*, p. 19.
58. Lasch, *The Minimal Self*, p. 98.
59. Fassin and Rechtman, *Empire of Trauma*, p. 20, emphasis added.
60. Polletta, *Fever*, pp. 60–2.
61. Canetti, *Crowds and Power*, p. 468.

62. E.W. Said, *Orientalism* (New York: Vintage, 1978).
63. I borrow from Walter Benjamin's well-known formulation on civilisation and barbarism in 'Theses on the Philosophy of History', trans. H. Zohn in *Illuminations*, ed. H. Arendt (New York: Schocken, 1969), p. 256 (Benjamin 1969).
64. The word 'shadow' is Edward Said's (*Orientalism*, p. 286).
65. T.J. Gunn, *Spiritual Weapons: The Cold War and the Forging of an American National Religion* (Westport: Praeger, 2009) (J. Gunn 2009).
66. G. Agamben, *Remnants of Auschwitz: The Witness and the Archive*, trans. D. Heller-Roazen (New York: Zone Books, 1999), p. 45; and see G. Anidjar, *The Jew, the Arab: A History of the Enemy* (Stanford, CA: Stanford University Press, 2003) (Agamben 1999; Anidjar 2003).
67. P. Levi, *Survival in Auschwitz*, trans. S. Woolf (New York: Simon and Schuster, 1996), p. 90 (Levi 1996).
68. C. Milosz, *The Captive Mind*, trans. J. Zielonko (London: Penguin, 2001), p. 57 (Milosz 2001).

References

Agamben, G. 1998. *Homo Sacer: Sovereign Power and Bare Life*, trans. Daniel Heller-Roazen. Stanford, CA: Stanford University Press.

———. 1999. *Remnants of Auschwitz: The Witness and the Archive*, trans. D. Heller-Roazen. New York: Zone Books.

Ahmad, I. 2009. *Islamism and Democracy in India: The Transformation of Jamaat-e-Islami*. Princeton, NJ: Princeton University Press.

Anidjar, G. 2003. *The Jew, the Arab: A History of the Enemy*. Stanford, CA: Stanford University Press.

Anidjar, G. 2007. When Killers Become Victims: Anti-Semitism and Its Critics. *Cosmopolis: A Review of Cosmopolitics* 3. http://agora.qc.ca/cosmopolis

———. 2012. Survival. *Political Concepts: A Critical Lexicon*, 2 (Winter). http://www.politicalconcepts.org/survival-gil-anidjar/

Asad, T. 2003. *Formations of the Secular: Christianity, Islam, Modernity*. Stanford, CA: Stanford University Press.

Balibar, E., et al. 2003. *Antisémitisme: l'intolérable chantage*. Paris: La Découverte.

Bearman, P.S., and K. Stovel. 2000. Becoming a Nazi: A Model for Narrative Networks. *Poetics* 27: 69–90.

Bender, C. 2010. *The New Metaphysicals: Spirituality and the American Religious Imagination*. Chicago: University of Chicago Press.

Benford, R.D., and D.A. Snow. 2000. Framing Processes and Social Movements: An Overview and Assessment. *Annual Review of Sociology* 26: 611–639.

Benjamin, Walter. 1969. Theses on the Philosophy of History. In *Illuminations*, ed. H. Arendt and trans. H. Zohn. New York: Schocken.

Blee, K.M. 1996. Becoming a Racist: Women in Contemporary Ku Klux Klan and Neo-Nazi Groups. *Gender and Society* 10(6) (December): 680–702

Brown, W. 1995. *States of Injury: Power and Freedom in Late Modernity*. Princeton, NJ: Princeton University Press.

Butler, J. 2004. *Precarious Life: The Power of Mourning and Violence*. London: Verso.

———. 2006. *Gender Trouble: Feminism and the Subversion of Identity*. New York: Routledge.

Canetti, E. 1984. *Crowds and Power*, trans. Carol Stewart. New York: Farrar, Straus and Giroux.

Cockburn, A., and J. St. Clair. 2003. *The Politics of Anti-Semitism*. Oakland, CA: AK Press and CounterPunch.

Davis, W.A. 2006. *Death's Dream Kingdom: The American Psyche since 9-11*. London: Pluto Press.

de Beauvoir, S. 2010. *The Second Sex*, trans. Constance Borde and Sheila Malovany-Chevallier. New York: Knopf.

Dirks, N.B. 2006. *The Scandal of Empire: India and the Creation of Imperial Britain*. Cambridge, MA: Belknap Press of Harvard University Press.

Esposito, R. 2010. *Communitas*, trans. Timothy Campbell. Stanford, CA: Stanford University Press.

Fassin, D., and R. Rechtman. 2009. *The Empire of Trauma: An Inquiry into the Condition of Victimhood*, trans. R. Gomme. Princeton, NJ: Princeton University Press.

Finkielkraut, A. 1994. *The Imaginary Jew*, trans. Kevin O'Neill and David Suchoff. Lincoln: University of Nebraska Press.

Goffman, E. 1986. *Frame Analysis: An Essay on the Organization of Experience*. Boston: Northeastern University Press.

Goodman, P. 1998. *Of One Blood: Abolitionism and the Origins of Racial Equality*. Berkeley: University of California Press.

Gunn, T.J. 2009. *Spiritual Weapons: The Cold War and the Forging of an American National Religion*. Westport: Praeger.

Haraway, D. 1989. *Primate Visions: Gender, Race, and Nature in the World of Modern Science*. New York: Routledge.

Hobbes, T. 1996. *Leviathan*, ed. Richard Tuck. Cambridge: Cambridge University Press.

Hofstadter, R. 1955. *Social Darwinism in American Thought*. Boston: Beacon Press.

James, C.L.R. 1989. *The Black Jacobins: Toussaint L'Ouverture and the San Domingo Revolution*. New York: Vintage.

Judaken, J. 2006. *Jean-Paul Sartre and the Jewish Question: Anti-antisemitism and the Politics of the French Intellectual*. Lincoln: University of Nebraska Press.

Kaviraj, S. 2010. *The Imaginary Institution of India: Politics and Ideas*. New York: Columbia University Press.

Lasch, C. 1984. *The Minimal Self: Psychic Survival in Troubled Times*. New York: W.W. Norton & Co..

Levi, P. 1996. *Survival in Auschwitz*, trans. S. Woolf. New York: Simon and Schuster.

Lindqvist, S. 1997. *The Skull Measurer's Mistake And Other Portraits of Men and Women Who Spoke Out Against Racism*, trans. Joan Tate. New York: New Press.

Mahmood, S. 2005. *Politics of Piety: The Islamic Revival and the Feminist Subject*. Princeton, NJ: Princeton University Press.

Mamdani, M. 1996. *Citizen and Subject: Contemporary Africa and the Legacy of Late Colonialism*. Princeton, NJ: Princeton University Press.

Marelli, J. 2006. Usages et maléfices du theme de l'antisémitisme en France. In *La république mise à nu par son immigration*, ed. Nacira Guénif-Souilamas. Paris: La Fabrique.

McAdam, D., S. Tarrow, and C. Tilly. 2004. *Dynamics of Contention*. Cambridge: Cambridge University Press.

Milosz, C. 2001. *The Captive Mind*, trans. J. Zielonko. London: Penguin.

Nader, L. 1974. Up the Anthropologist—Perspectives Gained from Studying Up. In *Reinventing Anthropology*, ed. Dell Hymes. New York: Vintage Books.

Orbuch, T.L. 1997. People's Accounts Count: The Sociology of Accounts. *Annual Review of Sociology* 23: 455–478.

Pagden, A. 1995. *Lords of all the World: Ideologies of Empire in Spain, Britain and France, c.1500–c.1800*. New Haven, CT: Yale University Press.

Polletta, F. 2006. *It Was Like a Fever: Storytelling in Protest and Politics*. Chicago: University of Chicago Press.

Scott, J.W. (ed). 1996. *Feminism and History*. Oxford: Oxford University Press.

Stern, A.M. 2005. *Eugenic Nation: Faults and Frontiers of Better Breeding in Modern America*. Berkeley: University of California Press.

U.S. Congress. 2004. Global Anti-Semitism Review Act, 29 September.

PART IV

Response

CHAPTER 9

Antisemitism, Islamophobia and the Search for Common Ground in French Antiracist Movements since 1898

Daniel A. Gordon

The dawn of the twenty-first century was a testing time for ideals of a united front against racism in France, witnessing sharp disagreement among antiracists about the relative importance of antisemitism and post-colonial racism, including Islamophobia. A flashpoint for this debate was in 2004, when France's best-known antiracist groups—the Mouvement contre le racisme et pour l'amitié entre les peuples (MRAP), the Ligue des droits de l'homme (LDH), the Ligue internationale contre le racisme et l'antisémitisme (LICRA) and SOS Racisme—publicly broke ranks over precisely such a fault-line. This chapter aims to set this acrimonious debate in a much longer-term historical context, by asking whether the opposing positions of what have been termed the 'Four

D.A. Gordon (✉)
Edge Hill University, Ormskirk, UK

J. Renton, B. Gidley (eds.), *Antisemitism and Islamophobia in Europe*, DOI 10.1057/978-1-137-41302-4_9

Sisters'[1] of French antiracism can be explained by truly irreconcilable approaches.

First, to outline the controversy of 2004. In May, the MRAP and the LDH pulled out of a demonstration against a rise in antisemitic attacks, arguing that the demonstration should be 'against antisemitism and against all racisms', not only 'against antisemitism'.[2] Conversely, in November, the organisers of the May demonstration, SOS Racisme and the LICRA, refused to take part in a demonstration 'against all racisms', claiming that some of the Muslim groups participating were insufficiently clear in their condemnation of antisemitism.[3] Certainly the November demonstration, as I experienced it, was far from achieving unity. Such polemics highlighted a struggle within the movements, in which patiently built alliances threatened to collapse. As Lynda Asmani of the Coordination des Berbères en France put it, this was a 'derisory debate, demobilising, even irresponsible when the urgency is to unite'.[4] These events were a classic example of what the newspaper *Libération* had described in 2003 as 'people who agree with each other 95% of the time to struggle against all racisms ending up by putting each other on trial for antisemitism or Islamophobia'.[5]

Yet was the controversy between antiracist organisations indicative of a deeper split? One way of understanding the disagreement of 2004 was, to put it in rather crude communitarian terms, as the breakdown of an earlier alliance of Jews and Muslims against the extreme right—under the weight of international events during the second Palestinian *intifada* and an irreconcilable divide over which racism to confront the most vigorously. The extremes of this debate can be summarised, for the sake of brevity and neutrality, as pole one and pole two. Those nearer pole one are primarily concerned with racism against colonial and post-colonial migrants. Because of the legacy of the Algerian war of independence, the most violent episode in the history of European decolonisation, this has the heaviest impacts on Arabs and Muslims. Islamophobia is therefore a central and pressing concern for those near pole one,[6] although the term only came into wide use in France recently, and remains very controversial.[7] Those nearer pole two are primarily concerned with antisemitism,[8] which, mirroring concern about Islamophobia, also has deep resonance in France because of the legacies of the Dreyfus Affair and Vichy France's collaboration with the Nazis. Exacerbating this split is the international dimension: pole one is broadly pro-Palestine, and pole two broadly pro-Israel; pole one thus accuses pole two of instrumentalising antisemitism, while pole two accuses pole one of seeking to cover it up. Indeed, both poles have faced accusations of seeking to import into France the Israeli–Palestinian conflict, which is thereby often framed as a contagious disease threatening France from outside.[9] Yet the conflict presents a very real trauma for French society,[10] not only because it has the largest Muslim and largest Jewish populations in Western Europe, but because, of the two most notorious human rights abuses committed by the French state in the twentieth century, one was against Jews and the other was against Muslims. Seemingly interminable debates around the highly sensitive issues of the deportation of Jewish people from France to extermination camps and the torture and extrajudicial executions of Algerian people—the former memory debate peaking from the 1970s to the mid-1990s,[11] and the latter during the early 2000s[12]—make the sheer weight of polemic surrounding either, let alone a combination of both multiplied by Israel–Palestine, so explosive.

It has therefore become a standard trope of republican discourse to complain of *communautarisme*: that French society is falling into separate communities, playing a highly dangerous game of competitive victimhood, in a turn to identity politics facilitated by the end of the more substantial ideological debates of the Cold War.[13] Outside observers also often share the view of an insuperable divide, whether taking sides or regretting that 'the two camps have remained entrenched in their respective positions'.[14] Michel Feher has nuanced this, pointing out that it is scarcely the case that Israel–Palestine arouses passions only among Jews and Muslims, with the rest of society looking on neutrally. Rather than a competition between minorities, he sees a broader argument between two different interpretations of contemporary history. One (corresponding, I would suggest, to pole two) sees the greatest danger in hatred against outsiders, carrying a threat of genocide; and the other (corresponding to pole one) sees the greatest danger in the domination of the poor by the rich. Feher pours cold water on the nostalgia for the world before 1989 often displayed by critics of *communautarisme* by observing continuities: although the poles are now divided more by which evil to combat than by which vision to struggle for, the philosophical bases of the disagreement have remained, rooted in the antitotalitarian and anti-imperialist discourses of the Cold War.[15] While this has the merit of looking beyond the immediate context, nevertheless Feher appears to share the widespread assumption of clearly defined and opposed poles.

Yet it is a well-trodden path among historians of migration in France to identify continuities between the intolerance of the 1930s and 1940s, most typically directed against Jews, and that of the 1980s, 1990s and 2000s, most typically directed against Muslims.[16] The specificity of anti-semitism in the former period and Islamophobia in the latter, though, is not always emphasised—for good reason, because both can also be seen as part of a long tradition of xenophobia in French society, also expressed virulently against fellow Catholic Belgians, Italians and Poles.[17] Moreover, important work by, among others, Paul Gilroy, Michael Rothberg and Max Silverman has questioned the idea that Holocaust memory and colonial memory have always been separate.[18] Part of the contemporary critique of what Gilroy calls 'camp thinking',[19] or what Jonathan Judaken calls 'ticket thinking',[20] is to question whether such divides are as immutable as they may appear. As Silverman points out, while Holocaust studies and post-colonial studies have tended to go their own ways in recent decades, this was not originally the case, and connections keep resurfacing in a variety of francophone texts.[21] Yet the contributions of Gilroy, Rothberg and Silverman remain at a broad-brush level of cultural analysis; arguably an empirical historical approach can add something further. Equally, recent historical works by Maud Mandel and Ethan Katz have pertinently questioned the popular notions that Muslim–Jewish relations in France either have always been characterised by enmity, or can be understood solely as a function of the Israeli–Palestinian conflict, or indeed only involved people whose primary self-identification was either Muslim or Jewish.[22] The main focus of Mandel and Katz's work is less, however, on the principal organisations of French antiracism than on other protagonists in such debates, at least for the period prior to the 1980s.[23] So, we might ask, does the idea of an unbridgeable divide hold up empirically over the longer term for the four organisations that clashed in 2004?

The 'Four Sisters' and Their Entangled Histories

Since the turn of the millennium, historiography on the oldest of those organisations, the LDH, has been given a new lease of life by a thesis by Cylvie Claveau, monographs by Simon Epstein and William Irvine, a series of articles by Norman Ingram and a book edited by Emmanuel Naquet and Gilles Manceron.[24] Much of this has been facilitated by the post–Cold

War restitution to France of the Ligue's pre–Second World War archives, confiscated to Berlin in 1940, when the organisation was banned during the Occupation, and then to Moscow in 1945.[25] Both the anglophone contributors to this debate, Irvine and Ingram, focus more on the Ligue's views on pacifism and Franco-German relations than on antiracism as such (though antisemitism is discussed in relation to international relations), and both see the history of the Ligue as to all intents and purposes at an end by 1940 at the latest; indeed, one of Ingram's articles is subtitled 'Who killed the Ligue des droits de l'homme?' It is true that the revived postwar Ligue never regained its pre-war six-figure membership of some 180,000, the figure oscillating between 6,000 and 10,000 since the late 1950s.[26] Nevertheless, as we shall see, reports of its death have been much exaggerated.[27] Bringing the Ligue's story up to date and within a comparative framework of other antiracist organisations may give us a *longue durée* perspective on different styles of campaigning.

Straight away, striking paradoxes emerge. The Ligue appeared in 2004 to fall near pole one, yet its history places it nearer pole two. The organisation would see any suggestion that it is soft on antisemitism as unthinkable given its origins. The oldest of the four, it was founded in 1898 by campaigners seeking to prove that Alfred Dreyfus was innocent, embuing the organisation thereafter with the moral mantle of 'Dreyfusard'. At its height during the inter-war period, when it was the largest human rights organisation in the world and larger than most French political parties, the Ligue developed a reputation as a doughty defender of the rights of Jewish and other immigrants. This reputation was enhanced during the Occupation, when many of its leading figures were killed by the Nazis—or by the French collaborationist militia the Milice, in the case of the Hungarian-born Victor Basch, the LDH's president from 1925 until his assassination in 1944 at the age of eighty. This is therefore an organisation highly conscious of its own past: many of its leaders, such as Madeleine Rebérioux, president from 1991 to 1995, have themselves been professional historians.[28]

However, this 'heroic version' is contested by Irvine, who argues that the Ligue was not a neutral human rights organisation, but a politicised left-wing one, close to the Radical and Socialist parties, containing many members who, out of pacifism, adopted a less than heroic approach to the rise of fascism in Europe.[29] Ingram takes a similarly critical position, engaging in historiographical hand-to-hand combat with Naquet over the presence or otherwise of a minority of Ligue members who could be described as antisemites.[30] Ingram even claims that in 1940, a small minority of local Ligue branches told Gestapo interrogators that they did not admit Jews—though the evidence is tricky to read, since those branches that did own up to having Jewish members were potentially putting them in great danger.[31] As Ingram notes, antisemitism is thus the key thread in this debate,[32] which has tended to pit French historians, often connected to the Ligue itself, against historians based outside France, particularly in Canada and Israel.

By contrast, the Ligue's attitude to colonialism, let alone Islamophobia, has tended to take a lower profile among most contributors to this debate. However, Claveau does give this significant attention, arguing that the Ligue was thoroughly caught up in the colonialist attitudes of the time, maintaining a hierarchical view of civilisations with Western Europe at the top.[33] *Ligueurs* in colonial Algeria argued that only assimilated Muslims should be represented in parliament; although Basch himself disagreed, arguing that assimilation should be freely chosen rather than forced, Claveau sees a contradiction, given that Basch also acted as an apologist for the colonialist general Lyautey.[34] Colonial abuses were denounced in a sensationalist and insensitive manner, with little thought for the dignity of victims, and in a way that tried to separate abuses from the principle of colonial rule.[35] Although there is little on this in Irvine's book, what there is similarly suggests ambivalence on colonialism: the Ligue majority denounced abuses while believing in a benevolent version of colonialism, and members in colonial North Africa could even be found expressing stereotypically racist views about Arabs. Ironically, the leader of the minority that did apparently firmly oppose colonialism, Félicien Challaye, was also extremely suspect, to say the least, on the question of antisemitism.[36] A caricature of pole one, perhaps—although Claveau uses a Saidian analysis to argue that Challaye, who openly praised aspects of the Vichy regime and was later involved in Holocaust denial, was not even a genuine anti-colonialist, but an Orientalist, at bottom just as racist as his opponents in his assumptions about the colonial Other.[37] However, the majority in the LDH remained opposed to Challaye's line, and his antisemitism. Some were still pro-French Algeria in the 1950s, and as late as 1967 the leadership was overtly pro-Israel—in the tradition of Victor Basch, who had been rather favourable to Zionism from as early as 1915, and Ferdinand Corcos, a leading LDH activist who in the 1930s had debated on Palestine against the Algerian nationalist leader Messali Hadj.[38] So, revisionism notwithstanding, it would still appear that the dominant position within the

LDH before decolonisation was closer to the notional pole two than pole one.

Given this, we might ask how the Ligue came to take the apparently pole one position of 2004. The answer seems to be that during the second half of the Algerian war of independence it evolved to an anticolonialist line, under the influence of dissident Socialists who formed the classically New Left Parti Socialiste Unifié, born out of this peak moment of anticolonialism. Some 18 members of the LDH's Central Committee between 1945 and 1975 became members of the PSU.[39] During the 1970s, what had been seen as a rather archaic relic of the Third Republic—'part of the furniture of the Republic'[40]—was joined by a new generation of activists.[41] This group of babyboomers, including Michel Tubiana, the Ligue's president from 2000 to 2005, was politically formed by the far Left of 1968. Tubiana, for example, passed through the Trotskyist Jeunesse communiste révolutionnaire.[42] Although its international politics would place it in anti-Zionist pole one, the JCR was far from insensitive to antisemitism and had many Jewish leading members,[43] like the LDH itself: Tubiana is of Algerian-Jewish origin. Another sign of this injection of *soixante-huitards* is that Tubiana joined the LDH on the initiative of his colleague Jean-Jacques Felice, the lawyer of choice for 1968-era radicals.[44] LDH activists were frequently involved in 1970s social movements, such as those in defence of immigrant workers. Something of a generational coup appears to have taken place at this point, for in 1976 half of the LDH's Central Committee were forced out by a rule change stipulating that Central Committee members had to be aged under seventy-five.[45] In 1977, out went the LDH's musty *Bulletin intérieur*, described in a damning internal report as in the style of a 'little Mass'—too inward looking, full of 'self-congratulation' and updates on the president's health.[46] In came *Hommes et Libertés*, a magazine designed for wider appeal, which Henri Noguères, president from 1975 to 1984 and a Ligue veteran since the 1930s, described as a 'wager on the future'.[47] During the 1990s, by which time the current orthodoxy among left-wing militants was moving rather nearer to pole one than pole two, the Ligue was joined by another new generation of activists. At the time of its hundredth anniversary in 1998, a study by academics at Sciences Po noted that the LDH had thereby avoided the crisis of militancy common to other organisations.[48] Today, older activists from the Algerian War generation often continue in the Ligue until lost to natural causes: it is not uncommon for the organisation's still active provincial branches to issue a humanist death notice in a local newspaper hailing the contributions of a recently deceased *ligueur de longue date* to 'the defence of the republican values of Liberty, Equality and Fraternity'.[49]

The Ligue thus remained anchored firmly on the Left, going so far as to endorse specific candidates at election time[50]—albeit usually only in the second ballot of presidential elections, at which only two candidates are present.[51] Arguably, this neatly sidesteps the issue of which of the many parties of the Left to endorse in the first ballot, while not concealing the broad allegiance of the Ligue to left-wing values that Irvine has noted for the interwar period, and which would still be hard to deny today. Although not a crude satellite of any one party, there is a sense in which the Ligue continues to play, on a much smaller scale, the role of a political club that Irvine and Ingram emphasise for the Third Republic—a kind of latter-day no-frills version of the Popular Front. Indeed, in 1977, Noguères explicitly justified the Ligue's support for the Union of the Left (an electoral pact between the Socialist, Communist and Left-Radical parties) with appropriate reference to French Left theology in the form of a 1937 speech by Léon Blum to the Ligue's Congress.[52] In 1998, it was estimated that around half of *ligueurs* vote Socialist, with smaller minorities voting Green, Communist or far Left.[53] Nevertheless, Ligue activists often tend to be those who left the 1968-era far Left because they had a greater attachment than their erstwhile comrades to individual liberty and 'bourgeois freedoms', for which LDH activism allowed a suitable and respectable outlet.[54] Again, this is a long-term continuity, paralleling the 1923 ban imposed by the Communist Party on its members joining the Ligue, regarded as petit-bourgeois, sentimental humanists. The sociological profile of *ligueurs* today appears to lend some support to this stereotype: the Sciences Po survey found that they are typically highly educated, upper-middle class and without religious beliefs.[55] As Eric Agrikiolansky argues, the Ligue often appeals to precisely those intellectuals who, whichever party of the Left they happen to be members of, resent having to follow the party line. Rebérioux, for example, was a dissident Communist, just as many of her contemporaries were dissidents from the Section française de l'internationale ouvrière, and later militants are sometimes dissidents within the Parti socialiste (PS).[56] Thus, while for Irvine the LDH was supposedly made irrelevant by the urban world of 1950s France,[57] its very archaism is today, in a world where modernisation is old hat, actually quite attractive to some people, radiating a kind of retro chic, tolerance of diversity and opportunities to continue the venerable battles of yesteryear.

It might be speculated, therefore, that the LDH's apparent flirtation with a pole one position by 2004 was an attempt to shake off its elitist image. This would be in some respects in keeping with its interwar history, when it sometimes combined a quasi-revolutionary rhetoric with a membership essentially composed of lawyers, teachers and civil servants, but might also present an opportunity to make up in retrospect for an anticolonialist fervour less apparent at the time. The Ligue has explicitly claimed that the dual heritage of its roles in the Dreyfus Affair and the Algerian War and subsequent campaigns for the rights of North African immigrants makes it 'doubly qualified to intervene' in favour of a two-state solution for Israel–Palestine.[58] Thus, it would be too caricatural to see the LDH as having shifted entirely to pole one, for its campaigning and memorial work is multi-faceted, cutting across our notional poles. The speakers at its 1998 centenary conference included, alongside the then President of the Republic Jacques Chirac and French Resistance heroes Lucie and Raymond Aubrac, one leftist speaker of Muslim background (the veteran Algerian opposition leader Hocine Aït Ahmed) and two leftist speakers of Jewish background (Alain Krivine of the Ligue communiste révolutionnaire and the Moroccan former political prisoner Abraham Serfaty).[59] They were chosen presumably because both Aït Ahmed and Serfaty are known for their long struggles, first against colonialism and then in favour of pluralism and minority rights in their home countries; antisemitic and anti-Sahraoui Arab nationalist arguments were used against Serfaty,[60] just as anti-Berber Arab nationalist arguments were against Aït Ahmed. The LDH is as vigorous in its memory activities on antisemitism—in 1998 it marked the centenary of the Dreyfus Affair by staging a reconstruction of Zola's trial,[61] and in 2000 it demanded disciplinary sanctions against an academic at the University of Lyon III accused of Holocaust denial[62]—as it is on colonialism. In 2001 Tubiana called for Paul Aussaresses, a general who wrote a book justifying the use of torture during the Algerian War, to be stripped of his medals,[63] while Tubiana's vice-president, the historian Gilles Manceron, has been active in promoting awareness of the police massacre of Algerian demonstrators in Paris on 17 October 1961.[64] In July–September 2004, the LDH's journal *Hommes et Libertés* devoted much of its issue to the question of antisemitism. While there may be an element of self-aggrandisement in the continual references to history, all this is quite fitting for an organisation that sees itself as 'the good memory and the bad conscience'[65] of the French Left, and the names of whose former leaders now adorn suburban bus stops.[66]

Of the four main protagonists in 2004, only one, the LICRA, could be consistently placed at putative pole two. Yet, just as the LDH was founded to fight for justice for a single individual victim of antisemitism, Alfred Dreyfus, so the LICRA can trace its origins directly to the defence of another, Simon Schwarzbard. In 1928 Schwarzbard, a Ukrainian-Jewish anarchist, shot dead Simon Petlioura, former head of the nationalist government of Ukraine, as he came out of a restaurant in Paris's Left Bank, on the grounds that Petlioura had been responsible for pogroms.[67] Thus, the Ligue internationale contre les pogroms, which soon became the Ligue international contre l'antisémitisme (LICA), was founded as part of a struggle against antisemitism, but also in defence of an act of terrorism.[68] The Schwarzbard affair was influential well beyond France. Henry Abramson has argued that much historiography on Ukrainian-Jewish relations follows similar arguments to those put at Schwarzbard's trial—the defence, as well as leftist, Soviet and Jewish sources, tended to admit that Schwarzbard killed Petlioura, but tried to justify this with reference to links between Petlioura and the pogroms, while the prosecution, and Ukrainian sources, tended to deny these links. However, the affair also demonstrates that the French anti-antisemitic milieu around the LICA was originally a broadly leftist one. Schwarzbard's lawyer, Henri Torrès, later became a member of the French Communist Party. Ukrainian sources, like the prosecution at the trial, further claim that Schwarzbard was not acting alone, but on behalf of Soviet intelligence, alleging links with an agent named Mikhail Volodin.[69] Yet the jury was not convinced, with its 'not guilty' verdict greeted by cries of 'Vive la France!' As Yosef Nedava points out, Schwarzbard was fortunate to have killed Petlioura in France rather than in England, where there was no precedent for someone to be acquitted of a murder purely on the grounds that the jury sympathised with their motives; his acquittal was a distinctively French phenomenon, for the political equivalent of a *crime de passion*. More recently, Schwarzbard has been the object of similar philosemitic attention from the French Left, since in 2010 his unpublished autobiographical writings were finally published in French translation as *Mémoires d'un anarchiste juif*, by the leftist publisher Syllepse in its neo-Bundist series Yiddishland.[70]

As for the LICA itself, if less famous than the LDH, it has also been portrayed as an energetic contributor to the struggle against antisemitism in 1930s France, mustering 50,000 members at its height.[71] Yet, just as Claveau, Irvine and Ingram have sought to challenge too heroic a view of the LDH, so Simon Epstein has offered a decidedly revisionist account of the LICA. Epstein's exploration of the role of former antiracists in collaboration after 1940 has identified a number of individuals with strange trajectories: Marcelle Capy, for example, a member of the LICA's Central Committee, went on to join the collaborationist Parti ouvrier et paysan français in 1941; while one of the most prominent fascists in wartime France, the former Communist Jacques Doriot, leader of the Parti populaire français, whose combination of anti-Jewish and pro-Palestinian positions would presumably place him at pole one, had in the early 1930s been known as a great friend of the LICA.[72] Epstein's account, though, by focusing narrowly on individuals' attitudes to the question of antisemitism, sometimes overstates the case by basing it on one criterion: he does not give much information on attitudes to colonial racism, and is influenced by a teleological assumption of the inherent futility of the struggle against antisemitism.[73]

So how might the LICA be analysed in terms of the double dimension of antisemitism and Islamophobia? The organisation was conscious of the need to broaden its appeal at least nominally: in 1979 the LICA became the LICRA by adding to its title the crucial words 'le racisme et', which it had also used between 1936 and 1948.[74] Indeed, recent ground-breaking works on the inter-war LICA/LICRA by Aomar Boum and Emmanuel Debono—made possible, similarly to work on the LDH, by the restitution of archives from Moscow—demonstrate that in the era of the Popular Front, the LICRA adopted a philo-Islamic position that arguably marks a stark contrast to its post-war reputation. Very active in colonial North Africa during this period, the LICRA argued for a Jewish–Muslim entente on the basis of a common defence against fascism, pursued through a variety of avenues, including donations to the poor during Ramadan. While remaining predominantly Jewish at national leadership level, some of the LICRA's local North African branches were headed by Muslims. In 1936–37, the organisation's founder, Bernard Lecache, initiated a rapprochement with both the Algerian nationalist leader Messali Hadj and the Algerian Muslim *ulema* leader Abdelhamid Ben Badis. Having begun his North African campaign at the beginning of 1936 in Tunisia, Lecache went on to visit Ben Badis' madrasa, and succeeded in attracting majority-Muslim audiences at public meetings.[75] One sheikh even claimed that 'The LICA is the true incarnation of the Islamic spirit'.[76] Such, albeit temporary, rapprochement was facilitated by the universalising discourse of French antiracism, for, as Boum notes, 'LICA conceptualised Jewish-Muslim relations in North Africa, and in Algeria in particular, through a political partnership with the Popular Front'.[77]

Yet long after the Second World War, antisemitism remained by far the top priority for the LICA. While framing calls for Jewish self-defence within a universalist discourse,[78] and even being criticised by the extreme Right for supposedly being too soft on Muslims in France, if not on Muslims in the Arab world,[79] it reverted close to the model of a pole two organisation. Although the LICA did denounce violence against North Africans in its journal *Le Droit de Vivre*, which described itself as 'the oldest antiracist newspaper in the world',[80] less prominence was given to this than to antisemitism,[81] and it was not always in a straightforward way. In 1971, for example, when a fifteen-year-old Algerian, Djellali Ben Ali, was killed by the husband of the concierge of his apartment block,[82] *Le Droit de Vivre* seemed to place an equivalence between the two parties.[83] Also in 1971, the LICA denounced the Association de solidarité franco-arabe for targeting only anti-Arab racism, arguing that 'Antiracism does not divide'.[84] Ironically, this was exactly the same argument used against the LICRA in 2004 for singling out antisemitism, except that it was being used by pole two against pole one rather than vice versa. Similarly, in 1981 the LICRA was accused in the immigrant-leftist magazine *Sans Frontière* of making an artificial distinction between racism and antisemitism, leading to a 'morbid competition between peoples so as to know which has the most deaths'.[85] So, typically of the broader discourse of French antiracism, both the LICRA in 1971 and its critics in 1981 and 2004 sought to claim the moral high ground of universalism for themselves, attributing baser, particularist motives to the opposing side. Certainly, the LICRA's consistently pro-Israel stance—its magazine carried advertisements for holidays in Israel and, ironically in view of the LICA's own origins, was a vociferous critic of the use of terrorism by the Palestine Liberation Organization (PLO)[86]—was a factor in the mutual suspicion of 2004, particularly since it was joined in its boycott of the November demonstration by the Conseil réprésentatif des institutions juifs de France (CRIF), whose then head, Roger Cukierman, was widely seen as an intransigent supporter of the government of Ariel Sharon.[87] However, there were other factors related to French domestic politics. As we have seen, the original LICA was on the Left: in the 1930s Jewish conservatives criticised young LICA activists for being too radical and leftist.[88] Yet, whereas LICA leaders had traditionally been Socialists, during the 1970s the organisation's then leader, Jean-Pierre Bloch, crossed to the Right, and his organisation was accused by left-orientated antiracists of supporting the anti-immigrant crack-downs of the Giscard d'Estaing government.[89] Patrick Gaubert, LICRA head between 1999 and 2010, had been a local councillor in the Gaullist Rassemblement pour la République (RPR), and in 2004 was elected as an MEP for the RPR's successor, the Union pour la majorité présidentielle. In the late 1980s and early 1990s, Gaubert had been put in charge of antiracism by Charles Pasqua,[90] the man for whom the journalistic cliché 'hard-line Interior Minister' should surely have been invented. Having been the '*Monsieur Racisme*' of Pasqua, more generally associated with repressive crackdowns on migrants, gave Gaubert an obvious credibility problem in the eyes of the Left, dominant in the other three groups.

A little like the LDH, the MRAP, although in 2004 seemingly central to pole one, has a history that stretches back into pole two—because it was founded by Jewish members of the French Resistance.[91] As late as the 1970s, it was much concerned with preventing a revival of neo-Nazism in West Germany[92] at a time when there was little prospect of this. So why, then, the animosity between the MRAP and the LICRA? Superficially it might be tempting to posit an ethnic explanation, since whereas the LICRA's leadership has consistently been Jewish, part of the MRAP's leadership in more recent times, notably Mouloud Aounit, who headed the organisation from 1989 to 2008, has come from an Algerian Muslim background. The MRAP's opponents accused Aounit and the MRAP of a '*dérive communautariste*',[93] and of covering up antisemitism within Muslim communities in order to avoid stigmatising them. Such accusations need to be viewed critically, since *communautarisme* has become an imprecise, catch-all term of abuse, used by both sides[94] to mean roughly 'you are making an illegitimate appeal to a particular ethnic group, whereas I am merely seeking to uphold the finest universal principles of the Republic'.

Rather than trying to pin everything on *communautarisme*, a more convincing explanation would be a longer-term political split dating back to the onset of the Cold War. The MRAP traces its lineage back to Mouvement national contre le racisme (MNCR), a clandestine wartime group set up by Communists to hide Jews and produce false papers.[95] Critics, however, reject this unproblematic filiation. In 2007, Maurice Winnykamen, a former MRAP militant living in retirement in Nice, wrote a book to denounce what he sees as the MRAP's 'insidious sliding to communitarianism'. Although Winnykamen, founder of the Association pour la mémoire des enfants juifs déportés des Alpes-Maritimes, sees his own MRAP activism as a way of giving something back to the MNCR for having saved his life (his mother, an MNCR activist, hid her eight-year-old son with a non-Jewish family), he accuses the MRAP of downplaying the role of Jews in the MNCR, and making a fundamental shift in 1977, when it changed its name from the Mouvement contre le racisme, l'antisémitisme et pour la paix to the Mouvement contre le racisme et pour l'amitié entre les peuples. Winnykamen portrays the name change as a deliberate strategy to focus exclusively on anti-Arab racism, out of a mistaken assumption that antisemitism had been defeated.[96] Those more sympathetic to the MRAP, on the other hand, see the name change not as abandoning the struggle against antisemitism, but as placing it within a wider struggle against racisms.

While Winnykamen's view on this appears to be overshadowed by more recent controversies, arguably the arguments at stake should also be seen in the context of the MRAP's historical relationship with the Parti communiste français (PCF). It is significant that Winnykamen is a disillusioned ex-Communist, and a disgruntled former employee of various companies linked to the PCF, who resigned from the party in 1978 because of its then leader George Marchais' dishonesty in pretending that such companies did not exist, following various grievances over the way the self-proclaimed party of the working class treated its own workers; Winnykamen is now a member of the PS.[97] Aounit's critics, such as Jean-Yves Camus,[98] tended to focus on an allegedly communitarian line and supposed softness on political Islam that, they claimed, enabled Aounit's election in 2004 as a regional councillor for the Seine-Saint-Denis, the northern suburbs of Paris heavily stigmatised for their large North African population. So was this simply the MRAP's president getting elected because of his Muslim background? While this may have played some role, to view it as the prime explanation does not adequately take into account the other well-known feature of the Seine-Saint-Denis, that this is the classic 'red belt', a historical bastion of the PCF. Although not a party member, Aounit was elected on a Communist-led list, supported the Communist candidate in the 2007 presidential elections, and often communicated through the pages of the Communist daily *L'Humanité*. Even within the anglophone academic literature on ethnicity in France, Alec Hargreaves, usually sceptical of overblown claims about *communautarisme*, assumes that the discrepancy between Aounit's 14%[99] in 2004 and the 6% in the Seine-Saint-Denis of the Communist candidate Robert Hue in the 2002 presidential election was evidence of an ethnic vote.[100]

However, ethnicity may be only one of several explanations, alongside more prosaic psephological ones: the enlargement in 2004 to a non-party list closer to grassroots civil society; the perennial tendency of the PCF, because of its combination of a legacy of local networks but weak credibility as a serious contender for the presidency, to perform significantly better in all other types of election than presidential ones; and that 2004 was the best ever regional election for the Left as a whole. Moreover, surely the 6%/14% statistic tells us as much about Hue's debacle in 2002—an extraordinary 'earthquake' election where left-wing voters deserted any candidate associated with the outgoing Plural Left government in favour of various protest votes or abstention—as it does about Aounit's success in 2004. The significant difference was a relative revival of the PCF at national level between these two dates, largely through being no longer in government and hence no longer a scapegoat for the failures of the Left in power. The extent of Aounit's impact is relativised by other elections in the Seine-Saint-Denis. In the European elections that were also held in 2004, the list headed by the PCF's Francis Wurtz reverted to a more respectable vote of over 11%.[101] Similarly, in the 2009 European elections, the PCF's Patrick Le Hyaric was elected an MEP with over 11% of the vote in the department, compared to less than 3% for the overtly communitarian campaign of Dieudonné's Liste antisioniste.[102] In the 2012 parliamentary elections, two pro-Communist candidates under the Front de gauche banner, hanging on to the coat-tails of Jean-Luc Mélenchon's high-profile presidential campaign, which gained some 17% support across the Seine-Saint-Denis, were re-elected to the National Assembly from the department, despite neither being from a Muslim or visible minority background; nor was any of the five Communist and allied deputies elected there in 2002, or any of the four elected in 2007.[103] This all suggests that Aounit's 14% was little different from what a PCF-supported local candidate, irrespective of ethnic origin, could expect to get in the Seine-Saint-Denis. Other than at times of total disaster for the party, the PCF and its satellites are capable of maintaining a non-negligible residual level of support in former red-belt areas of high migrant settlement, even when they do *not* present candidates of minority origin. To the extent that they have failed to maintain historical levels of support, there is no obvious explanation in terms of Jewish–Muslim antagonism. The Jewish background of Dany Cohn-Bendit, who headed the Green list, which with 17% of the vote received considerably more support in the Seine-Saint-Denis in the 2009 European elections than the PCF,[104] is well known. So is that of Daniel Goldberg, a former SOS Racisme activist who has been a PS deputy in that department since 2007, when he defeated not only the PCF candidate and two right-wing candidates of Maghrebi origin, but also Aounit himself (who this time, having failed to be selected as the Communist candidate, stood as an independent and got only 3%).[105] Yet neither was seemingly a deterrent to the electors of the Seine-Saint-Denis, even in a department that probably has one of the highest populations of Muslim background in France—suggesting that the broader political context remains determinant.

So, without wishing to posit a 'reds under the bed' conspiracy theory, the PCF's use of a candidate from the MRAP in 2004 was less communal politics than the recognisable behaviour of a satellite organisation, using figures from civil society to appeal to voters who do not directly identify with the party. Historiography on the Communist Party, when discussing these arm's-length satellite organisations, tends to give more space to the more obviously influential CGT trade union confederation[106]—where the informal relationship with the party was so close that it has been compared to that between Sinn Fein and the Irish Republican Army (IRA)—plus also sometimes, because it more blatantly served the interests of Soviet foreign policy, the Mouvement de la Paix, not to be confused with the similarly named MRAP.[107] Conversely, the only book on the MRAP in English, by Cathie Lloyd, while otherwise insightful, could have been more explicit on the Communist link, crucial in the organisation's origins. The foundation of the MRAP in 1949 resulted from the failure of the Alliance antiraciste, intended as a fusion of the MNCR with other antiracist groups including the LICA,[108] but to which the Cold War put paid. The MRAP retrospectively justified its own creation with reference to the inaction of the LICA, which it accused of having 'completely vanished into thin air the moment the Germans arrived in Paris'.[109] Ironically, the founder of the LICA, Bernard Lecache, was quite sympathetic to the Soviet Union and a founder member of the Communist Party, but had been expelled during the bolshevisation of the party in 1923 on the grounds that he was a kind of frivolous bourgeois journalist who was bringing the party into disrepute. Although during the Popular Front period of the mid-1930s the PCF had executed a U-turn on its previous policy of no participation in the LICA, from the 1949 split on we see the MRAP and the LICA going their separate ways.[110] The MRAP emphasised racism in the United States, in French colonies and towards colonial migrants to France.[111] Yet, as even the MRAP's own official history admits, it underplayed (until nuanced by the party's temporary flirtation with Eurocommunism)[112] human rights abuses in the Soviet Union, including antisemitism,[113] of which conversely the LICA and then the LICRA made much.[114] A telling example of the MRAP's satellite status was that in 1972 a MRAP delegation, including the Senegalese migrant workers' leader Sally N'Dongo, visited East Germany at the initiative of Association d'amitié RDA-France.[115] Certainly in recent decades ties between the MRAP and the PCF—like those between the CGT and the PCF—have loosened as the party has internally reformed and numerically declined, with Aounit the first MRAP president not to be a party member. Still, these links remain fundamental to understanding the organisation's history. In other words, the key point is that the MRAP–LICA rivalry was not originally between Muslims and Jews, but between two groups of Jews: one Communist, the other anti-Communist.

Nevertheless, the difference in emphasis did have an impact on the issues that we are examining in this book. Right from the time of the Occupation, the Communist leadership's increasingly French nationalist tone had made it less keen than its own MNCR affiliate to make Jews or antisemitism appear to be any kind of special case.[116] The MRAP, under Communist domination and desperate to avoid accusations of being simply a 'union of Jews',[117] therefore portrayed the LICA as remaining trapped in a one-dimensional struggle against antisemitism and failing to enlarge sufficiently into the struggle against anti-Arab racism. Although some within the MRAP were more favourable to Zionism, by the 1950s its predominant tendency was critical of the Israeli government, while drawing a distinction between the Israeli government and the Israeli people.[118] Tensions were apparent at the time of the Six Day War in 1967, when the MRAP was accused by the LICRA, and even some of its own members, of being 'pro-Arab'—although the MRAP's official position on the war was rather more neutral than these accusations might suggest, emphasising the need for each party to the Arab–Israeli conflict to understand the other's aspirations. Indeed, some MRAP members criticised their leadership for being too soft on Israel.[119] Simmering tensions on the MRAP–LICRA long-term front line—as it were, the Golan Heights or Checkpoint Charlie of French antiracism—resumed when a synagogue was bombed in Paris's rue Copernic in 1980. The LICRA accused the MRAP, portrayed as in league with the PLO, Colonel Gaddafi and the Soviet Union, of inaction in response to the attack,[120] whereas the MRAP claimed credit for having organised a mass demonstration of 300,000 people after Copernic, and pointed out that its own headquarters had been bombed a few months earlier.[121] However, the fact that the LICRA also used the Vitry 'bulldozer' affair of 1981, in which the Communist mayor of this Parisian suburb sent a bulldozer to demolish a hostel under construction for Malian workers, to attack the PCF and the MRAP[122] suggests that the main dividing line was not so much between different victims of racism, but a Communist/anti-Communist one, in which a variety of incidents could be instrumentalised to attack the other side politically. Moving to the 1990s, the fact that the MRAP claimed credit for the Gayssot law against Holocaust denial[123] and marked the hundredth anniversary of the Dreyfus Affair,[124] that Aounit went on demonstrations against NATO's war on Serbia[125] and chained himself to the railings of the Iranian embassy in defence of Iranian Jews accused of espionage,[126] together with the MRAP's extreme hostility to the Islamist movement in Algeria,[127] all suggest a classic French leftist and universalist line rather than any kind of communitarian one. Aounit's discourse on racism was a fairly old-fashioned socio-economic one[128] that is recognisably that of someone socialised in the red belt's municipal Communism. While he did place considerable emphasis on combating Islamophobia, he was also a personal friend of the leader of the Bureau national de vigilance contre l'antisémitisme, who issued a tribute following Aounit's death in 2012.[129] Even on the Middle East, while it is true that the Communist Party has increasingly tended to support the Palestinians, the positions of the MRAP and the PCF in favour of a two-state solution are more nuanced than their detractors suggest.[130] Finally, critics such as Winnykamen have cited Aounit's call in 2001 for Maurice Papon, the senior civil servant convicted of handing over Jews to the Nazis, to be released on grounds of ill health.[131] However, given Papon's role as police chief in 1961 in the deaths of hundreds of Algerian demonstrators, Papon was a hate figure in both Muslim and Jewish circles, so Aounit's statement could hardly have been made for populist communitarian reasons. Rather, it appears to have been made on the grounds of a universal principle: a MRAP communiqué clarified that if it was wrong to keep very old and ill people in prison, this should be applied universally—even to someone who had committed crimes as serious as Papon's, for which, it emphasised, he should never be forgiven.[132] In short, the thesis that the MRAP had espoused a pole one position by 2004 requires some nuancing.

Another piece of evidence undermining the thesis of total polarisation is that SOS Racisme, central to the notional pole two in 2004, is actually the only one of the four groups to have been founded in the context of post-colonial racism, during the mid-1980s at a time when antisemitism was not widely seen as a major current issue, but the scapegoating of black and North African immigrants by Jean-Marie Le Pen certainly was. The founding narrative of SOS Racisme, as propounded by its founder Harlem Désir, is that a student friend of his decided to return to Senegal following a racist incident on the metro when a woman falsely accused him of stealing her wallet. A multi-racial group of supposedly apolitical friends mounted a campaign and their slogan *Touche pas à mon pote* ('Hands off my mate') was a big success in the media. Given this, we might wonder why it played a pivotal role in pole two in 2004. While it would be far too simplistic to see SOS Racisme as a Jewish communitarian organisation, the division between it and other organisations has sometimes been seen in ethnic terms: with the exception of Malek Boutih, president from 1999 to 2003, SOS Racisme's leadership has more often tended to come from either an Afro-Caribbean or Jewish background than an Arab one. An important element of its original 1980s mobilisations were alliances forged with Jewish student groups, which may make this appear in retrospect as a golden age of Jewish–Muslim unity, but which attracted suspicion from some North African associations.[133] It has even been suggested that Désir's narrative appealed to mainstream opinion partly because the hero of the story was not North African[134] (at the time, representations of sub-Saharan Africans were less overwhelmingly negative, in part due to the legacy of a less violent decolonisation). SOS Racisme has thus long been sensitive to issues of antisemitism, while facing some credibility issues among Muslims. Nevertheless, its discourse highlighted similarities between Jews and Muslims: as Maud Mandel suggests, SOS Racisme's Jewish supporters 'expressly connected the two minorities in contradistinction to white European society'.[135] Dominique Sopo, president of SOS Racisme, explained the thinking behind the May 2004 demonstration thus: press releases were not enough, society had to react by marching, to avoid antisemitism returning to a space in public debate that it had not had since the Second World War. It had to be specifically about antisemitism to make this clear. He accused rival organisations of looking at society through 'the broken glasses of an exotico-victimist discourse' that wrongly assumed that people of immigrant origin would not mobilise against antisemitism alone.[136]

Yet again behind these arguments lies a deeper political context: namely, the divisions on the Left since the presidency of François Mitterrand. Many of the founders of SOS Racisme have gone on to careers as Socialist parliamentarians, with Désir rising by 2012 to follow in the footsteps of Mitterrand and François Hollande as the party's First Secretary. (Indeed, if one were to accept the premise that being elected by the voters of the Seine-Saint-Denis constitutes communitarianism, to be consistent one would need to apply this to Désir and Patrick Gaubert of the LICRA, who have both been MEPs for a constituency including it.) Critics of SOS Racisme therefore rebut the myth of the apolitical founding fathers and see the organisation, rather, as a Machiavellian creation of Mitterrand's inner circle aiming to defuse a preceding series of mobilisations by North African youth.[137] This was exacerbated by the suspicion with which the Socialist President was viewed in pro-Palestinian circles because of his role as Interior Minister in the early stages of the Algerian war of independence, and his relatively pro-Israeli foreign policy as President.[138] If we see SOS Racisme in a similar kind of satellite relationship to the PS as the MRAP to the PCF, then the antagonism of 2004 makes sense both within a longer historical pattern of Socialist support for Israel[139] versus radical Left support for the Palestinians, but also, perhaps more importantly, within wider divisions between the governing elites of the PS and the radical Left. Although it can to some extent be traced back further into the past, to the 1920 Congress of Tours and beyond, the fundamental rhetorical divide on the French Left between a 'Left of government' and a 'Left of the Left'—viewing each other with extreme suspicion as sold-out managers of capitalism or irresponsible demagogues, respectively—is substantially the same today as it has been since the 'austerity turn' of 1983–84, when a drastic U-turn in economic policy led to the departure of the Communist Party from government and a bitter sense of betrayal that stretches far beyond immigrant communities. Thus, in a mirror image of the MRAP's philocommunism, SOS instrumentalised anticommunism in accordance with the political line of the PS: in 1985 its journal *Touche pas à mon pote* carried a critical article about the situation of minorities in the Soviet Union, claiming that 'The Jews and Muslims of the USSR are also our mates'.[140] As Annie Kriegel pointed out at the time, it is significant that SOS appeared just when class was losing its purchase on mainstream politics.[141] SOS is a product of the slick and superficial age into which it was born, just as our other three organisations bear the imprints of their own founding eras. As Phillipe Juhem has argued, SOS's achievement was precisely its façade of an apparently non-partisan antiracism, tailor-made to attract media attention in the 1980s, but which would have been impossible during the more heavily politicised years before 1981. SOS activism appealed in part because it did not entail the personal sacrifices often demanded by traditional left-wing militancy.[142]

In fact, curiously, the 2004 demonstration could be seen as history repeating itself. In the autumn of 1985 there had also been two separate marches, by rival groups both claiming to be 'against racism and for equal rights'. Marches across France and through the capital were held by both SOS and a group of 'Franco-Arab supporters of autonomy',[143] who viewed SOS as an attempt to steal their thunder and the idea of a march as appropriating an idea already carried out in 1983 (the famous Marche pour l'Egalité, dubbed by the media the 'Marche des Beurs'[144]) and 1984. However, in 1985, the Jewish–Arab issue was mixed with wider concerns about the autonomy of minority youth from SOS's high-profile media approach: although aggravated by differences over the Middle East, the central issue at stake was a youth revolt being captured by someone else at the service of the PS.[145] As *Le Monde*'s correspondent Phillippe Bernard noted: 'The stake that they represent for the political parties of the Left and far Left is not helping to soften the divide.'[146] Thus, whereas SOS's culminating concert took place in the glamorous surroundings of the Place de la Concorde on 7 December, the rival event had taken place a week earlier in the symbolically charged inner-city immigrant district of Barbès, and had involved the participation of first-generation immigrant worker organisations, contrary to the dominant trend in 1980s France to forget these earlier struggles.[147] Yet, unlike in 2004, it is noteworthy that the MRAP hedged its bets by supporting both demonstrations,[148] suggesting that it had not yet fully chosen the pole it appeared to have joined nineteen years later. Something similar could be said of SOS Racisme, which at the time of the first 'headscarf affair' in 1989 was in favour of the right of Muslim girls to wear *hijab*[149] (a view it later changed) and in 1991 was against the first Gulf War, positions that might be more readily associated with pole one. As Maud Mandel argues, the late 1980s saw a subtle yet decisive shift away from ideas of a shared pluricultural public sphere associated with the early rhetoric of SOS Racisme, and towards ideas of dialogue between what were increasingly portrayed as essentially separate and particularistic communities.[150]

So ultimately an examination of the tangled history of all four protagonists has underlined my point about the relative fluidity of divisions in French antiracism. The impression, and reality, of divided marches in 2004 was not the result of an ethnic split. Rather, it was created by different political groups marching separately under their own banners and with their own agendas; hardly an unusual occurrence. This fits into a wider pattern of difficulty in uniting social movements in France, which have often been, as Ingram put it in relation to inter-war peace movements, both 'hale and hearty' and 'balkanized'.[151] For example, in 1997 *sans-papiers* groups had disagreed with SOS, the MRAP and the LICRA, this time all on the same side, over the date of a demonstration against a government immigration bill[152]; a banal tussle for control over demonstrations is not unknown on the French or any Left. And on the November 2004 demonstration, the Union juive française pour la paix and the Association des travailleurs maghrébins en France actually marched together under the same banner.[153] None of the four organisations involved in the 2004 controversy argued that antisemitism was unimportant; rather, the argument revolved around whether antisemitism had such specific features that the struggle against it needed to be pursued separately—or alternatively (the position of the LDH), that antisemitism indeed had specific features, but that the response to it and any racism had to be universal.[154] Looked at over the longer term, an examination of the main protagonists in the affair supports Jonathan Laurence and Justin Vaisse's claim that all 'Four Sisters' have a strong universalist outlook, and the differences are simply ones of strategy.[155]

Furthermore, even these differences become more complex once we look below leadership level. In 2004 there were many internal debates within the four organisations. For example, LDH activist Martine Cohen made a critique of the LDH's line in the Ligue's own journal, which Tubiana responded to at length.[156] Even within the Ligue's Central Committee, the journalist Antoine Spire admitted that there was an internal debate on whether there was a new antisemitism hiding behind anti-Zionism.[157] Meanwhile, the Lyons branch of SOS Racisme disobeyed the organisation's national stance by joining the November demonstration,[158] a reminder that splits over points of principle in Paris are sometimes regarded by provincial militants as a luxury that their smaller numbers cannot afford. And Aounit's leadership of the MRAP was heavily criticised from within the organisation, including by one of its best-known supporters, the philocommunist singer-songwriter Jean Ferrat.[159] Moreover, there is also some evidence of an overlap in membership between the different organisations.[160]

Thus, some of the more alarmist statements from both poles of the debate are not supported by the empirical evidence considered here; reports of the death of universalism are somewhat exaggerated. Indeed, French antiracism is often criticised for being too universalist, or for confusing the terms 'universal' and 'French'. Accusations that it subsumes the specificities of particular racisms into a vague republican or antifascist discourse have been made both in relation to antisemitism[161] and in relation to colonial racism, as well as in antifascist mobilisations.[162] It is true, as Libby Saxton points out, that analogies between different racisms can sometimes be used to pit victims against each other—as with the lawyer Jacques Vergès, whose position was pole one at its most crude, since his main line of defence at the 1987 trial of his client Klaus Barbie, the Gestapo 'Butcher of Lyons', was to argue that France had no right to put Germans on trial for doing to Jews what France had itself done to Africans and Arabs.[163] Yet analogies can also be used to create mutual solidarity.

An Alternative History—of Mutual Solidarity

I would like to end this chapter by opening up some possibilities for exploring this hidden history, which it is to be hoped will be more fully researched by scholars in future. Some such moments, once marginalised, are already beginning to be rediscovered. Firstly, it is significant that one quite widespread reaction to the killing of at least 130 Algerian Muslim demonstrators on 17 October 1961 by Parisian police under the command of Maurice Papon was to compare it to wartime brutality against Jews. This may appear obvious in retrospect, given Papon's 1998 conviction for complicity in crimes against humanity for his earlier role in the deportation of Jews from Bordeaux. Yet, although the facts about Papon's wartime activities were not made public until 1981, the antisemitism/Islamophobia link was already being made in 1961. Thus, the PSU newspaper *La Tribune Socialiste* illustrated its issue denouncing the curfew that applied only to Muslims with a photograph of German soldiers and Jews with their hands up. It emphasised that the measures were a betrayal of the wartime struggle against racism: 'When will there be a star (yellow or green), when will there be ghettos, when will there be the final solution of the Algerian problem in France?'[164] Similar sentiments were expressed by Elie Kagan, the photographer whose pictures of 17 October have been widely reproduced, and who had himself had to wear a yellow star as a child.[165] Likewise, Jacques Panijel's film *Octobre à Paris* culminated in a declaration that not only juxtaposed racist terms for Jews and Algerians respectively, but also united both within a universalising frame: 'So what more needs to be done to make everyone understand that everyone is a *youpin*, that everyone is a *bicot*? Everyone.'[166] These striking words would be repeated by Jean-Luc Einaudi, the author of the most influential investigations in France about the events of October 1961, both at Papon's trial and at the 1999 trial where Papon unsuccessfully attempted to sue Einaudi for libel.[167]

While it was not the only historical comparison made in 1961—the Peterloo Massacre of 1819,[168] the suppression of the Paris Commune in 1871,[169] the SS massacre of the village of Oradour-sur-Glane in 1944[170] and the 1957 Battle of Algiers[171] also featured—the antisemitism comparison was widespread enough that the term 'pogrom' was used by more than one publication. Notably, Jean-Paul Sartre's *Les Temps Modernes* wrote: 'Pogrom: until now, the word has not been translated into French.'[172] The power of this word was enough to have the issue in which this sentence appeared seized by police, even though the facts it reported were the same as in other freely available journals. The statement was not literally true: 'pogrom' had been used in French to denote a massacre of Jews in Tsarist Russia since 1903 and to denote a racist massacre of any group of people since 1926.[173] Such an extension of usage was controversial, though, with the most acceptable extension in a colonial context being to denote the killing of European settlers by the indigenous population.[174] Yet at least in activist circles, *Les Temps Modernes*'s usage stuck: K.S. Karol's account in the *New Statesman* of 'what are officially called "Moslem manifestations", but which those who watched can only describe as a new type of pogrom, carried out by heavily armed police against a defenceless section of the civilian population' was entitled simply 'The Paris Pogrom'.[175]

A potentially problematic element of the comparison was that Algerian nationalist discourse has sometimes, such as in a declaration by the then President of the National Council of the Algerian Revolution, Benyoucef Ben Khedda,[176] banded around too freely the word 'genocide', perhaps in an attempt to play on European guilt about the fate of the Jews, knowing how sensitive a theme this was for French society.[177] This was an exaggeration: however brutal the war, there was never any state policy to kill every Algerian. Nevertheless, exterminationist sentiments were not absent from the minds of some policemen. One Algerian victim of police violence, Abdelkader Khannous, testified that while being beaten with iron bars at a police station, the police shouted 'what the fuck are you doing there in our country; if you stay there, we'll kill you all'.[178] And though the ultimate ends differed, the means bore some similarity. Because Papon's role in the Holocaust was not then public knowledge, the element sometimes singled out for comparison was not Papon's activities in Bordeaux, but the roundup of Parisian Jews by French police at the Vélodrome d'Hiver cycling stadium in July 1942.[179] This element of the comparison was appropriate on two grounds: the selection of detainees essentially on racist grounds, and the sheer scale of the roundup (around 13,000 arrests in 1942 and 11,500 in 1961); it was the first time since 1942 that the police, lacking enough police vans, had requisitioned public transport buses for the purpose. In 1958 Papon had masterminded a smaller roundup of Algerians actually at the Vél d'Hiv, where seven years previously there had also been a police operation to prevent an Algerian nationalist rally, but the stadium had since been demolished.[180] Indeed, the Holocaust was sometimes alluded to by the police themselves, with threats made to send Algerians to 'the gas chambers' being one of a number of factors cited by Jim House as explaining false rumours that circulated among Algerians of gassings at the Palais des Sports, another holding centre.[181] Moreover, the antisemitism comparison was also made by Jewish participants in the debate, including the Auschwitz Deportees Association.[182] For Daniel Mayer, LDH president from 1958 to 1975 and a former Socialist government minister, the comparison that sprang to mind was Krystallnacht.[183] Mindful of parallels with intolerance that they themselves suffered, Jewish community organisations from across the spectrum, including the Union des sociétés juives de France, the Communist-supporting Union des juifs pour la résistance et l'entraide and the chief rabbi, Jacob Kaplan, all issued denunciations, and the Union des étudiants juifs held a protest meeting.[184]

Thus, the position of Vergès, who was to argue at an anniversary conference at the Algerian Cultural Centre in Paris that the atrocities of French colonialism were 'far more serious, far more current and far more frightening for the future' than 'a crime committed by some Nazis against some Jews',[185] was not typical of reactions at the time, and twenty-five years later remained in a minority even on the radical Left. Vergès was denounced by other prominent 17 October campaigners, including Einaudi, Didier Daeninckx and Pierre Vidal-Naquet, who accused him of antisemitism and attempting to minimise the Holocaust.[186] There has been some debate about the validity of comparison, with *Les Temps Modernes*'s statement that 'we refuse to differentiate between the Algerians crammed into the Palais des Sports waiting to be "deported" and the Jews locked up at Drancy before Deportation' criticised at the time by the left-Catholic *Esprit*, Madeleine Réberioux, the liberal Gaullist René Capitant, and in retrospect by Vidal-Naquet and Einaudi.[187] Yet the vast majority of those making such comparisons in the heat of the moment appear to have done so not in order to belittle the Holocaust, but to attract the sympathy of

French public opinion, assumed to be already convinced of the wrongness of antisemitism, towards the plight of the Algerians within familiar universalising terms of reference (perhaps because, as House suggests, there was less awareness in French opinion of the other potential reference frame of previous colonial violence).[188] Vidal-Naquet, who had himself signed *Les Temps Modernes*'s petition that he later criticised, also later tried to put in a historical atlas the words 'Anti-Algerian pogrom in Paris' against the date 17 October 1961, even though his publisher at first refused on the grounds that the word only applied to Jews.[189]

A second example of this universalising tendency within French antiracism is the memory of the sheltering by the Paris Mosque of North African Jews from the Holocaust, for it was not only the Jews and Christians of the MNCR who saved the lives of Jews in France. In recent years the previously little-known role of the Mosque in saving Jews—which some have preferred to overlook because it fits uneasily into narratives of eternal Muslim–Jewish enmity—has been investigated by authors including Robert Satloff and Ethan Katz, and depicted in a feature film by the Franco-Moroccan filmmaker Ismaël Ferroukhi.[190] Revolving around the relationship between an Algerian Muslim who joins the French Resistance and the Algerian Jewish singer Salim Halali, *Les hommes libres* breaks multiple taboos about the marginalised role of immigrants in the Resistance, homosexuality among North Africans and, crucially, Jewish–Muslim relations. As Ferroukhi's film correctly portrays, Halali's life was saved by passing as a Muslim, thanks to false papers supplied by the Mosque's Rector, Si Kadour Benghrabit (providing an alibi relying on similarities between the most intimate of Jewish and Muslim rituals, to circumvent the Germans and Vichy police's practice of inspecting men's penises to check for circumcision). While not contesting the veracity of such real examples where the Mosque did save Jewish lives, and while highlighting the role played by memorialisation of this episode in various efforts to foster better relations between Jews and Muslims in contemporary France, Katz seeks to nuance historical understanding of the Mosque's role. He suggests that Benghrabit's actions were, rather than straightforwardly heroic, characterised by a complex mixture of resistance, accommodation *and* collaboration (as there were also examples where Benghrabit appears to have declined opportunities to pass off individual Jews as Muslims).[191] Since something very similar could be said of many non-Muslim, non-Jewish French notables—one of the central thrusts of the past four decades of historiography on France during the 'Dark Years' has been to break down simplistic moral binaries[192]—this approach has the merit of steering the debate beyond the hero/villain dichotomies that so often characterise international discussions of antisemitism and Islamophobia.

While the Paris Mosque's café, if not its wartime history, has long been firmly on the Parisian tourist trail, *Les hommes libres* also foregrounds a much more marginalised site of memory, the Muslim cemetery at Bobigny, in the heart of the Seine-Saint-Denis,[193] where a false tombstone was put up to convince the Germans that Halali's father was a Muslim. Although Halali died in obscurity in a care home in 2005, his ashes scattered at the cemetery overlooking Nice, he appears to be enjoying posthumous recognition. A compilation of his recordings, released in 2009, was in 2012 on sale at an exhibition about the Jews of Algeria at Paris's Musée d'art et d'histoire du judaïsme.[194] The exhibition itself—which emphasised Algeria's historical role as a refuge from European antisemitism, and stressing antisemitism not, until late on, by indigenous Algerians but rather by European colonists[195]—was evidence of a certain universalist antiracism among museum curators, with visitors invited on a *parcours croisé* ('mixed journey') encompassing both the exhibition on the Jews of Algeria and a simultaneous exhibition at the Cité national de l'histoire de l'immigration (CNHI) on the experience of (Muslim) Algerians in France during the Algerian war of independence.[196]

The CNHI has also tried to bring together these apparently divergent histories. In 2009 it hosted *Ma Proche Banlieue*, an exhibition of photographs by the French-Jewish photographer Patrick Zachmann, predominantly about people of colonial migrant ancestry in the 'near suburbs'—the economically deprived periphery of Marseilles and Paris. Zachmann aimed to present a more human portrait of everyday life there in contrast to sensationalist images of rioting, by juxtaposing photographs taken in 1984 with more recent ones of the same people. Yet the way in which he framed the exhibition also sought to bring together the experiences of Jews with those more conventionally viewed by French society as 'immigrants', by including *La Mémoire de mon père* (1998), a video interview with his own father, whose parents were killed in the Holocaust. The accompanying book revealed how the Paris region locations were also 'near suburbs' in the sense of being close to Zachmann personally, since he himself grew up in the *banlieue*, and the publication featured photographs by him of Jewish life there. He also noted that in the original 1984 project all bar one of his interviewees, although mostly of Muslim background, had said that they cared nothing for religion or for the differences in origin between themselves and him. He acknowledged more recent difficulties, but sought to avoid a reductive focus on violence as the dominant image of the *banlieue*.[197] Zachmann's work is thus a reminder of the hidden history of quiet Jewish–Muslim co-existence in the *banlieue*. Although today this tends to attract far less attention than the reverse, as late as 1995 the makers of the influential film *La Haine* could feature a Jewish character as one of a multi-ethnic trio of protagonists in an apparently realist drama.[198] Indeed, twenty-first-century French and francophone cinema has seen a wider resurgence of representations of Jewish–Arab relations, at least some of which offer an optimistic vision of co-existence.[199]

So whereas reportage in anglophone media tends to strongly emphasise inter-ethnic discord in contemporary France, attempts to shed a more positive light on the history of Muslim–Jewish relations are something of a growth industry in France itself. Various entangled histories, entwined in the universalist discourse of French antiracism, continue to make waves today. In 2013, for example, Abdelwahab Meddeb and Benjamin Stora received rather more publicity for their *Histoire des relations entre juifs et musulmans des origines à nos jours* than might typically be expected of a co-edited academic book running to over a thousand pages.[200] The book was accompanied by a television documentary broadcast on the Franco-German channel Arte.[201] And just before 11 September 2001, Marc Cheb Sun, a journalist from the Seine-Saint-Denis of half-Egyptian, half-Italian origin, had decided to launch the magazine *Respect Mag*, now claiming a readership of 40,000, which describes itself as 'urban, social and mixed', aiming to promote a 'fusion of civilisations'. While predominantly aimed at post-colonial minorities, in 2012 *Respect Mag* published a special issue on 'Jews of France', the final of a trio of issues also including 'Muslims of France' and 'Blacks of France'. Arguably, this can be seen as a bold attempt to tackle the thorny issue of bringing Jews into multi-culturalism: Cheb Sun claimed that Jews were the forgotten part of 'mixed France'. The 'Jews of France' special issue featuring a multiplicity of exchanges between Jews and other minorities featuring content by, among others, the rapper Arabian Panther and the Jewish historian Esther Benbassa—and a dialogue between Antoine Beaufort, president of the youth wing of the LICRA, and Samy Debah, president of Collectif contre l'islamophobie en France. In this, Debah was critical of the LICRA, but also of the MRAP and the LDH, for what he perceived as a failure to recognise the term Islamophobia. Yet the issue also included a speech by Bariza Khiari, a Socialist senator of Muslim origin, affirming that 'Islamophobia and anti-semitism are two sides of the same coin'.[202]

Thus, while antisemitism and Islamophobia have both been realities in modern France, so have struggles against them. Moreover, those struggles do not have to be considered as zero-sum games. Rather, at their best, they have gone together. Broadly this study supports Hargreaves' contention that, recent narrow factionalism notwithstanding, ethnic separatism is weak in France.[203] A strong universalist thread runs through the histories of many of the movements considered here. In particular, there is a strong emphasis within many of these histories on altruism, rather than self-interested identity politics. While it is easy to criticise this universalist approach as paternalistic and, as we have seen, often open to political manipulation, we might also ask: Is there not something admirable in its refusal to see the world in ethnic categories? What, in short, is wrong with altruism?

Acknowledgements I would like to thank the organisers of and audiences at the Antisemitism and Islamophobia in Europe Conference in 2008, the Society for the Study of French History's Conference at Trinity College, Dublin in 2009, and an Erasmus Masters seminar at Oldenburg University in 2014, as well as Mary Horbury, William Horbury and Ethan Katz, for their comments on earlier versions.

Notes

1. Jonathan Laurence and Justin Vaisse, *Integrating Islam: Political and Religious Challenges in Contemporary France* (Washington, 2006), pp. 69–71 (Laurence and Vaisse 2006).
2. Dominique Sopo, *SOS Antiracisme* (Paris, 2005), pp. 67–8 (Sopo 2005).
3. *Libération*, 8 November 2004. Attention focused on the Union des organisations islamiques de France, generally seen as more hard line than other groups. Critics (e.g. Caroline Fourest, *Le choc des préjugés: le l'impasse des postures sécuritaires et victimaires* [Paris, 2007], pp. 27–9 [Fourest 2007]) see it as a Trojan horse for fundamentalism, although other sources indicate that its leadership has made condemnations of antisemitism and held meetings with Jewish leaders: Laurence and Vaisse, *Integrating Islam*, pp. 104–7.
4. *Le Droit de Vivre*, June 2004.
5. *Libération*, 8 December 2003.
6. For example, Vincent Geisser, *La nouvelle islamophobie* (Paris, 2003) (Geisser 2003).
7. Using 'islamophobie' as a search term on the Bibliothèque Nationale de France's catalogue gives only twenty-eight entries, of which only fifteen refer to items in French published in France, all published since

2003, whereas using 'antisémitisme' gives 1,937 entries: 'BnF catalogue général', http://catalogue.bnf.fr/servlet/ListeNotices?host=catalogue [consulted 26 May 2013]. A number of high-profile figures, including the former prime minister Manuel Valls and the assassinated *Charlie Hebdo* editor Charb, have strongly opposed use of the term 'islamophobie': *Libération*, 21 September 2013; Charb, *Lettre aux escrocs de l'islamophobie qui font le jeu des racistes* (Paris, 2015). For a defence of the validity of the concept of Islamophobia, see Marwan Mohammed and Abdellali Hajjat, *Islamophobie. Comment les élites francaises fabriquent le "problème musulman"* (Paris, 2013) (Charb 2015; Mohammed and Hajjat 2013).

8. For example, Pierre-André Taguieff, *La nouvelle judéophobie* (Paris, 2002). The apparently polarised nature of this debate is symbolised by the fact that the title of Geisser's book was a direct riposte to that of Taguieff (Taguieff 2002).
9. For example, *Libération*, 14 January 2009.
10. See Samir Kassir and Farouk Mardam-Bey, *Itinéraires de Paris à Jérusalem*, 2 vols (Paris, 1992); Denis Sieffert, *Israël-Palestine, une passion française: la France dans le miroir du conflit israélo-palestinien* (Paris, 2004); Pascal Boniface (ed.), *La société française et le conflit israélo-palestinien*, special issue of *La Revue internationale et stratégique*, no. 58 (Summer 2005); Michel Feher, 'Le Proche Orient hors les murs. Usages français du conflit israélo-palestinien', in *De la question sociale à la question raciale? Représenter la société française*, eds. Didier Fassin and Eric Fassin (Paris, 2009), pp. 91–105; *Le Moyen-Orient, une passion française?*, special issue of *Matériaux pour l'histoire de notre temps*, no. 96 (2009); *La Palestine en débat 1945–2010*, special issue of *Confluences Méditerranée*, no. 72 (Winter 2009–2010); Nathalie Debrauwere-Miller (ed.), *Israeli-Palestinian Conflict in the Francophone World* (London, 2010); Keith Reader, ed., 'France and the Middle East', special issue of *Modern and Contemporary France*, 22, no. 1 (February 2014); Maud Mandel, *Muslims and Jews in France: History of A Conflict* (Princeton, 2014); Ethan Katz, *The Burdens of Brotherhood: Jews and Muslims from North Africa to France* (Cambridge, MA, 2015) (Kassir and Mardam-Bey 1992; Sieffert 2004; Boniface 2005; Feher 2009; *Le Moyen-Orient, une passion française?* 2009; *La Palestine en débat 1945–2010* 2009–2010; Debrauwere-Miller 2010; Reader 2014; Mandel 2014; Katz 2015).

11. See Henry Rousso, *The Vichy Syndrome: History and Memory in France since 1944* (Cambridge, MA, 1994); Joan B. Wolf, *Harnessing the Holocaust: The Politics of Memory in France* (Stanford, 2004) (Rousso 1994; Wolf 2004).
12. See Benjamin Stora and Mohammed Harbi (eds.), *La Guerre d'Algérie: 1954–2004 la fin de l'amnésie* (Paris, 2004); Raphaëlle Branche, 'The State, Historians and Memories: The Algerian War in France, 1992–2002', in *Contemporary History on Trial: Europe Since 1989 and the Role of the Expert Historian*, eds. Harriet Jones, Kjell Ostberg and Nico Randeraad (Manchester, 2007), pp. 159–73; Aïssa Kadri, Moula Bouaziz and Tramor Quemeneur (eds.), *La guerre d'Algérie revisitée: nouvelles générations, nouvels regards* (Paris, 2015) (Stora and Harbi 2004; Branche 2007; Kadri et al. 2015).
13. For example, *Le Nouvel Observateur*, 11–17 December 2003. Cf. Gérard Noiriel, *Immigration, antisémitisme et racisme en France (XIXe-XX siècles). Discours publics, humiliations privées* (Paris, 2007), p. 639 (*Le Nouvel Observateur* 2003; Noiriel 2007).
14. Peace, 'Antisémitisme nouveau', p. 112.
15. Feher, 'Proche Orient'.
16. For example, Olivier Milza, 'La gauche, la crise et l'immigration (années 1930–années 1980)', *Vingtième Siècle*, 7 (1985), 127–40; Ralph Schor, *Français et immigrés en temps de crise (1930–1980)* (Paris, 2004); Daniel A. Gordon, 'The Back Door of the Nation State: Expulsions of Foreigners and Continuity in Twentieth Century France', *Past and Present*, 186 (February 2005), 201–32; Noiriel, *Immigration, antisémitisme et racisme*. There is a symmetry in historiography, with Ralph Schor's *L'Opinion française et les étrangers (1919–1939)* (Paris, 1985) giving rise to his student Yvan Gastaut's *L'immigration et l'opinion en France sous la Ve République* (Paris, 2000) (Milza 1985; Schor 1985, 2004; Gordon 2005; Gastaut 2000).
17. See Gérard Noiriel, *The French Melting-Pot: Immigration, Citizenship and National Identity* (Minneapolis, 1996). As late as the early 1970s, it was not only North Africans but also Portuguese who were living in shanty towns: Daniel A. Gordon, *Immigrants and Intellectuals: May '68 and the Rise of Antiracism in France* (Pontypool, 2012) (Noiriel 1996; Gordon 2012).

18. Paul Gilroy, *Between Camps: Nations, Cultures and the Allure of Race* (London, 2004); Max Silverman, 'Interconnected Histories: Holocaust and Empire in the Cultural Imaginary', *French Studies*, LXII, no. 4 (2008), 417–28; Michael Rothberg, *Multidirectional Memory: Remembering the Holocaust in the Age of Decolonization* (Stanford, 2009) (Gilroy 2004; Silverman 2008; Rothberg 2009).
19. Gilroy, *Between Camps*, p. 82.
20. Jonathan Judaken, 'So What's New? Rethinking the "New Antisemitism" in a Global Age', *Patterns of Prejudice*, 42, nos 4–5 (2008), p. 559 (Judaken 2008).
21. Silverman, 'Interconnected Histories'; Max Silverman, 'Race, Memory and the Image in Jean-Luc Godard's *Histoire du cinéma*', paper to Ethnicity, 'Race' and Racism Seminar, Edge Hill University, 22 February 2013; Max Silverman, *Palimpsestic Memory: The Holocaust and Colonialism in French and Francophone Fiction and Film* (Oxford, 2013) (Silverman 2013a, b).
22. Mandel, *Muslims and Jews*; Ethan Katz, 'Tracing the Shadow of Palestine: The Zionist-Arab Conflict and Jewish-Muslim Relations in France 1914–1945', in *Israeli-Palestinian Conflict*, ed. Debrauwere-Miller, pp. 25–40; Katz, *Burdens of Brotherhood* (Katz 2010).
23. However, Johannes Heuman is currently researching a monograph on the attitudes of French antiracist organisations to Jewish–Muslim relations.
24. Cylvie Claveau, *L'Autre dans les Cahiers des droits de l'homme, 1920–1940. Une selection universaliste de l'alterité à la Ligue des droits de l'homme et du citoyen* (PhD thesis, McGill University, 2000); Simon Epstein, *Les Dreyfusards sous l'Occupation* (Paris, 2002); Willliam Irvine, *Between Justice and Politics: The Ligue des droits de l'homme, 1898–1945* (Stanford, 2007); J.P. Daughton, Review of Irvine, *H France Review*, 7, no. 81 (July 2007); Joel Blatt (ed.), 'Roundtable Review: *Between Justice and Politics: The Ligue des droits de l'homme, 1898–1945*', *H-Diplo Roundtable Reviews*, VIII, no. 11 (2007), 1–44; Norman Ingram, 'Defending the Rights of Man: The Ligue des droits de l'homme and the Problem of Peace', in *Challenge to Mars: Essays on Pacifism from 1918 to 1945*, eds. Peter Brock and Thomas Socknat (Toronto, 1999); Norman Ingram, 'Selbstmord or Euthanasia? Who Killed the Ligue des droits de l'homme?', *French History*, 22, no. 3

(September 2008), 337–57; Norman Ingram, 'A la Recherche d'une guerre gagnée: The Ligue des droits de l'homme and the War Guilt Question (1918–1922)', *French History*, 24, no. 2 (June 2010), 218–35; Norman Ingram, 'La Ligue des droits de l'homme et le problème allemand', *Revue d'histoire diplomatique*, 124, no. 2 (June 2010), 119–31; Emmanuel Naquet and Gilles Manceron (eds.), *Etre Dreyfusard, hier et aujourd'hui* (Rennes, 2009) (Claveau 2000; Epstein 2002; Irvine 2007; Daughton 2007; Blatt 2007; Ingram 1999, 2008, 2010a, b; Naquet and Manceron 2009).

25. Sonia Combe, *Retour de Moscou: Les archives de la Ligue des droits de l'homme 1898–1940* (Paris, 2004) (Combe 2004).
26. Eric Agrikiolansky, *La Ligue française des droits de l'homme et du citoyen depuis 1945: sociologie d'un engagement civique*, pp. 80, 206–7 (Agrikiolansky 2004).
27. To put this in an international comparison, in May 2009 Shami Chakrabati, head of Liberty, the nearest that Britain has to an equivalent of the Ligue, claimed with some triumph at the organisation's seventy-fifth anniversary conference that it had never before had as many as its current 10,000 members, in a country with a similar population to France. Liberty lacks, it might be added, the network of local branches that Ingram and Irvine rightly stress as essential to the Ligue's importance to wider French society during the Third Republic. Liberty's Council is so dominated by London-based lawyers that any candidates for Council who fall outside this socio-geographical category often use the fact as a central plank of their election addresses. While a similar story can be told of the Ligue's Parisian leadership, it also maintained some three hundred local branches: *Hommes et Libertés*, June–August 1998.
28. Michel Vovelle, 'La tradition des historiens de la Révolution ligeurs', in *1898–2004: une mémoire pour l'avenir*, supplement to *Hommes et Libertés*, October/November/December 2004, p. 43; Agrikiolansky, *Ligue française des droits de l'homme*, pp. 33–6, 59–63; Ellen Crabtree, 'A Historian's Passion for Politics: Madeleine Rebérioux and the Ligue des Droits de l'Homme', paper to Society for the Study of French History Conference, Durham University, 12 July 2014 (Vovelle 2004; Crabtree 2014).
29. Irvine, *Between Justice and Politics*.

30. Naquet and Manceron, *Etre Dreyfusard*; Ingram, 'A la Recherche d'une guerre gagnée', pp. 219–20; Ingram, 'Ligue des droits de l'homme et le problème allemand', pp. 119–24.
31. Ingram,'Selbstmord or Euthanasia?', pp. 347–51.
32. Ingram, 'Ligue des droits de l'homme et le problème allemand', p. 123.
33. Claveau, *L'Autre dans les Cahiers*, p. 210.
34. Claveau, *L'Autre dans les Cahiers*, pp. 196–7.
35. Claveau, *L'Autre dans les Cahiers*, pp. 168, 199.
36. Irvine, *Between Justice and Politics*, pp. 144–5, 212–13; Gilles Manceron, *Marianne et les colonies: une introduction à l'histoire coloniale de la France* (Paris, 2003), pp. 236–51 (Manceron 2003).
37. Claveau, *L'Autre dans les Cahiers*, pp. 206–8.
38. Irvine, *Between Justice and Politics*, pp. 212–13; Stéphane Bunel, *De la Palestine à la France: l'antiracisme en question*, p. 42; Catherine Fhima and Catherine Nicault, 'Victor Basch et la judéité', in *Victor Basch 1863–1944: un intellectuel cosmopolite*, eds. François Basch, Liliane Crips and Pascale Gruson (Paris, 2000), pp. 199–236; Katz, 'Shadow of Palestine', pp. 32–3 (Fhima and Nicault 2000).
39. Gilles Morin, 'La Ligue des droits de l'homme et la guerre d'Algérie', in *Mémoire pour l'avenir*, pp. 60–1; Agrikiolansky, *Ligue française des droits de l'homme*, pp. 50–1. On the PSU, see Tudi Kernalegenn, François Prigent, Gilles Richard and Jacqueline Sainclivier (eds.), *Le PSU vu d'en bas. Réseaux sociaux, mouvement politique, laboratoire d'idées (années 1950–années 1980)* (Rennes, 2009); Daniel A. Gordon, 'A "Mediterranean New Left"? Comparing and Contrasting the French PSU and the Italian PSIUP', *Contemporary European History*, 19, no. 4 (November 2010), 309–30 (Morin 2004; Kernalegenn et al. 2009; Gordon 2010).
40. Gilles Perrault, 'Enquête sue un centenaire au-dessus de tout soupçon', in *Mémoire pour l'avenir*, p. 7 (Perrault 2004).
41. Agrikiolansky, *Ligue française des droits de l'homme*, pp. 68–99.
42. *Qu'est-ce que la Ligue des droits de l'homme?* (Paris, 2003), pp. 14–16 (*Qu'est-ce que la Ligue des droits de l'homme?* 2003).
43. Jean-Paul Salles, *La Ligue communiste révolutionnaire (1968–1981). Instrument du Grand Soir ou lieu d'apprentissage?* (Rennes, 2005), pp. 307–12 (Salles 2005).

44. *Hommes et Libertés*, July–September 2008; Michelle Zancarini-Fournel, *Le moment 68. Une histoire contestée* (Paris, 2008), pp. 121–2 (Zancarini-Fournel 2008).
45. Agrikiolansky, *Ligue française des droits de l'homme*, p. 119.
46. 1978 report, quoted in *Hommes et Libertés*, May–June 1987.
47. *Hommes et Libertés*, June 1977.
48. *Le Monde*, 10–11 May 1998.
49. For example, *Ouest France Ille-et-Villaine*, 12 April 2013. Appropriately enough, the deceased, Yves Quéau, was the head teacher of the Lycée Emile Zola in Rennes, whose hall had hosted Dreyfus' second trial in 1899: Association pour la mémoire du lycée et du collège de Rennes, 'Du collège Saint-Thomas au lycée Emile Zola', http://www.citescolaire-emile-zola-rennes.ac-rennes.fr/spip.php?article53 [accessed 12 June 2013].
50. Which, in contrast, Liberty carefully avoids. Irvine, *Between Justice and Politics*, makes a similar contrast with the American Civil Liberties Union.
51. *Mémoire pour l'avenir*, pp. 73, 77, 79, 81, 94.
52. *Hommes et Libertés*, June 1977.
53. *Le Monde*, 10–11 May 1998.
54. *Qu'est-ce que la Ligue des droits de l'homme?*, p. 18.
55. *Le Monde*, 10–11 May 1998.
56. Agrikiolansky, *Ligue française des droits de l'homme*, pp. 95–9, 229–40.
57. Irvine, *Between Justice and Politics*, p. 221.
58. *Hommes et Libertés*, October–November 1988.
59. *Le Monde*, 10–11 May 1998.
60. Abraham Serfaty, *Le Maroc, du noir au gris* (Paris, 1998) (Serfaty 1998).
61. *Le Monde*, 24 February 1998.
62. *Le Monde*, 15 December 2000.
63. *Le Monde*, 5 May 2001.
64. Gilles Manceron, 'La triple occultation d'un massacre', in *Le 17 octobre des Algériens*, eds. Marcel and Paulette Péju (Paris, 2011), pp. 111–85 (Manceron 2011).
65. Henri Leclerc, Ligue president from 1995 to 2000, speaking at a memorial ceremony for his predecessor Yves Joffa in 1999: Henri Leclerc, 'Yves Jouffa', in *Mémoire pour l'avenir*, p. 99 (Leclerc 2004).

66. 'Francis de Pressensé', named after the Ligue's president from 1903–14, features alongside 'Léon Blum' and 'Résistance' as stops on one bus route through the southern Paris suburb of Châtenay-Malabry, while there is a square in Paris named after Victor and Hélène Basch.
67. Jean Maîtron, *Dictionnaire biographique du mouvement ouvrier français*, vol. 41 (Paris, 1992), pp. 191–2 (Maîtron 1992).
68. Ralph Schor, *L'opinion française et les étrangers*, pp. 484–9.
69. Patricia Kennedy Grimsted, 'The Odyssey of the Petliura Library and the Records of the Ukrainian Nationalist Republic during World War II', *Harvard Ukrainian Studies*, XXII (1998), 181–208; Henry Abramson, *A Prayer for the Government: Ukrainians and Jews in Revolutionary Times 1917–1920* (Harvard, 1999), pp. 168–78 (Grimsted 1998; Abramson 1999).
70. Yosef Nedava, 'Some Aspects of Individual Terrorism: A Case Study of the Schwarzbard Affair', in *Terrorism in Europe*, eds. Yonah Alexander and Kenneth Myers (London, 1982), pp. 29–39; Simon Schwarzbard, *Mémoires d'un anarchiste juif* (Paris, 2010), p. 262 (Nedava 1982; Schwarzbard 2010).
71. Ralph Schor, *L'Antisémitisme en France dans l'entre-deux-guerres* (Paris, 2005), pp. 212–16 (Schor 2005).
72. Simon Epstein, *Un paradoxe français: antiracistes dans la Collaboration, antisémites dans la Résistance* (Paris, 2008), pp. 118–19, 221–7 (Epstein 2008).
73. Epstein, *Paradoxe français*, p. 384.
74. LICRA, 'Histoire de la LICRA', http://www.licra.org/histoire-licra [accessed 24 June 2013]; Catherine Lloyd, *Discourses of Antiracism in France* (Aldershot, 1998), p. 61, n. 61 (Lloyd 1998).
75. Emmanuel Debono, *Aux origines de l'antiracisme. La LICA (1927–1940)* (Paris, 2012), pp. 8–9, 262–9; Aomar Boum, 'Partners against Anti-Semitism: Muslims and Jews Respond to Nazism in French North African Colonies, 1936–1940', *Journal of North African Studies*, 19, no. 4 (2014), 554–70 (Debono 2012; Boum 2014).
76. Debono, *Aux origines de l'antiracisme*, p. 268.
77. Boum, 'Partners against Anti-Semitism', p. 565.
78. Lloyd, *Discourses*, p. 104.

79. Annie Kling, *La France licratisée: enquête au pays de la Ligue internationale contre le racisme et l'antisemitisme* (Strasbourg, 2007), pp. 156–62 (Kling 2007).
80. *Le Droit de Vivre*, December 1971, September–October 1973.
81. Gastaut, *L'immigration et l'opinion*, p. 183; for examples of continuing emphasis on antisemitism, *Le Droit de Vivre*, January 1973, February 1973 and March 1973.
82. Gordon, *Immigrants and Intellectuals*, pp. 122–6; Amit Prakash, 'Murder in La Goutte d'Or: The Mouvement des Travailleurs Arabes, Surveillance and Counter-Surveillance', paper to Society for French Historical Studies 59th Annual Meeting, Harvard University/MIT, 4–7 April 2013 (Prakash 2013).
83. Le *Droit de Vivre*, December 1971.
84. *Le Droit de Vivre*, December 1971.
85. *Sans Frontière*, 27 November 1981.
86. *Le Droit de Vivre*, May 1980; *Sans Frontière*, 27 November 1981.
87. In 2002, for example, a polemic took place between Cukierman and Michael Tubiana of the LDH, who took Cukierman to task for ignoring proposals for joint demonstrations against antisemitic attacks and instead linking them to political support for Israel. Reminding Cukierman of the LDH's origins, Tubiana argued: 'You are making a mistake in denying the universality of the struggle against racism and in mixing the fight against antisemitism with unilateral support for a state. In refusing to open your demonstration to organisations other than those which make up the CRIF, you make a necessarily universal fight into a communitarian step, thus denying the principles of the Republic': *Qu'est-ce que la Ligue des droits de l'homme?*, pp. 13–14.
88. Schor, *L'antisémitisme en France*, pp. 212–14.
89. Gastaut, *L'immigration et l'opinion*, p. 183; *Sans Frontière*, 27 November 1981.
90. Hélène Emeret (ed.), *Combattre le racisme: les combats de la LICRA 1927–2002* (Paris, 2002), p. 91 (Emeret 2002).
91. See Lloyd, *Discourses.*
92. *Droit et Liberté*, February 1973, September 1976, October 1976, November–December 1976, June 1979.
93. For example, *Marianne*, 2–8 February 2008.

94. For example, Sieffert, *Israël-Palestine*, p. 217 accuses the CRIF, LICRA and SOS Racisme of *communautarisme*, just as Tubiana did against Cukierman.
95. Albert Lévy et al., *Chronique d'un combat inachevé: 50 ans contre le racisme* (Paris, 1999), p. 9 (Lévy et al. 1999).
96. Winnykamen, *Grandeur et misère*.
97. Maurice Winnykamen, 'Biographie de Maurice Winnykamen', http://www.ajpn.org/personne-Maurice-Winnykamen-2669.html [accessed 4 June 2013].
98. Jean-Yves Camus, 'The French Left and Political Islam: Secularism versus the Temptation of an Alliance', *Engage Journal*, no. 3 (September 2006) (Camus 2006).
99. Whereas Hargreaves gives this figure for the whole department, Camus maximises his polemical case by citing only higher figures from those individual towns that gave the highest vote to Aounit.
100. Alec Hargreaves, *Multi-Ethnic France: Immigration, Politics, Culture and Society* (New York/London, 2007), p. 135 (Hargreaves 2007).
101. *Le Monde*, 15 June 2004.
102. *Le Monde*, 9 June 2009.
103. *L'Humanité*, 19 June 2012; *Libération*, 18 June 2002; *Le Monde*, 19 June 2007.
104. *Le Monde*, 9 June 2009.
105. *Le Monde*, 19 June 2007.
106. For example, Jean Montaldo, *La France Communiste* (France, 1978), pp. 67–80; George Ross, *Workers and Communists in France: From Popular Front to Eurocommunism* (Berkeley, 1982) (Montaldo 1978; Ross 1982).
107. For example, David Caute, *Communism and the French Intellectuals 1914–1960* (London, 1964), pp. 44, 187–90; Montaldo, *France Communiste*, pp. 111–12; Irwin Wall, *French Communism in the Era of Stalin: The Quest for Unity and Integration, 1945–1962* (Westport, 1983), p. 129; Tony Judt, *Past Imperfect: French Intellectuals, 1944–1956* (Berkeley, 1992), pp. 222–5 (Caute 1964; Wall 1983; Judt 1992).
108. Lloyd, *Discourses*, p. 61, n. 2.
109. *Droit et Liberté*, May 1979.
110. Jean Maîtron, *Dictionnaire biographique du mouvement ouvrier français*, vol. 34 (Paris, 1989), pp. 48–50; Epstein, *Paradoxe français*, pp. 194–6, 222 (Maîtron 1989).

111. Lévy, *Chronique*, p. 12.
112. Lévy, *Chronique*, pp. 62–3; *Droit et Liberté*, February 1973.
113. Lévy, *Chronique*, p. 25.
114. For example, *Le Droit de Vivre*, October 1980.
115. *Droit et Liberté*, December 1972.
116. Karen Adler, *Jews and Gender in Liberation France* (Cambridge, 2004), p. 149 (Adler 2004).
117. Lévy, *Chronique*, p. 12.
118. Johannes Heuman, 'Antiracism and Zionism after the Liberation of France', paper to Society for the Study of French History Conference, Chichester University, 4 July 2016 (Heuman 2016).
119. Yvan Gastaut, 'La Guerre des Six jours et la question du racisme en France', *Cahiers de la Méditerranée*, no. 71 (2005), citing letters in the MRAP's archives (Gastaut 2005).
120. *Le Droit de Vivre*, October 1980.
121. Lévy, *Chronique*, pp. 94–5.
122. *Le Droit de Vivre*, January 1981; 'De Copernic à Vitry', special supplement to *Le Droit de Vivre*, February 1981 (De Copernic à Vitry 1981).
123. *L'Humanité*, 9 July 2001.
124. Lloyd, *Discourses*, p. 67.
125. Cécile Amar, *Les nouveau communistes*, p. 171.
126. A photograph of which appears, ironically, in the official history of LICRA: Emeret, *Combattre le racisme*, p. 42.
127. Lévy, *Chronique*, pp. 124–5.
128. For example, Mouloud Aounit, 'Vous avez dit intégration?', in *Le Grand livre contre le racisme* (Paris, 2007) (Aounit 2007).
129. Bureau national de vigilance contre l'antisémitisme, 'Le BNVCA attristé par la decès de Mouloud Aounit présente ses condoléances à sa famille et ses collaborateurs', 10 August 2012, http://www.sosantisemitisme.org/communique.asp?ID=715 [accessed 3 June 2013] (Bureau national de vigilance contre l'antisémitisme 2012).
130. *L'Humanité*, 9 July 2001; Marie-George Buffet, 'La connaissance de l'Autre: un chemin pour la paix', in *Société française*, ed. Boniface, pp. 16–22. Sieffert, *Israël-Palestine*, pp. 151–3 suggests that good relations between PCF and PLO were established only after the latter's acceptance of a two-state solution, because in 1947–48 the PCF had followed the Soviet line in supporting the

United Nations' partition plan and the creation of Israel (Buffet 2005).

131. Winnykamen, *Grandeur et misère*, pp. 116–22.
132. *L'Humanité*, 8 February 2001.
133. Adil Jazouli, 'Le dialogue judéo-arabe', *Migrations Société*, 1, no. 2 (April 1989), 19–22; Alec Hargreaves, 'The Political Organisation of the North African Immigrant Community in France', *Ethnic and Racial Studies*, 14, no. 3 (1991), 361 (Jazouli 1989; Hargreaves 1991).
134. Paul Yonnet, *Voyage au centre du malaise française: l'antiracisme et le roman national* (Paris, 1993), pp. 101–10 (Yonnet 1993).
135. Mandel, *Muslims and Jews in France*, p. 132.
136. Sopo, *SOS Antiracisme*, pp. 55–6, 66–8.
137. For example, Saïd Bouamama, *Dix ans de marche des Beurs* (Paris, 1994), p. 131. For a nuanced view, see Phillippe Juhem, 'SOS-Racisme, histoire d'une mobilisation "apolitique". Contribution à une analyse des transformations des représentations politiques après 1981', doctoral thesis, Université de Paris 10—Nanterre, 1998, http://juhem.free.fr [accessed 15 July 2016]; Daniel A. Gordon, 'French and British Antiracists Since the 1960s: A *rendez-vous manqué?*', *Journal of Contemporary History*, 50, no. 3 (July 2015), 606–31 (Bouamama 1994; Juhem 1998; Gordon 2015).
138. Although there was some ambiguity in his policy (Mitterrand was the first Western head of state to call for a Palestinian state, and sent the French navy to rescue Yasser Arafat from Lebanon), in 1982 he also became the first French head of state to address the Knesset: Sieffert, *Israël-Palestine*, pp. 165–75.
139. French Socialists have had good relations with the Israeli Labour Party and its predecessors ever since the 1930s: Sieffert, *Israël-Palestine*, pp. 89–110.
140. *Touche pas à mon pote*, October 1985.
141. Yonnet, *Malaise française*, pp. 112–13, 120–1.
142. Juhem, 'SOS Racisme', Chapter 9.
143. *Le Monde*, 1 November 1985.
144. See *1983 La Marche pour l'égalité et contre le racisme*, special issue of *Migrance*, no. 41 (2013) (*1983 La Marche pour l'égalité et contre le racisme* 2013).

145. *Le Monde*, 1 and 30 November 1985; Bouamama, *Dix ans*, pp. 132–51.
146. *Le Monde*, 30 November 1985.
147. Gordon, *Immigrants and Intellectuals*, pp. 216–23.
148. *Le Monde*, 30 November 1985.
149. Gastaut, *L'immigration et l'opinion*, p. 591.
150. Mandel, *Muslims and Jews in France*, pp. 125–52.
151. Norman Ingram, *The Politics of Dissent: Pacifism in France, 1919–1939* (Oxford, 1991), pp. 1, 2 (Ingram 1991).
152. *Le Monde*, 1 March 1997.
153. *L'Humanité*, 8 November 2004.
154. Statements made by the different organisations in Benoît Hervieu-Léger, 'Antisémitisme: une brèche dans la lutte antiraciste', *Réforme*, 13 January 2005 (Hervieu-Léger 2005).
155. Laurence and Vaisse, *Integrating Islam*, pp. 69–71.
156. *Hommes et Libertés*, January–March 2005.
157. *Hommes et Libertés*, July–September 2004.
158. *L'Humanité*, 8 November 2004.
159. *Le Figaro*, 11 August 2012.
160. Agrikiolansky, *Ligue française des droits de l'homme*, p. 51.
161. Lloyd, *Discourses*, pp. 70, 85; Adler, *Jews and Gender*.
162. Lloyd, *Discourses*, p. 101; Jim House and Neil MacMaster, *Paris 1961: Algerians, State Terror and Memory* (Oxford, 2006), p. 236 (House and MacMaster 2006).
163. Libby Saxton, 'Terms of Engagement: Algeria, France and the Middle East in Barbet Schroeder's *L'Avocat de la terreur* and Philippe Faucon's *Dans la vie*', *Modern and Contemporary France*, 19, no. 2 (May 2011), 216 (Saxton 2011).
164. *La Tribune Socialiste*, 14 October 1961.
165. Association Sortir du Colonialisme (ed.), *Le 17 octobre 1961 par les textes de l'époque* (Paris, 2011), p. 27 (Association Sortir du Colonialisme 2011).
166. Bibliothèque de documentation internationale contemporaine, Nanterre (BDIC), O PIECE 380 RES, 'Le Comité Maurice Audin et *Vérité-Liberté* présentent "Octobre à Paris"'; excerpt played in *Secret History: Drowning By Bullets*, first broadcast Channel 4, 13 July 1992.
167. Jean-Luc Einaudi, *Octobre 1961: un massacre à Paris* (Paris, 2011), pp. 48, 93 (Einaudi 2011).

168. Pierre Vidal-Naquet, *Torture: Cancer of Democracy* (Harmondsworth, 1963), p. 118 (Vidal-Naquet 1963).
169. *Le Communiste*, November 1961; *Fédération de France du FLN, Section universitaire, Bulletin d'information*, 30 December 1961 (*Fédération de France du FLN* 1961).
170. BDIC 4 DELTA RES, no. 2: Action catholique ouvrière, 'Après les manifestations algériennes de Paris les 17 et 18 octobre 1961', 30 October 1961; Actualité de l'Emigration (ed.), *17 octobre 1961: mémoire d'une communauté* (Paris, 1987), pp. 29–33; Maurice Papon, *Les chevaux du pouvoir* (Paris, 1988), p. 183 (Actualité de l'Emigration 1987; Papon 1988).
171. BDIC 4 DELTA 143 RES, 'Déclaration du FLN sur la répression et les mesures policières dans la région parisienne', 17 octobre 1961 and 'Appel au peuple français', 18 October 1961; *Vérité-Liberté*, November 1961; Papon, *Chevaux*, p. 181 (BDIC 4 DELTA 143 RES 1961a, b).
172. Quoted in *Vérité-Liberté*, November 1961.
173. *Trésor de la langue française*, vol. 13 (Paris, 1988), p. 641; *Grand Larousse de la langue française*, vol. 5 (Paris, 1976), p. 4393 (*Trésor de la langue française* 1988; *Grand Larousse de la langue française* 1976).
174. For example, *L'Algérienne*, 1 October 1961.
175. *New Statesman*, 27 October 1961.
176. *Le Temps du Niger*, 26 October 1961; cf. Einaudi's critique of the Algerian pro-government newspaper *El Moudjahid*'s use of the term in 2000: Einaudi, *Octobre 1961*, pp. 117–19.
177. Daniel A. Gordon, *The Paris Pogrom, 1961* (MA dissertation, University of Sussex, 1998), p. 44; Jim House, 'Memory and the Creation of Solidarity During the Decolonisation of Algeria', *Yale French Studies*, no. 118–19 (2010), 28 (Gordon 1998; House 2010).
178. BDIC O PIECE 557 RES: République Algérienne, Ministère de l'Information, 'Les manifestations algériennes d'octobre 1961 et la répression colonialiste en France', p. 49.
179. For example, *Vérité-Liberté*, November 1961.
180. Jean-Luc Einaudi, *La bataille de Paris* (Paris, 1991), p. 113; *Le Monde*, 17 October 1997; House, 'Memory and the Creation of Solidarity', p. 21 (Einaudi 1991).
181. House, 'Memory and the Creation of Solidarity', p. 29.

182. House, 'Memory and the Creation of Solidarity', p. 26.
183. Einaudi, *Bataille de Paris*, p. 242.
184. Einaudi, *Bataille de Paris*, pp. 210, 238, 241.
185. Actualité de l'Emigration, *17 octobre 1961*, p. 65.
186. Pierre Vidal-Naquet, *Assassins of Memory: Essays on the Denial of the Holocaust* (New York, 1992), pp. 132, 190; Einaudi, *Octobre 1961*, pp. 117–19 (Vidal-Naquet 1992).
187. Jim House and Neil MacMaster, *Paris 1961* (Oxford, 2006), p. 224; Vidal-Naquet, *Assassins of Memory*, p. 128; House, 'Memory and the Creation of Solidarity', p. 27.
188. House and MacMaster, *Paris 1961*, p. 224; House, 'Memory and the Creation of Solidarity', p. 24.
189. Actualité de l'Emigration, *17 octobre 1961*, pp. 45–6.
190. Robert Satloff, *Among the Righteous: Lost Stories from the Holocaust's Long Reach into Arab Lands* (Jackson, 2007); *Les hommes libres* (2011), dir. Ismaël Ferroukhi; Ethan Katz, 'Did the Paris Mosque Save Jews? A Mystery and Its Memory', *Jewish Quarterly Review*, 102, no. 2 (Spring 2012), 256–87 (Satloff 2007; Katz 2012).
191. Katz, 'Did The Paris Mosque Save Jews?'.
192. Julian Jackson, *France: The Dark Years, 1940–1944* (Oxford, 2001) (Jackson 2001).
193. See Marie-Ange Adler, *Le cimetière musulman de Bobigny* (Paris, 2005) (Adler 2005).
194. Salim Halali, *Trésors de la chanson judéo-arabe/Jewish-Arab Song Treasures* (Buda Musique, 2009) (Halali 2009).
195. This point can be summed up by a sign on one beach: 'Forbidden to Arabs, dogs and Jews': Pierre Daum, *Ni valise ni cercueil: les pieds-noirs restés en Algérie après l'indépendance* (Arles, 2012), p. 181; see also Benjamin Stora, *Les trois exils. Juifs d'Algérie* (Paris, 2011) (Daum 2012; Stora 2011).
196. *Juifs d'Algérie*, Musée d'art et d'histoire du judaïsme, Paris, 28 September 2012–27 January 2013.
197. *Ma proche banlieue*, Cité national de l'histoire de l'immigration, 26 May–11 October 2009; Patrick Zachmann, *Ma proche banlieue* (Paris, 2009) (Zachmann 2009).
198. Silverstein, 'Context of Antisemitism and Islamophobia', p. 19.
199. See Carrie Tarr, 'Jewish-Arab Relations in French and Maghrebi Cinema', in *Studies in French Cinema: UK Perspectives 1985–2010*,

eds. Will Higbee and Sarah Leahy (Bristol, 2011), pp. 321–36; Alyssa Goldstein Sepinwall, 'Jewish-Muslim Romance with a French Twist: Jean-Jacques Zilbermann's *He's My Girl*', *Fiction and Film for French Historians*, 6, no. 6 (April 2016) (Tarr 2011; Sepinwall 2016).
200. For example, *Le Nouvel Observateur*, 10–16 October 2013.
201. *Juifs et musulmans, si loin, si proches* (2013), dir. Karim Miské.
202. *Respect Mag*, no. 36, October–November–December 2012.
203. Hargreaves, *Multi-Ethnic France*, pp. 126–7.

References

1983 La Marche pour l'égalité et contre le racisme. Special issue of *Migrance*, no. 41, 2013.

Abramson, Henry. 1999. *A Prayer for the Government: Ukrainians and Jews in Revolutionary Times 1917–1920*. Harvard.

Actualité de l'Emigration. 1987. *17 octobre 1961: mémoire d'une communauté*. Paris.

Adler, Karen. 2004. *Jews and Gender in Liberation France*. Cambridge.

Adler, Marie-Ange. 2005. *Le cimetière musulman de Bobigny*. Paris.

Agrikiolansky, Eric. 2004. *La Ligue française des droits de l'homme et du citoyen depuis 1945: sociologie d'un engagement civique*. Paris.

Aounit, Mouloud. 2007. Vous avez dit intégration? In *Le Grand livre contre le racisme*. ed. Jacquard, Albert et al. Paris.

Association Sortir du Colonialisme. 2011. *Le 17 octobre 1961 par les textes de l'époque*. Paris.

BDIC 4 DELTA 143 RES. 1961a. Déclaration du FLN sur la répression et les mesures policières dans la région parisienne, 17 October.

———. 1961b. Appel au peuple français, 18 October.

Blatt, Joel (ed). 2007. Roundtable Review: *Between Justice and Politics: The Ligue des droits de l'homme, 1898–1945. H-Diplo Roundtable Reviews VIII*(11): 1–44.

Boniface, Pascal (ed). 2005. *La société française et le conflit israélo-palestinien*. Special issue of *La Revue internationale et stratégique*, no. 58, Summer.

Bouamama, Saïd. 1994. *Dix ans de marche des Beurs*. Paris.

Boum, Aomar. 2014. Partners Against Anti-Semitism: Muslims and Jews Respond to Nazism in French North African Colonies, 1936–1940. *Journal of North African Studies* 19(4): 554–570.

Branche, Raphaëlle. 2007. The State, Historians and Memories: The Algerian War in France, 1992–2002. In *Contemporary History on Trial: Europe Since 1989 and the Role of the Expert Historian*, ed. Harriet Jones, Kjell Ostberg, and Nico Randeraad. Manchester.

Buffet, Marie-George. 2005. La connaissance de l'Autre: un chemin pour la paix. In *La société française et le conflit israélo-palestinien*. Special issue of *La Revue internationale et stratégique*, ed. Pascal Boniface, no. 58, Summer.

Bureau national de vigilance contre l'antisémitisme. 2012. Le BNVCA attristé par la decès de Mouloud Aounit présente ses condoléances à sa famille et ses collaborateurs. 10 August. Accessed 3 June 2013, http://www.sosantisemitisme.org/communique.asp?ID=715

Camus, Jean-Yves. 2006. The French Left and Political Islam: Secularism Versus the Temptation of an Alliance. *Engage Journal*, no. 3, September.

Caute, David. 1964. *Communism and the French Intellectuals 1914–1960*. London.

Charb. 2015. *Lettre aux escrocs de l'islamophobie qui font le jeu des racistes*. Paris.

Claveau, Cylvie. 2000. *L'Autre dans les Cahiers des droits de l'homme, 1920–1940. Une selection universaliste de l'alterité à la Ligue des droits de l'homme et du citoyen*. PhD thesis, McGill University.

Combe, Sonia. 2004. *Retour de Moscou: Les archives de la Ligue des droits de l'homme 1898–1940*. Paris.

Crabtree, Ellen. 2014. A Historian's Passion for Politics: Madeleine Rebérioux and the Ligue des Droits de l'Homme. Paper to *Society for the Study of French History Conference*, Durham University, 12 July.

Daughton, J.P. 2007. Review of Irvine. *H-France Review* 7(81) (July).

Daum, Pierre. 2012. *Ni valise ni cercueil: les pieds-noirs restés en Algérie après l'indépendance*. Arles.

De Copernic à Vitry. Special supplement to *Le Droit de Vivre*, February 1981.

Debono, Emmanuel. 2012. *Aux origines de l'antiracisme. La LICA (1927–1940)*. Paris.

Debrauwere-Miller, Nathalie (ed). 2010. *Israeli-Palestinian Conflict in the Francophone World*. London.

Einaudi, Jean-Luc. 1991. *La bataille de Paris*. Paris.

———. 2011. *Octobre 1961: un massacre à Paris*. Paris.

Emeret, Hélène (ed). 2002. *Combattre le racisme: les combats de la LICRA 1927–2002*. Paris.

Epstein, Simon. 2002. *Les Dreyfusards sous l'Occupation*. Paris.

———. 2008. *Un paradoxe français: antiracistes dans la Collaboration, antisémites dans la Résistance*. Paris.

Fédération de France du FLN, Section universitaire, Bulletin d'information, 30 December 1961.

Feher, Michel. 2009. Le Proche Orient hors les murs. Usages français du conflit israélo-palestinien. In *De la question sociale à la question raciale? Représenter la société française*, ed. Didier Fassin and Eric Fassin. Paris.

Fhima, Catherine, and Catherine Nicault. 2000. Victor Basch et la judéité. In *Victor Basch 1863–1944: un intellectuel cosmopolite*, ed. François Basch, Liliane Crips, and Pascale Gruson. Paris.

Fourest, Caroline. 2007. *Le choc des préjugés: le l'impasse des postures sécuritaires et victimaires*. Paris.

Gastaut, Yvan. 2000. *L'immigration et l'opinion en France sous la Ve République*. Paris.

———. 2005. La Guerre des Six jours et la question du racisme en France. *Cahiers de la Méditerranée*, no. 71.

Geisser, Vincent. 2003. *La nouvelle islamophobie*. Paris.

Gilroy, Paul. 2004. *Between Camps: Nations, Cultures and the Allure of Race*. London.

Gordon, Daniel A. 1998. *The Paris Pogrom, 1961*. MA dissertation, University of Sussex.

———. 2005. The Back Door of the Nation State: Expulsions of Foreigners and Continuity in Twentieth Century France. *Past and Present* 186 (February): 201–232.

———. 2010. A "Mediterranean New Left"? Comparing and Contrasting the French PSU and the Italian PSIUP. *Contemporary European History* 19(4) (November): 309–330.

———. 2012. *Immigrants and Intellectuals: May '68 and the Rise of Antiracism in France*. Pontypool.

———. 2015. French and British Antiracists Since the 1960s: A *rendez-vous manqué*? *Journal of Contemporary History* 50(3): 606–631.

Grand Larousse de la langue française, vol. 5. Paris, 1976.

Grimsted, Patricia Kennedy. 1998. The Odyssey of the Petliura Library and the Records of the Ukrainian Nationalist Republic during World War II. *Harvard Ukrainian Studies* XXII: 181–208.

Halali, Salim. 2009. *Trésors de la chanson judéo-arabe/Jewish-Arab Song Treasures*. Buda Musique.

Hargreaves, Alec. 1991. The Political Organisation of the North African Immigrant Community in France. *Ethnic and Racial Studies* 14(3): 361.

———. 2007. *Multi-Ethnic France: Immigration, Politics, Culture and Society*. New York/London.

Hervieu-Léger, Benoît. 2005. Antisémitisme: une brèche dans la lutte antiraciste. *Réforme*, 13 January.

Heuman, Johannes. 2016. Antiracism and Zionism after the Liberation of France. Paper to *Society for the Study of French History Conference*, Chichester University, 4 July.

House, Jim. 2010. Memory and the Creation of Solidarity During the Decolonisation of Algeria. *Yale French Studies*, no. 118–119: 28.

House, Jim, and Neil MacMaster. 2006. *Paris 1961: Algerians, State Terror and Memory*. Oxford.

Ingram, Norman. 1991. *The Politics of Dissent: Pacifism in France, 1919–1939*. Oxford.

———. 1999. Defending the Rights of Man: The Ligue des droits de l'homme and the Problem of Peace. In *Challenge to Mars: Essays on Pacifism from 1918 to 1945*, ed. Peter Brock and Thomas Socknat. Toronto.

———. 2008. Selbstmord or Euthanasia? Who Killed the Ligue des droits de l'homme? *French History* 22(3) (September): 337–357.

———. 2010a. A la Recherche d'une guerre gagnée: The Ligue des droits de l'homme and the War Guilt Question (1918–1922). *French History* 24(2) (June): 218–235.

———. 2010b. La Ligue des droits de l'homme et le problème allemand. *Revue d'histoire diplomatique* 124(2) (June): 119–131.

Irvine, Willliam. 2007. *Between Justice and Politics: The Ligue des droits de l'homme, 1898–1945*. Stanford.

Jackson, Julian. 2001. *France: The Dark Years, 1940–1944*. Oxford.

Jazouli, Adil. 1989. Le dialogue judéo-arabe. *Migrations Société* 1(2) (April): 19–22.

Judaken, Jonathan. 2008. So What's New? Rethinking the "New Antisemitism" in a Global Age. *Patterns of Prejudice* 42(4–5): 559.

Judt, Tony. 1992. *Past Imperfect: French Intellectuals, 1944–1956*. Berkeley.

Juhem, Phillippe. 1998. SOS-Racisme, histoire d'une mobilisation "apolitique". Contribution à une analyse des transformations des représentations politiques après 1981. Doctoral thesis, Université de Paris 10—Nanterre. http://juhem.free.fr

Kadri, Aïssa, Moula Bouaziz, and Tramor Quemeneur (eds). 2015. *La guerre d'Algérie revisitée: nouvelles générations, nouvels regards*. Paris.

Kassir, Samir, and Farouk Mardam-Bey. 1992. *Itinéraires de Paris à Jérusalem*, 2 vols. Paris.

Katz, Ethan. 2010. Tracing the Shadow of Palestine: The Zionist-Arab Conflict and Jewish-Muslim Relations in France 1914–1945. In *Israeli-Palestinian Conflict in the Francophone World*, ed. Nathalie Debrauwere-Miller. London: Routledge.

———. 2012. Did the Paris Mosque Save Jews? A Mystery and Its Memory. *Jewish Quarterly Review* 102(2) (Spring): 256–287.

———. 2015. *The Burdens of Brotherhood: Jews and Muslims from North Africa to France*. Cambridge, MA.

Kernalegenn, François Prigent, Gilles Richard, and Jacqueline Sainclivier (eds). 2009. *Le PSU vu d'en bas. Réseaux sociaux, mouvement politique, laboratoire d'idées (années 1950—années 1980)*. Rennes.

Kling, Annie. 2007. *La France licratisée: enquête au pays de la Ligue internationale contre le racisme et l'antisemitisme*. Strasbourg.

La Palestine en débat 1945–2010. Special issue of *Confluences Méditerranée*, no. 72, Winter 2009–2010.

Laurence, Jonathan, and Justin Vaisse. 2006. *Integrating Islam: Political and Religious Challenges in Contemporary France*. Washington.

Le Moyen-Orient, une passion française? Special issue of *Matériaux pour l'histoire de notre temps*, no. 96, 2009.

Le Nouvel Observateur, 11–17 December 2003.

Leclerc, Henri. 2004. Yves Jouffa. In *1898–2004: une mémoire pour l'avenir*, supplement to *Hommes et Libertés*, October/November/December.

Lévy, Albert, et al. 1999. *Chronique d'un combat inachevé: 50 ans contre le racisme*. Paris.

Lloyd, Catherine. 1998. *Discourses of Antiracism in France*. Aldershot.

Maîtron, Jean. 1989. *Dictionnaire biographique du mouvement ouvrier français*, vol. 34. Paris.

———. 1992. *Dictionnaire biographique du mouvement ouvrier français*, vol. 41. Paris.

Manceron, Gilles. 2003. *Marianne et les colonies: une introduction à l'histoire coloniale de la France*. Paris.

———. 2011. La triple occultation d'un massacre. In *Le 17 octobre des Algériens*, ed. Marcel and Paulette Péju. Paris.

Mandel, Maud. 2014. *Muslims and Jews in France: History of A Conflict*. Princeton.

Milza, Olivier. 1985. La gauche, la crise et l'immigration (années 1930–années 1980). *Vingtième Siècle* 7: 127–140.

Mohammed, Marwan, and Abdellali Hajjat. 2013. *Islamophobie. Comment les élites francaises fabriquent le "problème musulman"*. Paris.

Montaldo, Jean. 1978. *La France Communiste*. Paris.

Morin, Gilles. 2004. La Ligue des droits de l'homme et la guerre d'Algérie. In *1898–2004: une mémoire pour l'avenir*, supplement to *Hommes et Libertés*, October/November/December.

Naquet, Emmanuel, and Gilles Manceron (eds). 2009. *Etre Dreyfusard, hier et aujourd'hui*. Rennes.

Nedava, Yosef. 1982. Some Aspects of Individual Terrorism: A Case Study of the Schwarzbard Affair. In *Terrorism in Europe*, ed. Yonah Alexander and Kenneth Myers. London.

Noiriel, Gérard. 1996. *The French Melting-Pot: Immigration, Citizenship and National Identity*. Minneapolis.

———. 2007. *Immigration, antisémitisme et racisme en France (XIXe-XX siècles). Discours publics, humiliations privées*. Paris.

Papon, Maurice. 1988. *Les chevaux du pouvoir*. Paris.

Perrault, Gilles. 2004. Enquête sue un centenaire au-dessus de tout soupçon. In *1898–2004: une mémoire pour l'avenir*, supplement to *Hommes et Libertés*, October/November/December.

Prakash, Amit. 2013. Murder in La Goutte d'Or: The Mouvement des Travailleurs Arabes, Surveillance and Counter-Surveillance. Paper to *Society for French Historical Studies 59th Annual Meeting*, Harvard University/MIT, 4–7 April.

Qu'est-ce que la Ligue des droits de l'homme? Paris, 2003.
Reader, Keith. 2014. France and the Middle East: The Israeli–Palestinian Conflict in Contemporary France. Special issue of *Modern and Contemporary France* 22(1) (February).
Ross, George. 1982. *Workers and Communists in France: From Popular Front to Eurocommunism.* Berkeley.
Rothberg, Michael. 2009. *Multidirectional Memory: Remembering the Holocaust in the Age of Decolonization.* Stanford.
Rousso, Henry. 1994. *The Vichy Syndrome: History and Memory in France since 1944.* Cambridge, MA.
Salles, Jean-Paul. 2005. *La Ligue communiste révolutionnaire (1968–1981). Instrument du Grand Soir ou lieu d'apprentissage?* Rennes.
Satloff, Robert. 2007. *Among the Righteous: Lost Stories from the Holocaust's Long Reach into Arab Lands.* Jackson.
Saxton, Libby. 2011. Terms of Engagement: Algeria, France and the Middle East in Barbet Schroeder's *L'Avocat de la terreur* and Philippe Faucon's *Dans la vie. Modern and Contemporary France* 19(2) (May): 216.
Schor, Ralph. 1985. *L'Opinion française et les étrangers (1919–1939).* Paris.
———. 2004. *Français et immigrés en temps de crise (1930–1980).* Paris.
———. 2005. *L'Antisémitisme en France dans l'entre-deux-guerres.* Paris.
Schwarzbard, Simon. 2010. *Mémoires d'un anarchiste juif.* Paris.
Sepinwall, Alyssa Goldstein. 2016. Jewish-Muslim Romance with a French Twist: Jean-Jacques Zilbermann's *He's My Girl. Fiction and Film for French Historians* 6(6) (April).
Serfaty, Abraham. 1998. *Le Maroc, du noir au gris.* Paris.
Sieffert, Denis. 2004. *Israël-Palestine, une passion française: la France dans le miroir du conflit israélo-palestinien.* Paris.
Silverman, Max. 2008. Interconnected Histories: Holocaust and Empire in the Cultural Imaginary. *French Studies LXII*(4): 417–428.
———. 2013a. Race, Memory and the Image in Jean-Luc Godard's *Histoire du cinéma.* Paper to *Ethnicity, 'Race' and Racism Seminar*, Edge Hill University, 22 February.
———. 2013b. *Palimpsestic Memory: The Holocaust and Colonialism in French and Francophone Fiction and Film.* Oxford.
Sopo, Dominique. 2005. *SOS Antiracisme.* Paris.
Stora, Benjamin. 2011. *Les trois exils. Juifs d'Algérie.* Paris.
Stora, Benjamin, and Mohammed Harbi (eds). 2004. *La Guerre d'Algérie: 1954–2004 la fin de l'amnésie.* Paris.
Taguieff, Pierre-André. 2002. *La nouvelle judéophobie.* Paris.
Tarr, Carrie. 2011. Jewish-Arab Relations in French and Maghrebi Cinema. In *Studies in French Cinema: UK Perspectives 1985–2010*, ed. Will Higbee and Sarah Leahy. Bristol.

Trésor de la langue française, vol. 13. Paris, 1988.
Vidal-Naquet, Pierre. 1963. *Torture: Cancer of Democracy*. Harmondsworth.
———. 1992. *Assassins of Memory: Essays on the Denial of the Holocaust*. New York.
Vovelle, Michel. 2004. La tradition des historiens de la Révolution ligeurs. In *1898–2004: une mémoire pour l'avenir*, supplement to *Hommes et Libertés*, October/November/December.
Wall, Irwin. 1983. *French Communism in the Era of Stalin: The Quest for Unity and Integration, 1945–1962*. Westport.
Wolf, Joan B. 2004. *Harnessing the Holocaust: The Politics of Memory in France*. Stanford.
Yonnet, Paul. 1993. *Voyage au centre du malaise française: l'antiracisme et le roman national*. Paris.
Zachmann, Patrick. 2009. *Ma proche banlieue*. Paris.
Zancarini-Fournel, Michelle. 2008. *Le moment 68. Une histoire contestée*. Paris.

CHAPTER 10

The Price of an Entrance Ticket to Western Society: Ayaan Hirsi Ali, Heinrich Heine and the Double Standard of Emancipation

David J. Wertheim

Current public debates on Islam mostly revolve around the question of whether, and to what extent, Muslim immigrants should be required to adjust to the values of the modern societies in which they live. Such debates echo similar debates concerning the 'Jewish Question' that were conducted during the processes of Jewish emancipation in many European countries throughout the nineteenth century. In both cases there were/are controversies dealing with religious dress, circumcision, ritual slaughter, the language of sermons and the content of religious texts. In such debates, apologists for Muslims or Jews frequently criticise far-reaching calls for infringements on the right of religious minorities to profess their religion and culture as veiled forms of antisemitism or Islamophobia. In some cases they could be right, in others they could be wrong. What I would like to

D.J. Wertheim (✉)
Menasseh ben Israel Institute, Amsterdam, The Netherlands

J. Renton, B. Gidley (eds.), *Antisemitism and Islamophobia in Europe*, DOI 10.1057/978-1-137-41302-4_10

discuss here, however, is the limited understanding of Islamophobia and antisemitism that is implied in these debates. Understood as the motivation behind calls for assimilation, antisemitism and Islamophobia may easily be taken to be phenomena whose central—and unacceptable—feature is the demand for Jews or Muslims not to be who they really are—to compromise their inner selves. Such an understanding, however, does no justice to the complexity and venom behind forms of antisemitism and Islamophobia that are related to processes of emancipation. In this chapter, the stories of two figures whose works and lives were closely connected to their emancipatory journey, Heinrich Heine and Ayaan Hirsi Ali, will serve to reveal this complexity. Both of their lives were devoted in large part to their expressed wish to acquire an entrance ticket into a non-Jewish or non-Muslim society to which they felt a true affinity, and to an extent (Hirsi Ali more than Heine) rejected the religious traditions from which they came. Yet both became, I will argue, victim to a certain kind of stereotyping that had everything to do with their attempts at emancipation. My main point will be that this was a stereotyping that not only asked them not to be themselves, but—ironically and impossibly—also expected them to remain true to their despised otherness, to be their supposedly authentic selves. I will call the cause of their predicament 'the double standard of emancipation'. The significance of uncovering this double standard lies not only in its benefits for our understanding of antisemitism and Islamophobia, helping to debunk its stereotypes, but also in the way in which it helps us to make sense of the sometimes captivating ways in which those targeted by it, like Hirsi Ali and Heine, behave and act.

Ayaan Hirsi Ali, a Somali-born former Dutch politician and writer now resident in the United States, is one of the most intriguing figures in the fierce and sometimes outright violent debates on the immigration and integration of Muslim immigrants that have been dominating Dutch public discourse for over a decade. Her public image combines the fragility of a beseeching refugee with the self-assuredness of an overconfident combatant, and her significance for Dutch public debate can hardly be exaggerated. Many of her interviews and actions made headlines. She has been either loved or hated, but for a time her opinions had great influence and they were difficult, if not impossible, to ignore. It was Hirsi Ali who forced the debate's direction, towards a critical examination of Islam and the ultimate consequences of secularism.

However, there was also another side to Hirsi Ali's public image. In a surprisingly short time span, her reputation in the Netherlands deteriorated and she lost much of her grasp on the debates that she herself had initiated. It has been fascinating to see the downfall of her public image from a role model for successful integration to an untrustworthy, spoiled diva. Three parliamentary debates on Hirsi Ali may serve to exemplify this demise. The first two were held in May and June 2006.[1] During these debates the Dutch parliament went to great lengths to solve problems that Hirsi Ali had with her Dutch nationality. Members of parliament (MPs) proposed solutions that would never apply to other immigrants in her position. The sheer time devoted to this issue—the second debate lasted until after sunrise—attested to the importance parliament attached to Hirsi Ali, as did the fact that the debates resulted in the fall of the Dutch government. The third debate was held just over a year later, when Hirsi Ali had moved to the United States, and dealt with the Dutch government's decision to stop paying for Hirsi Ali's protection outside the Netherlands.[2] By this point, the willingness among MPs to help Hirsi Ali had dramatically plummeted. Only 7 of the 150 MPs were willing to do something against this decision. Even the MPs who belonged to her political party, the Volkspartij voor Vrijheid en Democratie (People's Party for Freedom and Democracy; VVD), and MPs who, like her, lived under the daily threat of Muslim fundamentalists, did not cast a vote that could help her out.

The political question of who was wrong and who was right in these debates will not be my concern here. Nevertheless, Hirsi Ali's case raises important issues that reach further: How could this have happened? Who or what was responsible for this demise? She herself, because of her actions, or the harsh climate in the Netherlands against Muslim immigrants? (After all, she too was an immigrant from a Muslim country.) And, most importantly, what processes were at play here that are inherent in the acceptance and integration of minorities?

My intention here will be to shed light on these questions by a making a comparison between Hirsi Ali and the renowned German-Jewish poet, essayist and writer Heinrich Heine (1797–1867), whose struggle to find his place as a Jew in German society bears remarkable resemblance to the battles of Hirsi Ali's conducted in the Netherlands of the twenty-first century. However, before we come to this comparison it will be necessary to discuss more extensively Hirsi Ali's public life in the Netherlands: her opinions, her actions, the political affairs in which she came to be involved and the precise way in which her public image deteriorated.

Hirsi Ali's biography reads as the perfect example of integration and assimilation. She was born a Muslim in Somalia, but migrated to the

Netherlands as a political refugee in 1992. There she learned to speak immaculate Dutch—she initially became a translator—and left Islam. She also became a member of the think-tank of the Dutch labour party, Partij van de Arbeid (PvdA), where she published her first articles and her career as a public figure in the Netherlands started. After a number of years, however, she became frustrated with the PvdA, whose ideas in her eyes were too politically correct when it came to Islam. Then, right before the 2003 elections, she caused her first political stir, when she crossed over from the PvdA to the more right-wing and conservative VVD. She was drawn to this party for its open concern that the growth of the Muslim minority posed a problem to Dutch society, and its favouring of compulsory integration and restrictions on immigration. In return, the VVD made her an MP. There, as spokeswoman on integration, she continued to provoke debate, arguing, for example, that Muhammad had been a pervert and for the closure of certain Islamic schools.[3] Not long after her switch to the VVD, the murder of the film-maker Theo van Gogh made Hirsi Ali instantly world famous. She had collaborated with van Gogh on a ten-minute film called *Submission*, which was broadcast on Dutch public television on 29 August 2004.[4] The film features a woman wearing a veil obscuring her face, and a transparent cloth revealing her naked body. The woman delivers a monologue about the way in which the Qur'an commands her to subordinate to sexual and other abuse by men. Closeups show quotations from the Qur'an written on her naked body, which is marked with bruises.

Many Muslims considered the movie blasphemous and some started to threaten those behind it. In September 2004, van Gogh was murdered by a Muslim fundamentalist, who shot him eight times, stabbed him and put a knife into his body with a letter threatening Hirsi Ali. Ever since, she has needed extremely heavy protection. Living with such protection, Hirsi Ali continued her political work as a MP in her party, and both friends and enemies admired her for her courage in doing so. It was the peak of her popularity. Van Gogh had praised her before his death for her courage 'to play the role of heretic'.[5] Deputy Prime Minister Gerrit Zalm said of her, 'She is completely authentic, although she is friendly and nice, she does not save anyone, myself included.'[6] One of Hirsi Ali's greatest devotees, the philosopher Herman Phillipse, wrote her a letter, which he published in a pamphlet. He praised her for her refusal to compromise on her beliefs: 'What I admire in you is that even though you are a politician, you say what you mean and believe what you say.'[7] With these words, van Gogh,

Zalm and Phillipse gave an insightful explanation of the secret behind Hirsi Ali's popularity. What they show is that she was primarily admired for her authenticity. The fact that she did not weaken her tone despite the terrifying reality of threats aimed at her further reinforced this image. Whereas the average politician was perceived to be willing to trade his or her innermost convictions for just a crumb of power, here was someone who did not compromise. This virtue went beyond the opinions that people could have over her particular ideas, and was cause for admiration even by her opponents.

However, as it turned out, her much admired consistency and authenticity also became her weak point. In 2006, the makers of *Zembla*, a Dutch documentary series, made a film about Hirsi Ali that was intended to attack her integrity. It was sarcastically entitled 'The Holy Ayaan'.[8] And with some facts and many insinuations it tried to prove that Hirsi Ali was not the person she pretended to be. The documentary-makers had tried but failed to prove that Hirsi Ali's claim that she had been circumcised was wrong,[9] and now insinuated that Hirsi Ali had not escaped from a forced marriage, but used her husband to flee to the West. In the documentary, Hirsi Ali was interviewed and recounted—as she had publicly done before[10]—that in the process of her asylum request she had not disclosed the whole truth to the authorities and had told some untruths as well. She admitted that she had pretended to be a political refugee, whereas in fact she had fled to the Netherlands to escape a forced marriage. For this she had, out of fear of the possibility that her family would trace her, not given her true name and date of birth. In the context of the insinuations in this documentary, this old news turned out to be explosive. It put her party, the VVD, in an awkward position. The party's policy was to restrict immigration, and it strongly favoured measures taken against immigrants who had not been honest during the procedures of their asylum requests. The policy referred to the precedent of other immigrants who had lost their Dutch citizenship for faking names and dates of birth. Now, it turned out, an MP for their own party had done the same. However, the bomb truly exploded when the responsible Minister for Integration and Immigration, Rita Verdonk (also from the VVD), took the step of issuing a letter to Hirsi Ali stating that she was withdrawing Hirsi Ali's Dutch passport.[11] It was realised immediately that this letter would have grave consequences. First of all, it became uncertain whether the Dutch government would continue taking care of Hirsi Ali's protection, as it was doing at that time. Many feared, therefore, that Verdonk's course of action endangered Hirsi

Ali's personal safety. Second, without her Dutch nationality, Hirsi Ali was not formally allowed to be a member of the Dutch Parliament. To some, the ministerial reasoning that Hirsi Ali's Dutch citizenship had always been invalid even questioned the validity of all the parliamentary decisions made during Hirsi Ali's term of office. No one was willing to consider this suggestion seriously, but it was clear that Hirsi Ali could not remain a MP. She herself then decided to resign. The decision was made easier by the fact that she had already been planning to leave parliament to become a member of the conservative American think-tank the American Enterprise Institute. But the resignation now had to be sooner than planned, and was therefore a painful step.

Yet Hirsi Ali's withdrawal from politics did not mark the end of the affair. For her new job, she needed her Dutch passport, and therefore parliament demanded that Verdonk find some kind of creative solution that would enable Hirsi Ali to retain her Dutch nationality. The solution came six weeks later with a statement that Hirsi Ali signed after lengthy negotiations that she and her lawyer had conducted with Verdonk. In that statement, she argued that she had not lied about her name, because the name she had used, which she had later called her false name—Hirsi Ali—was, according to Somalian law, in fact her real name, even though her father was called Hirsi Magan.[12] She also stated that she regretted the confusion that she had caused, taking all of the blame onto herself and clearing the minister of any responsibility. This statement immediately raised many eyebrows, and in an interview the next day Hirsi Ali, holding the passport in her hand, bluntly declared that she had only signed the statement to keep her passport.[13] A subsequent parliamentary debate revealed that the minister had misused her power by demanding this statement in return for the passport, and it was this fact that eventually brought down the government.[14] Hirsi Ali had admitted freely that she had lied about her name, and that she had regretted doing so.

I think it is here that the demise of her public image truly began. From a legal perspective, what she did was perfectly understandable, but it affected the root of her popularity. The affair undermined her image as a figure driven by a persistent unwillingness to compromise, including her habit of saying what she believed, the very quality for which she was so admired. She brilliantly demonstrated this trademark one last time during the dramatic press conference in which she announced her resignation from the Dutch parliament. Dressed in a white shirt that accentuated her integrity, she impressed many observers with a declaration intended to end once and for all the doubts about her name, but which was also an eloquent testimony to the backgrounds that formed her:

> You may ask, what is my name? I am Ayaan, the daughter of Hirsi, who is the son of Magan, the son of Isse, the son of Guleid, who was the son of Ali, who was the son of Waiáys, who was the son of Muhammad, of Ali, from Umar, from the lineage of Umar, the son of Mahamus. I am from this clan. My primordial father is Darod, who left Somalia eight hundred years ago from Arabia and founded the great tribe of the Darod. I am a Darod, a Macherten, an Osman Mahamud and a Magan. Last week there was still some confusion about my name. What is my name? You now know my name.[15]

But for all her efforts to be open, in the end, as we saw, she had found herself in a situation in which it had simply become impossible for her to keep to the norms of complete honesty, and she signed a false declaration. As a result, less noble public sentiments surfaced against her. The admiration for Hirsi Ali had always mirrored the suspicion towards immigrants who were believed to be dishonest in their loyalty to the Netherlands, to liberal democracy, and to the motivations they gave in their efforts to acquire refugee status. In that sense, she was held up as the positive exception among immigrants from Muslim backgrounds. She proved that immigrants who were not willing to go as far in their criticism of Islam as she had done were not truly committed to the Western values of liberal democracy. Hirsi Ali also demonstrated that it was possible for a person of Muslim background to become an admired and reliable Dutch citizen in a Netherlands ruled by a government that prided itself for taking a hard line on integration and immigration. But now, more and more people began to look at Hirsi Ali in a different way. She became a self-asserting 'femme fatale of politics', a 'diva' whose biography was ridden with lies.[16] Although it was rarely expressed in so many words, it is my belief that her switching from one political party to another, her befriending of prominent and influential figures, the extravagant ways in which she dressed and her misrepresentations about her name were not seen as the traits of an exceptional Muslim, but were viewed as the embodiment of the shrewdness of the average Muslim immigrant.

It is here that the comparison with the writer Heinrich Heine becomes relevant. Heine, of course, lived in a different age, and in a different place. He was born in Düsseldorf, grew up and was educated in the German fashion, and German was his native tongue. Yet Heine was Jewish. And as a Jew, his position in German society was not self-evident. He lived from 1797 to 1856 in an age when the place of Jews in German society was an issue of constant debate. His birth took place during the Napoleonic occupation of that city. The Napoleonic armies exported the ideals of the French Revolution, including the emancipation of the Jews. However, after these armies were defeated and had withdrawn, discriminatory measures against Jews were installed once again. Most notably, Jews were barred from certain prominent careers such as the military or academia.[17] Heine, like many Jews of his generation, decided to surmount this obstacle by converting to Christianity. He was well aware that this decision had nothing to do with his religious convictions. Like Hirsi Ali, he too felt the need to compromise to be able to belong in the world that he wished to join. In 1825, he became a Christian for his career. Like Hirsi Ali, he did not try to hide his pragmatism. He commented on this act in what is perhaps his most famous aphorism: 'The baptismal certificate is the entrance ticket to European culture.'[18] We may say, therefore, that what her Dutch passport was to Hirsi Ali, his baptismal certificate was to Heine. Interestingly enough, here too a name change was involved, from Harry Heine to Johann Christian Heinrich Heine.[19]

Heine's aphorism meant that to be taken seriously in Europe as a Jew, it was necessary to barter with what was then considered one of the most important elements of one's inner self: one's religious beliefs. That Heine was well aware of this is shown by the fact that, like other Jewish converts, he tried to circumvent the problem by deliberately choosing conversion to the Lutheran state religion, which did not require him to attest to one's beliefs in the dogmas of Christianity. Later, Heine could therefore argue that his Christianity meant nothing more than the mere fact that he was 'parading as an Evangelical Christ in a Lutheran Church book' and was of no consequence to his writing.[20] Still Heine came to regret and feel shame for his decision, suffering the insincerity of his conversion in spite of this way out.

The way in which Heine had to compromise reflected a much wider demand that Jews were untrue to themselves. As Jews, they were considered a foreign element in German culture with different values and a different God, and in order to deserve their equality, it was generally assumed, Jews needed to adjust themselves to German values. To what extent they needed to adjust was an issue for debate. It could involve a wide range of acts, including becoming Christian, renouncing the Jews' election by God, praying in German and/or abandoning hopes for a rebuilt Jerusalem. Heine was not the only Jew to find himself in such a quandary. Many Jews of his generation followed the same path.

Yet assimilation almost necessarily involves an element of hiding or even betraying one's identity. After all, assimilation entails the abandonment of central aspects of a person's background, such as language, name, values and manners, in favour of those of the society to which they wish to belong. Such changes would not be problematic if the consequences of such actions for one's identity were considered a private matter, and if the notion of the authentic self was a non-issue. However, in the late nineteenth and early twentieth centuries, the authenticity of the self became an increasingly significant question in society. The great upheavals of modernity in the nineteenth century created an obsession in German conservative intellectual circles with rootedness and authenticity. Romanticists not only searched for authenticity in German history, philosophy, music and myths, they also started to *extol authenticity itself*. Fidelity to one's inner nature became a moral requirement in itself. This demand then backfired on the Jews, who, for their very adjustments to society, were now easily taken to represent superficiality and inauthenticity.

The stereotypes of modern antisemitism show well how Judaism was equated with such vices. Antisemitic stereotypes continuously stressed the inauthenticity that was taken to be the nature of the German Jew. Jews were considered wolves in sheep's clothing, parasites who did not create anything, but utilised everything of value to their personal economic benefit. Jews were blamed for never producing authentic goods, they were considered incapable of producing authentic art, and their obsession with money, which epitomised inauthenticity, only showed their lack of interest in anything that was of non-monetary value. Cultural critics believed that the reason German culture was deteriorating or even degenerating could be traced to the prominence of Jews in the cultural world. They were, it was constantly emphasised, only interested in misusing it as a source of income.[21]

Thus, Jews were subject to two different and opposing kinds of social pressure: on the one hand, emancipation asked them to assimilate and abandon their original way of life; on the other hand, it made them vulnerable to the charge of not being authentic, the direct consequence of their desired assimilation. They were, therefore, faced with what I will call a double standard of emancipation: a double standard that desires the near-impossible combination of assimilation and authenticity.

Heine was not an authentic Jew. He never aspired to be a traditional Jew. He grew up within a family that was already quite assimilated, his knowledge of Hebrew and Jewish religious matters was limited, and his Jewish faith probably even more so. He attended Jewish but also Catholic schools and tried—to a certain extent he also succeeded—in establishing himself as a German poet, writing in German. Yet even though he took the ultimate step and converted to Christianity, his baptismal certificate would never become the entrance ticket to European culture that he had expected it to be. For one, he did not get his academic career. However, it was also as a poet that he encountered resistance. Notwithstanding his conversion, Heine remained, in the eyes of antisemitic critics, a Jewish poet—a negative label that had nothing to do with the use of any Jewish subjects or with the Jewish religion, and everything to do with his presumed lack of truthfulness. Heine's style was considered inauthentic, pretentious, not genuine art. As the literary critic Wolfgang Menzel wrote:

> [Heine] lacked the profound seriousness of Byron, and in particular Byron's nobility. Already the first products of his mind showed his Jewish style [sein Jüdeln], his showing off, display, less with what is beautiful than with the Gold, which he secured by publishing in prose and verse.[22]

An even better example is Richard Wagner's discussion of Heine in his notorious antisemitic pamphlet *Jewry in Music*:

> At the time when our poetry came a lie ... then was the office of a highly gifted poet-Jew to bare with fascinating taunts that lie, that bottomless aridity and Jesuitical hypocrisy of our Versifying which still would give itself the airs of true poesis.[23]

Becoming a Christian had backfired in the end, therefore. Instead of doing away with the suspicion that Jews suffered as an alien element in a Christian society, it only added to it, confirming the inauthenticity that was presumed to be inherent in the Jewish mind.

Does the comparison between Hirsi Ali and Heine extend this far? Is authenticity at present the obsession it was for German romantics and neoromantics of the nineteenth century? Perhaps not, but we can certainly see its importance increasing in reference to debates over integration. The attack of *Zembla* on the integrity of Hirsi Ali, and the deterioration of her public image, were key signs. Since then, its importance has become more explicit, mainly due to the rise of the politician Geert Wilders, who once sat together with Hirsi Ali for the VVD in the Dutch parliament, and then left to start his own very successful right-wing populist party, Partij voor de Vrijheid (Freedom Party; PVV). As leader of this party Geert Wilders has made a big issue of the dual citizenship of Dutch Moroccans and Turks in the Netherlands, and proposed that they should not hold government office because their loyalty is not to be trusted. He also introduced the concept of Taqiyya into Dutch public debate, a not very commonly known Islamic notion commanding Muslims to hide their faith in the face of persecution. To Wilders, this notion entails that some Muslims behave in a liberal way, while in their heart they think differently. He concludes from the existence of this notion that it is impossible to know which Muslims actually are 'committing Taqiyya'.[24] This obviously implies that all Muslims need to be suspected.

It should not be overlooked that Hirsi Ali was not only a victim of a modern version of this double moral standard of emancipation; she herself was instrumental in establishing its modern version.[25] Basic to her ideas are not only the typical liberal values of secularism, women's rights, tolerance of sexual diversity and freedom of speech, but also those of being consistent, true to oneself and unwilling to lie. These values permeate her entire criticism of Islam, which bears many resemblances to that of Geert Wilders, whom she called her 'pal' when he left the VVD. The honesty imperative was basic already to her decision to leave Islam, which she made after the 9/11 attacks. For her, the bombing of the World Trade Center showed that Islam was a religion that led to terrible things. She believed that Osama bin Laden interpreted the Qur'an accurately, and therefore concluded that to be a moderate Muslim was self-deception. In her autobiography, she states this quite simply; she writes that she left Islam because she did not want to lie any more.[26]

This idea persisted in her writings on Islam after this point. Her criticism is informed by what I would like to call a strong either/or perspective. In Hirsi Ali's eyes, someone is either Muslim or liberal—there is nothing in between. As a consequence, she has very little understanding of attempts to find reconciliation between Islam and the liberal state. The headscarf, in her eyes, is always suppression, even if women who wear it argue that they do so to express their identity. She talks about the dangers inherent in 'pure Islam',[27] implying that in the faith of a Muslim, a man who does not want to beat his wife is impure. She once asked Muslim schoolchildren in front of a television camera to decide what was more important, the Qur'an or the Dutch constitution.[28] It was clear that for her there was no middle way, and that Muslims should commit blasphemy or open themselves to the charge of not being a reliable citizen. From this way of thinking came her provocations. In an interview, she called Muhammad a pervert according to modern-day opinion, said that Islam was a backward culture and defended 'the right to offend'.[29] She did not see a reason to hold back so as not to estrange 'moderate' Muslims. According to Hirsi Ali's reasoning, if they deplored such use of freedom of speech they only proved that they were not moderate. A corollary of these views was that attempts to reconcile Islam with liberal and democratic values were necessarily flawed, as they were bound to end in hypocrisy. Finally, she attached such value to the truth that in her biography, published in 2006 soon after the affair with her passport, she admits to having been in the wrong in lying about her asylum story, since in the Netherlands she had 'discovered that it is wrong not to tell the whole truth'.[30] It is an astonishing admission, since she could not have written half that book without that very lie. After all, her Dutch passport was basic to everything she has become. Once again, we see the workings of the double standard of emancipation in how Hirsi Ali's wish to comply with the norms that Dutch society expected of her had to end in a contradiction.

Heine too struggled with this double standard. At times, he shared the romanticist sentiment that valued authenticity. Heine looked for it in his Jewish background. He romanticised about Jewish history in some of his writings, such as the *Rabbi of Bacherach*. He also took part in the Verein für Wissenschaft des Judenthums, an organisation whose members tried to take pride in their Jewish origin in a non-religious way by establishing a so- called science of Judaism. And he even tried to prove the genuineness of his convictions by challenging his enemies to duels, and actually fought them quite regularly. But in the end, he also embraced his own hypocrisy, and decided to be authentic in his inauthenticity. Heine's talent was to make this into his creative drive. That is why he could make jokes about his conversion. I believe that on a deeper level it is also what is behind his trademark irony, which both ridicules and cherishes. This elusiveness is what makes his poetry. What Heine understood is that when it comes to processes of integration and assimilation, it does not help to be too strict on authenticity, and I think it is here that Hirsi Ali's criticism of Islam misses the target. Yet their stories also show that for an immigrant pressed to integrate into a foreign society, it is easier to become a poet than an activist.

Notes

1. *Debat over het oordeel van de minister voor Vreemdelingenzaken en Integratie over de nationaliteit van mevrouw A. Hirsi Ali (383 kB) 24-05-2006 | Handelingen 2005–2006, nr. 78, pag. 4836–4884 session date: 16-05-2006 | Tweede Kamer*; Debat over het kabinetsbesluit inzake de nationaliteit van mevrouw Hirsi Ali, en over de brief van de minister voor Vreemdelingenzaken en Integratie d.d. 27 juni 2006 over de naturalisatie van mevrouw Ayaan Hirsi Ali (30559, nr. 6) *17-07-2006 | Handelingen 2005–2006, nr. 96, p. 5976–6025 session date: 28-06-2006 | Tweede Kamer.*
2. Debat over de beveiliging van mevrouw Hirsi Ali (234 kB) *02-11-2007 | Handelingen 2007–2008, nr. 10, p. 646–666 session date: 09-10-2007 | Tweede Kamer.*
3. Ali Ayaan Hirsi and Carla Benink, *Mijn vrijheid: de autobiografie* (Amsterdam: Augustus, 2007), p. 372; ibid., p. 379 (Ayaan Hirsi and Benink 2007).
4. 'Submission: Part I', *IMDb*, http://www.imdb.com/title/tt0432109/ (accessed 6 May 2011); The co-producer of Van Gogh's Column Productions has withdrawn the film from distribution and refused to give anyone permission to screen it. However, it can easily be found and viewed on YouTube.
5. Anke Manschot, 'Zij heeft de importantie van Aletta Jacobs'; Harriet Freezerring voor Ayaan Hirsi Ali' in: Opzij December 1, 2004 (Manschot 2004).
6. Anke Manschot, 'Zij heeft de importantie van Aletta Jacobs'; Harriet Freezerring voor Ayaan Hirsi Ali' in: Opzij December 1, 2004.
7. Herman Philipse, *Verlichtingsfundamentalisme? Open brief over Verlichting en fundamentalisme aan Ayaan Hirsi Ali, mede bestemd voor Piet Hein Donner* (Amsterdam: Bert Bakker, 2005), p. 14 (Philipse 2005).
8. 'De Heilige Ayaan', *Vara*, http://bit.ly/iTIBdx (accessed 6 May 2011).
9. Ayaan Hirsi and Benink, *Mijn vrijheid: de autobiografie*, p. 419.
10. Ibid., p. 366.
11. Ibid., p. 418.
12. 'Verklaring van mevrouw Ayaan Hirsi Ali', *kamerstukken*, http://bit.ly/lKRWTQ (accessed 6 May 2011).

13. 'Ayaan Hirsi Ali over besluit minister Rita Verdonk', *NOVA*, http://bit.ly/iFKBM9 (accessed 10 May 2011).
14. Debat over het kabinetsbesluit inzake de nationaliteit van mevrouw Hirsi Ali, en over de brief van de minister voor Vreemdelingenzaken en Integratie d.d. 27 juni 2006 over de naturalisatie van mevrouw Ayaan Hirsi Ali (30559, nr. 6) *17-07-2006 | Handelingen 2005–2006, nr. 96, p. 5976–6025 session date: 28-06-2006 | Tweede Kamer.*
15. 'Perconferentie Hirsi Ali 16 mei 2006', *Opzij*, http://bit.ly/lLMkmf (accessed 6 May 2011).
16. For an example of this: 'Femme Fatale van de Politiek', *De Telegraaf*, 1 July 2006, p. 5 (Femme Fatale van de Politiek 2006).
17. Jeffrey L. Sammons, *Heinrich Heine: A Modern Biography* (Princeton, NJ: Princeton University Press, 1979), p. 108 (Sammons 1979).
18. Hartmut Kircher, *Heinrich Heine und das Judentum* (Bonn: Bouvier, 1973), p. 121 (Kircher 1973).
19. Ibid., p. 119.
20. Heinrich Heine and Renate Francke, *Uber Deutschland, 1833–1836: Aufsätze über Kunst und Philosophie* (Berlin and Paris: Akademie-Verlag; CNRS, 1972), p. 17. I thank Professor Gilman for alerting me to this reference (Heine and Francke 1972).
21. One of the best examples of this trend was the hugely popular book Einem Deutschen [Julius Langbehn], *Rembrandt als Erzieher* (Leipzig: C.L. Hirschfeld, 1890) (Langbehn 1890).
22. Kircher, *Heinrich Heine und das Judentum*, p. 131.
23. Paul R. Mendes-Flohr and Jehuda Reinharz, *The Jew in the Modern World: A Documentary History* (New York: Oxford University Press, 1980), p. 330 (Mendes-Flohr and Reinharz 1980).
24. Debat over de regeringsverklaring en de algemene politieke beschouwingen naar aanleiding van de Miljoenennota voor het jaar 2011 (32500) 05-11-2010 | Handelingen 2010–2011, nr. 13, pag. 7–78 session date: 26-10-2010 | Tweede Kamer.
25. Of this there are also examples in the history of Jewish integration in nineteenth-century Germany, most notably Karl Marx, who, a Jew himself, equated Judaism with the vices of commerce in his pamphlet 'On the Jewish Problem': Karl Marx and Hartwig Brandt, *Zur Judenfrage die antisemitischen Schriften von Karl Marx* (Berlin: Reichmann, 1982) (Marx and Brandt 1982).

26. Ayaan Hirsi and Benink, *Mijn vrijheid: de autobiografie*, p. 423.
27. 'Hirsi Ali: Islam kern terreurprobleem', *nos*, http://goo.gl/VOdOu, Pure islam (accessed 10 May 2011).
28. Maria Benbrahim, 'Hirsi Ali en Moslimscholen 2', *Trouw.nl*, http://goo.gl/KMNOV (accessed 10 May 2011).
29. Ayaan Hirsi Ali, 'Het recht om te beledigen', *Liberales*, http://goo.gl/zuBmQ (accessed 10 May 2011).
30. Ayaan Hirsi and Benink, *Mijn vrijheid: de autobiografie*, p. 423.

References

Ayaan Hirsi, Ali, and Carla Benink. 2007. *Mijn vrijheid: de autobiografie*. Amsterdam: Augustus.

Femme Fatale van de Politiek. *De Telegraaf*, 1 July 2006.

Heine, Heinrich, and Renate Francke. 1972. *Uber Deutschland, 1833–1836: Aufsätze über Kunst und Philosophie*. Berlin and Paris: Akademie-Verlag; CNRS.

Kircher, Hartmut. 1973. *Heinrich Heine und das Judentum*. Bonn: Bouvier.

Langbehn, Julius. 1890. *Rembrandt als Erzieher*. Leipzig: C.L. Hirschfeld.

Manschot, Anke. 2004. Zij heeft de importantie van Aletta Jacobs. *Opzij*, 1 December.

Marx, Karl, and Hartwig Brandt. 1982. *Zur Judenfrage die antisemitischen Schriften von Karl Marx*. Berlin: Reichmann.

Mendes-Flohr, Paul R., and Jehuda Reinharz. 1980. *The Jew in the Modern World: A Documentary History*. New York: Oxford University Press.

Philipse, Herman. 2005. *Verlichtingsfundamentalisme? Open brief over Verlichting en fundamentalisme aan Ayaan Hirsi Ali, mede bestemd voor Piet Hein Donner*. Amsterdam: Bert Bakker.

Sammons, Jeffrey L. 1979. *Heinrich Heine: A Modern Biography*. Princeton, NJ: Princeton University Press.

CHAPTER 11

The Impact of Antisemitism and Islamophobia on Jewish–Muslim Relations in the UK: Memory, Experience, Context

Yulia Egorova and Fiaz Ahmed

The relationship between the Jewish and Muslim communities of Europe is often constructed by public discourse as polarised due to the Israel–Palestine conflict. Indeed, in the summer of 2014, the mass media presented numerous reports suggesting that the relations between Europe's Jews and Muslims were deteriorating following the military action between Israel and Gaza. At the same time, it has been argued by social scientists and humanities scholars that the discussion of Jewish–Muslim relations needs to be situated in the wider context of the position of 'minority' communities in Europe. In this chapter, we will adopt the same approach in focusing on the case of the UK and will use ethnographic analysis to highlight the context-dependent nature of Jewish–Muslim relations.

Y. Egorova • F. Ahmed (✉)
Durham University, Durham, UK

J. Renton, B. Gidley (eds.), *Antisemitism and Islamophobia in Europe*, DOI 10.1057/978-1-137-41302-4_11

The topic of Jewish–Muslim relations in the modern world has produced substantial literature stemming from a wide range of disciplines.[1] A number of studies demonstrated the importance of taking into consideration the broader socio-historical context of European colonialism, as well as local experiences of the two communities when considering Jewish–Muslim relations in Europe.[2] In this chapter, we will suggest that Jewish–Muslim relations in the UK offer fertile ground for applying the same socio-historical approach and will focus on one particular aspect of this relationship revealed in our ethnographic work with both groups—the importance of the local context for the formation of mutual attitudes towards and perceptions of the two communities.

A number of sociological studies and surveys examined the opinions of British Muslims about Jews and Judaism to suggest that negative attitudes towards the Jews are more common among British Muslims than among the general population,[3] but none of the studies published so far has provided a detailed ethnographic analysis of the mutual perceptions of the two communities in the UK or explored the attitudes of British Jews towards Muslims. In 2013–15 we conducted a study that involved in-depth interviews with forty British Jews and British Muslims and participant observation of the meetings of two initiatives in Jewish–Muslim dialogue. The overall study revealed that community members demonstrate a wide range of views regarding each other and that their relationship provides an important example of sizeable groups of Jews and Muslims often living side by side and successfully negotiating different types of mutual perceptions and understandings. What we would like to focus on in this chapter is one particular aspect of these relations—the way they are shaped by and, at the same time, reflect wider public British attitudes towards 'minority communities' in general and towards Jews and Muslims in particular. This chapter will contribute to the main theoretical themes explored in this volume by suggesting that the attitudes that the two communities exhibit towards each other intersect with wider public discourses about Jews and Muslims, and 'minority' groups in the UK, and that their perceptions of the 'other' community often reflect their own sense of in/security and experiences of discrimination in British society. In the following two sections we will present data from our interviews and participant observation, focusing first on the Jewish and then on the Muslim respondents. In the final section, we will return to the wider problematics of the context-dependent nature of Jewish–Muslim relations.

History, Memory, Experience

> So, you are studying the relationship between Jews and Muslims? Very interesting! Though, I must say, I could never understand Muslims very well. Why would anyone want to become a suicide-bomber? It is very hard for a British person to understand it!

This is how Miriam,[4] one of Yulia's Jewish interviewees, responded to her description of our project's objectives. In a little while Miriam added, 'I should not be prejudiced though. I have met so many lovely Muslim people who have been so helpful. But you hear so much in the news about all these terrorists. And then, of course, if you look at what they say about Israel …'

Miriam's comments do not by any means reflect the whole complexity of the way in which British Jews relate to their Muslim neighbours, but they do highlight the concern and hesitation about Jewish–Muslim relations that are present among some members of British Jewish communities. In this section we will suggest that what appears to account for this hesitation is both the general negative stereotyping of Muslims that originates in the mainstream mass media and public discourse, and the antisemitic discrimination that Jewish communities have experienced in the UK.

The question of Jewish–Muslim relations looms large in the debate about what became to be known as the 'new antisemitism'. The emergence of this debate in Jewish communities followed the intensification of antisemitic violence that resurfaced in the past two and a half decades and appeared to correlate with events in the Middle East.[5] As Matti Bunzl has discussed at length in his seminal essay on the topic, some commentators in the Jewish constituencies have called attention to those cases of antisemitic violence where the perpetrators are Muslims.[6]

The responses that we received from our Jewish interviewees about their experiences of interactions with British Muslims were positive, although almost every respondent talked about the concern present in their congregations. This is how Michael, a Jewish man in his sixties, attempted to explain this concern to us: 'Unfortunately there is a fear [among British Jews] of being overtaken, and I think it's the numbers, because we're such a small number and Muslims are a relatively large number … and then there is of course the question of Israel.'

Our Muslim participants reported examples of attitudes revealed towards them that were similar to those described by Michael. For instance, Sayyid, an activist of inter-faith dialogue, told Fiaz how once he and his Jewish counterpart tried to organise an event that was supposed to bring together Jewish and Muslim schoolchildren. The event fell through because the Jewish parents felt that it would not be safe enough for their children to be in contact with a Muslim group.

This example, as well as the quote with which we started the section, highlight how wider societal discourses about Islam appear to affect the way in which some British Jewish people relate to British Muslims. It is clear that some of their hesitation stems from the rhetoric of the 'war on terror' that is common in the mainstream mass media and public discourse, and is not at all limited to the Jewish constituency.[7] At the same time, it appears that in addition to the general context of mainstream British/ European discourse about Islam, the way in which British Jews relate to their Muslim neighbours is mediated by their own historical memories and personal experiences of antisemitism.

Many of our Jewish interviewees, similarly to Michael, were very critical of what they saw as anti-Muslim prejudices in their congregations, but they also put them in the broader context of the history of antisemitism in general, and of their communities' experiences. For instance, when Yulia asked Baruch, an Orthodox Jewish man, who lives in a neighbourhood that is home to a sizeable Jewish congregation, what, in his opinion, the attitudes of British Muslims towards his community were, he said that many problems stemmed from the fact that Muslims, as well as the general population, tended to associate all Jews with Israel without knowing enough about Jewish history. He noted that, for instance, young people in general and young Muslims in particular did not realise the role that the pogroms in Russia had played in providing the context for Zionism becoming more popular among European Jews. Had British Muslims known about the history of antisemitism, their attitude towards Zionism and Israel might have been more positive, he argued.

When asked about the way in which Muslims were perceived in his congregation, he said that though most of his neighbours had never experienced any open conflicts with them, they were constantly concerned about the possibility of such a confrontation.

> You sometimes can just see it, how young *haredi*[8] men walk down the street, they see an Asian looking man, assume that he is Muslim and simply turn white and start shaking with fear. It is very sad indeed, but one has to understand that being *haredi* means being visibly Jewish, which means that these

> boys must have experienced antisemitic abuse before even if it had nothing to do with the Muslims.

This quote raises important questions about the way in which in the case of British Muslims and *haredi* Jews, race and religion co-constitute and co-produce each other in ways that lead to stereotyping and discrimination, which could be explored at length in a separate paper. The experiences of the secular, Reform and modern Orthodox Jewish people may be different, but their answers to our questions about Jewish–Muslim relations also contain accounts of historical and recent discrimination in the UK, which again go beyond the Muslim context.

As Miriam's quote suggests, she strongly identifies herself (and, probably, by implication her community) as British and constructs Muslims as quintessentially non-British. However, her words may also be hiding a story of assimilation that some of our Jewish respondents described as quite problematic.

Speaking about the experiences of their co-religionists in the UK in comparison with those of British Muslims, all Jewish respondents observed that British Jews were a very well-assimilated community who (possibly with the exception of *haredi* Jews) did not stand out the way that British Muslims did. Their responses thus construct the same juxtaposition between Jews and Muslims that Miriam attempted, although several of them also noted that the reason why British Jews were so successful in assimilating was because they were forced to do so by the mainstream society in the past. For instance, Edith stressed that she felt envious of British Muslims because, she argued, they were represented in the UK in larger numbers and (in her perception) they were not under the same pressure to assimilate. Another Jewish respondent, Simon, a man in his early seventies, observed that today British Jews were well respected in the country, because they had succeeded in secular education, but to do so they had 'to change their ways a lot'.

Indeed, as Keith Kahn-Harris and Ben Gidley point out, '[t]he British Jewish community emerged during a time of monoculturalism, in which it was difficult for minorities publically to articulate their concern'.[9] Interestingly, similar observations were made by some of our Muslim respondents too, who in their interviews referred not only to the history of Jewish assimilation in the UK, but also to the wider history of anti-Jewish prejudice. As Omar, a Muslim man in his thirties, put it when we asked him to compare the position of the two communities in Britain:

> People are ok with the Jewish community now because it had to pay the price for it. It [their acceptance] didn't happen overnight, it took hundreds of years and it took millions of people dying in the Holocaust to get to where they are—so yes, the Muslim community might look and say 'they're normalised and no one seems to have a problem with them', but it came at a price.

Omar's words echo the intervention made by anthropologist Jonathan Boyarin in relation to 'the interrelated problematics of movements such as Third World liberationism, feminism, and the struggle to reinvent Jewishness'.[10] In exploring this problematics, Boyarin suggests that we should pay attention to the post-colonial condition of the Jewish people. Boyarin warns against conflating the condition of the Jewish people with the politics of the Jewish state, and argues that though Israel is not normally described as a post-colonial state—quite the opposite, it is often seen as a First World power on a mission to extend its territory—we should not make an assumption that 'Jews can't be in a postcolonial situation'.[11] Instead, he advocates an approach that calls attention not only to the spatial, but also to the temporal dimension of anti-imperial struggle, and argues that in deciding whether the Jewish people could be seen as a subaltern group, one should consider not only their contemporary condition, but also their history.[12]

We suggest in this respect that in reflecting on the position of their community in the UK in relation to that of British Muslims, many of our Jewish respondents revealed a degree of subaltern self-perception stemming from their historical memories of persecution. In addition, some of them put the topic of Jewish–Muslim relations in the context not only of the history of antisemitism, but also of their own experiences in the UK. For instance, David, a religiously observant Jewish man in his fifties, shared the view that we had heard from Omar. Like most of our Jewish respondents, he suggested that what the Muslim communities, which represent mostly relatively recent immigrants, were going through in the UK now was what Jewish people had experienced in the past, and that the reason why the levels of overt antisemitism in the UK, in his view, were comparatively lower than those of anti-Muslim prejudice was because Jews had already been forced to assimilate. 'If you're getting a lot of hassle, if you can't live in a society because of the way you look, sometimes that can force people to abandon things quicker', he said. Reflecting on his own experiences, he told us how when he was an adolescent growing up in the suburbs, he would regularly receive verbal antisemitic abuse.

Michael, whom we quoted earlier, drew a direct link between his concern about antisemitism in the UK and the question of Jewish–Muslim relations. When we asked him what, in his opinion, the main issues in these relations were, he stated that all the problems between local Jews and Muslims were stemming from the conflict in the Middle East. However, though he was critical of Israel, it was hard for him to stop supporting it because of the possibility of antisemitic persecution in the UK:

> I have to say I can see both sides of this because I've dealt with a lot of Muslims ... and the prejudice begins and starts really with Israel and Palestine, that's where the prejudice is, it's nothing to do with [what we do] on a day-to-day basis—in other words we can get on well ... I am not the greatest Zionist in the world ... I'm not a person that says Israel is wonderful and it's faultless. But I've got to survive, you see, so if it became bad here I'd go there and that's my survival kit.

This quote implies that the attachment that Michael feels to the state of Israel, a country whose politics he also considers to be a major stumbling block in the development of Jewish–Muslim relations, stems from his concern about the security of Jewish people in the UK. Attitudes towards Israel, current among contemporary British Jewish communities, are extremely diverse,[13] and it is not at all our suggestion that all British Jews see Israel as key to their survival. However, Michael's words are indicative of a sense of insecurity that some British Jewish people have that might explain an attachment to Israel. When we asked Michael if he had personally experienced antisemitism, he did not reference any traumatic encounters with Muslims. However, he did recount how once, back in the 1970s, when he was a student at a prestigious UK university, he shared with a fellow student doing a law degree that he was Jewish. On hearing this the student, who, according to Michael, was of solid upper-middle-class background, expressed his surprise at the way Michael looked—'I thought Jewish people had horns', he said.

As we later discovered in an interview with Sayyid, this time-old antisemitic discourse was still circulating in the UK today. After Sayyid told Fiaz about the failed inter-faith sports event, he added that while British Jews appeared to have unsettling concerns about British Muslims, the latter were susceptible to negative stereotyping too. Sayyid said:

> Once I visited a Jewish museum, and a lady who worked there told me that recently they had a group of Muslim kids on a school trip. One of the kids, a nine or ten year old girl asked her to kneel down, because she wanted to see her horns. 'To see what?!' the lady said. 'Your horns. My parents say that Jews have horns.'

This archaic anti-Jewish imagery, which associates Jews with Satan's consorts, goes back to the Middle Ages, when emerging evangelising Christianity started a relentless anti-Jewish campaign in an attempt to differentiate 'true Israel' (Christianity) from ancient Israel and its Torah.[14] As Michael's and Sayyid's quotes show, this imagery (whether its proponents believed in its literal meaning or not) has stayed in circulation until the turn of the twenty-first century, creating a circuit of stereotyping practices, which in these cases were exhibited by an upper-middle-class man of secular or Christian background and the Muslim parents of a schoolgirl. As both examples suggest, this circuit is proving to be doubly damaging for British Jewish–Muslim relations, because it both spreads anti-Jewish attitudes among local Muslims and contributes to the overall sense of insecurity among British Jewish communities, which then, combined with the general negative stereotyping of Muslims propagated by the mass media, interpellates their perceptions of their Muslim neighbours. As we will argue in the following section, what adds yet another dimension to this circuit is British Muslims' own experience of discrimination.

Images of Difference

> You want to know why I came to this Forum? I wanted to meet Jewish people and learn from them how to become more successful. We, Muslims, are so much behind British Jews, and they are so much ahead. We need to learn from them and catch up.

This is how Raza, a Muslim student in his early twenties, explained to Yulia why he had decided to start attending a forum that brought together young people from local Muslim and Jewish communities. Like Miriam's quote with which we started the previous section, Raza's words hardly do justice to the wide-ranging approaches that his community has to Jewish–Muslim dialogue. However, as we saw in a number of other interviews, the trope of Jewish people being 'successful' and 'ahead', which in other contexts would have read as an antisemitic stereotype, was used by our Muslim respondents to describe their own condition of discrimination.

As we already mentioned in the previous section, many of our Muslim interviewees noted that while Jewish people might be in a stronger position in the UK now and were subjected to less prejudice than Muslims, that was only because they had already had to face a lot of discrimination in the past, which had led to their assimilation. In commenting on the contemporary and historical experiences of the two groups, a significant number of our Muslim respondents suggested that the position of European Muslims now was comparable to that of European Jews in the first half of the twentieth century. For instance, this is how Ibrahim, a man in his thirties, put it when we asked him to comment on the position of Jews and Muslims in Europe:

> They [Jewish people] have been through everything that we're starting to go through now, they've been through that whole cycle ... Because this is how it started, if you look at the Jewish history in Europe, it started off with little things in the media and ended up with what happened, Hitler and the Second World War, right? So there is this mild sort of, simmering, concern [among British Muslims] ... Are we the next Jewish community in World War time? And is that the future? And do we all just have to be ...what are the answers? We're sort of looking for the answers I suppose ... I don't know whether you'll hear that a lot, but it's there ... Because it's horrific what happened [in the history of the Jewish people], and how it started with little things like this.

Some of our Muslim respondents put forward to us the idea that their co-religionists should know Jewish history better to understand their own position in Europe. As Tarik, a coordinator of an inter-faith forum, put it:

> Unfortunately, there are a lot of Muslims in the UK who don't know about the history of the Jews in Europe, otherwise, they would see the similarities in their experiences. A lot of Muslims in Britain today do not know about the history of European Jews, so sometimes my brothers make statements [about the Jews] which are very hurtful, because they don't know the history—because if you know the history of Jews in Europe, then there are experiences which can repeat themselves, you know?

For Tarik, tragic events in Jewish history have thus become an important reference point both in understanding the history of European Muslims and in seeking ways to develop Muslim–Jewish dialogue. At the same time, many Muslim respondents stated that at the moment their community was receiving less legal support and facing more prejudice than their Jewish counterparts. For instance, this is how Ibrahim continued to answer our question about the position of Jews and Muslims in Europe:

> Right now I think various things affect both communities, but the Jewish community is very well protected through laws, and through their lobbying, and through the support that they have through the government, and through MPs and so forth, whereas the Muslim community doesn't have that ... You can say what you want about Muslims and you can do what you want to Muslims, and no one really is there to fight your corner, that's how I see it. You know, I have been in a [job] interview, and because of the way I was dressed they actually said to me at the end of the interview that 'you're clearly from a particular faith and that's difficult for us'. Now, if you said that to a Jewish person ... that organisation would be in serious trouble, but you can do it to a Muslim person. So I think the adversity affects both communities, but the quality of the advocacy and support is differential, and it doesn't exist for the Muslim community as it does for the Jewish community.

Issues that Ibrahim raised in this interview have become the subject of a dynamic academic discussion among scholars of Judaism and Islam who have explored the relationship between antisemitism and Islamophobia in Europe. Matti Bunzl has argued that while overt political antisemitism still exists in contemporary Europe, it is embraced mainly by the extreme Right, which occupies only a marginal position on the political horizon of the continent. When it comes to Muslims, Bunzl argues, Islamophobic political and mass media agendas appear to be much more mainstream.[15] A vivid example of this problematics was discussed in academic and media sources in the aftermath of the *Charlie Hebdo* events of January 2015. For instance, Didier Fassin has highlighted how in France freedom of speech laws put Muslims at a disadvantage, and how *Charlie Hebdo* itself had put limits to its application of free speech when they fired one of its cartoonists for writing an antisemitic piece, but ignored contributions that were clearly Islamophobic.[16] Writing specifically about the effect that anti-Muslim discrimination in France has had on French Jewish–Muslim relations, Silverstein has pointed out that young European Muslims tend to present their situation as consonant with other oppressed and racialised groups, including the Palestinians in the West Bank. This, combined with the situation where Islamophobic statements are tolerated more under freedom of speech laws than antisemitic remarks, leads young Muslims to view Jews as 'fully integrated European insiders and indeed iconic of all which is intolerable in their own lives', Silverstein argues.[17]

What has become a major contributing factor to tensions in Jewish–Muslim relations in France was how local Jewish and Muslim communities coming from North Africa were differentially treated by the state.[18] British Jews and British Muslims do not share the same region of origin, as most British Jews are the descendants of immigrants from continental Europe, while the majority of British Muslims are connected to South Asia.[19] However, like in France, the context of their past histories plays an important role in the formation of their mutual perceptions. As we noted in the previous section, Baruch invoked the history of pogroms to explain why he was disappointed with the position that many British Muslims took on Israel. Some of our Muslim respondents put their views of Jewish–Muslim dialogue in the context of their ancestors' histories too. For instance, when we asked Fatimah, who had stated that the conflict in the Middle East was the main issue in Jewish–Muslim relations, what Palestine meant to her, she replied that the history of the Palestinians reminded her of the history of South Asian Muslims—the community from which she had come—under British rule.

In addition, it appears that in contemporary Britain, like in France, Jews and Muslims are often juxtaposed by the political Right. For instance, Meer and Noorani in their discussion of antisemitism and Islamophobia in the UK point out a *Daily Telegraph* article that advised British Muslims to follow the example of British Jews who, the author of the article argued, managed to integrate better into British society because all branches of Judaism accepted civil law.[20] It would not be at all surprising if such generalisations that essentialise both traditions, but construe Judaism as more compatible with life in the UK than Islam, were detrimental to Jewish–Muslim relations.

What appears to be equally problematic for the development of Jewish–Muslim dialogue is the political Right portraying Jewish people as potential 'allies' of European Christians in the fight against the 'Islamisation' of Europe,[21] and the mass media constructing Muslims as the 'natural' enemies of the Jews and levelling blanket accusations of antisemitism at the entire Muslim community. As Silverstein has argued, European mass media often portray Muslims as the main victimisers of European Jews, and present the attacks on Jewish persons and property where perpetrators were Muslim as further evidence of Muslim immigrants' susceptibility to extremist ideologies and failure to integrate.[22] We saw examples of the negative effect that this rhetoric can have on Jewish–Muslim dialogue in our ethnography. One of our respondents, Daniel, who was involved in an initiative that brings together young Jews and Muslims, told Yulia that it was often hard for him to recruit Muslim participants. When he tried to explore what was preventing them from taking part in these events, he was told that some young Muslims felt that it would not be legally safe for them to engage in these activities. They were worried that they might be asked about their position on Israel and, if they made any negative comments, they would be arrested, they told Daniel.

Making negative remarks about Israel does not constitute a criminal offence in the UK. However, we suggest that the scepticism of these young people should not be dismissed as sheer delusion. The episode that Daniel described is starkly reminiscent of the episode that Sayyid related, when a group of Jewish parents refused to let their children take part in an interfaith event that was going to include Muslims. In both cases the response was based on what on the face of it looks like groundless suspicion—suspicion that any interaction with Muslims was dangerous for Jews, and that Jews were so well protected by the legal system that making any negative comments about the Jewish state could lead to an arrest, particularly if these comments were made by a Muslim person whom society already expected to become a perpetrator of anti-Jewish violence. In both cases the suspicion was hardly based on evidence, but it is important to consider it as a symptom of a more general feeling of insecurity experienced in both communities, the topic that we will address in the following section.

Discussion

Baruch's words about some of his co-religionists' reaction to 'Asian-looking men' were brought home to us later that same day when Yulia interviewed a Muslim respondent who lived in the neighbouring area. Amir, a young man of Pakistani descent, told her that he admired his local Jewish community and felt very lucky to live close to what he described as a Jewish neighbourhood. When Yulia asked Amir if he had ever faced discrimination or prejudice, he recounted the following episode:

> Just a few days ago I was jogging in the park close to where I live. I had my headphones on and was just jogging. There was a man walking down the street who saw me and started walking in the opposite direction. That was unpleasant, and this was not an isolated incident.

The brief episodes that Baruch and Amir described illuminate some of the key themes in the overall problematics of Jewish–Muslim relations in the UK and probably in the Diaspora more widely. As was discussed in the previous section, some members of the Jewish communities have strong security concerns that hark back to the long and tragic history of anti-Jewish violence, and in the political climate of contemporary Europe are more often than not associated with Muslims. The latter, in their turn, are struggling to shed the image of foreign, racialised others, who are seen as the victimisers of Jews and a constant security threat. In the case of Amir, anti-Muslim/racist discrimination does not appear to have affected his attitude towards British Jews. However, in other cases, such as those described by Silverstein, Mandel and Katz,[23] it can lead to Muslims seeing their Jewish compatriots as 'luckier' and more powerful citizens.

We have shown throughout this chapter that though the role that the conflict in the Middle East plays in shaping Jewish–Muslim relations is difficult to ignore, the way in which the minutiae of these relations develop is determined by the local experiences of British Jews and Muslims, as well as, to a degree, by their collective historical memories. As we argued earlier, how the Israel–Palestine conflict itself is understood in both groups is to a large extent mediated by these local experiences. It appears that for many British Jews and British Muslims, antisemitism and Islamophobia constitute a significant factor that determines their place in the vexed picture of Jewish–Muslim relations in the UK, and we argue that the social hesitation and fear that some British Jews and British Muslims have against each other is a symptom of wider problems in the way in which 'minority' groups are perceived and treated in the UK.

As we showed in the previous sections, both personal and historical experiences of discrimination were frequently referred to in our respondents' accounts of their view of Jewish–Muslim relations and of their perception of the 'other' group. In the case of the Jewish communities, the prior existing sense of insecurity, combined with exposure to public and mass-media discourses that construct Muslims as a security threat in general, and a threat to Jewish persons and organisations in particular, forces some members of the Jewish constituency to view Muslims with suspicion. Similarly, the post-colonial experiences of Muslims in Europe, particularly after the events of 9/11 and 7/7, have conditioned some of them to expect discrimination and unlawful prosecution, be doubtful about the future of Muslims in Europe and ask whether Europe is not on the brink of another, this time anti-Muslim, Holocaust. When it comes specifically to their relations with their Jewish counterparts, this overall feeling of insecurity sometimes intersects both with time-old anti-Jewish stereotypes and conspiracy theories, and with a reaction to public discourses that 'other' both communities, but position Jews and Muslims differently, placing the latter at the bottom of the British 'hierarchy of minorities'.

We argue that these experiences, combined with exposure to wider anti-Jewish and anti-Muslim discourses, produce mistrust that, to build on John Jackson's insight, 'translate fear into social action',[24] such as when Jewish parents prevent their children from participating in an inter-faith event out of fear that they will become victims of a terrorist attack, or when young Muslims refuse to engage in dialogue with their Jewish counterparts out of fear that they will be accused of making anti-Israeli statements and arrested. The example of Jewish–Muslim relations in the UK could be usefully referred to in exploring Jewish–Muslim relations in the Diaspora worldwide, as we expect the local conditions to be an important factor of these relations in many parts of the world. To go beyond the context of Jewish–Muslim relations, our case study highlights how the expectations and social fears that different communities have about the way they will be treated by others are an important indicator of the overall state of inter-communal relations in the UK and the level of socio-political comfort that they experience not just in relation to each other, but also in relation to the mainstream society.

Notes

1. Illuminating analysis of Jewish–Muslim relations in the Middle East and North Africa is offered in, for instance, Bernard Lewis, *The Jews of Islam* (Princeton, NJ: Princeton University Press, 1984), Ron Nettler and Suha Taji-Farouki, *Muslim-Jewish Encounters: Intellectual Traditions and Modern Politics* (Amsterdam: OPA, 1998), Tudor Parfitt (ed.), *Israel and Ishmael. Studies in Muslim-Jewish Relations* (Richmond: Curzon Press, 2000), Norman Stillman, *The Jews of Arab Lands in Modern Times* (Philadelphia, PA: Jewish Publication Society, 2003), Aomar Boum, *Memories of Absence: How Muslims Remember Jews in Morocco* (Stanford, CA: Stanford University Press, 2013). Muslim attitudes towards Jews and Israel in different parts of the world were explored in Moshe Ma'oz (ed.), *Muslim Attitudes to Jews and Israel: The Ambivalences of Rejection, Antagonism, Tolerance and*

Cooperation (Brighton, Portland, Toronto: Sussex Academic Press, 2010). Abdelwahhab Meddeb and Benjamin Stora co-edited the formidable *A History of Jewish-Muslim Relations, From the Origins to the Present Day* (Princeton, NJ: Princeton University Press, 2013) (Lewis 1984; Nettler and Taji-Farouki 1998; Parfitt 2000; Stillman 2003; Boum 2013; Ma'oz 2010; Meddeb and Stora 2013).

2. For instance, anthropologist Paul Silverstein has argued in his ethnographic research of Jewish–Muslim relations that in France and North Africa, Muslim populations have demonstrated an ability both to reject and empathise with their Jewish compatriots, and that in France the negativity that some local Muslims exhibit towards Jews can be seen as a response to the state oppression directed at North African immigrants and their children (Paul Silverstein, 'The Fantasy and Violence of Religious Imagination: Islamophobia and Anti-Semitism in France and North Africa', in *Islamophobia/Islamophilia: Beyond the Politics of Enemy and Friend*, ed. Andrew Shryock [Bloomington: Indiana University Press, 2010], pp. 141–71 [Silverstein 2010]). The historian Maud Mandel's recent monograph convincingly argues that daily interactions between French Jews and French Muslims are diverse and go far beyond the polarisation over the Israel–Palestine issue. Mandel also stresses the importance of paying attention to 'the way global dynamics, both in the Middle East and in French North Africa, came together with national and even local factors to shape Muslim-Jewish relations in postcolonial France' (Maud Mandel, *Muslims and Jews in France: History of a Conflict* [Princeton, NJ: Princeton University Press, 2014], p. 3 [Mandel 2014]). Similarly, Ethan Katz concludes in his monograph exploring the histories of Jews and Muslims from North Africa in France that 'interactions between Jews and Muslims in France were not a binary but a decidedly triangular affair', as the state played a key role in defining these interactions from their very inception, and because Jews and Muslims themselves understood and expressed these attitudes and relations through the prism of their relationship to the French (Ethan Katz, *The Burdens of Brotherhood: Jews and Muslims from North Africa to France* [Cambridge, MA: Harvard University Press, 2015], pp. 24–5 [Katz 2015]).

3. Gunther Jikeli, 'Antisemitism Among Young Muslims in London', *International Study Group Education and Research on Antisemitism. Colloquium 1. Aspects of Antisemitism in the UK*, 5 December 2009, London (Jikeli 2009).
4. For the purposes of protecting our respondents' anonymity their names have been changed.
5. Matti Bunzl, *Anti-Semitism and Islamophobia: Hatreds Old and New in Europe* (Chicago: Prickly Paradigm Press, 2007); Keith Kahn-Harris and Benjamin Gidley, *Turbulent Times: The British Jewish Community Today* (London: Continuum, 2010), p. 140; Brian Klug, 'Interrogating "New Anti-Semitism"', *Ethnic and Racial Studies*, 36, no. 3 (2013), 468–82 (Bunzl 2007; Kahn-Harris and Gidley 2010; Klug 2013).
6. Bunzl, 'Anti-Semitism and Islamophobia: Hatreds Old and New', p. 25.
7. See, for instance, Didier Fassin, 'In the name of the Republic: Untimely Meditations on the Aftermath of the *Charlie Hebdo* Attack', *Anthropology Today*, 31, no. 2 (2015), 4–5 (Fassin 2015).
8. Haredi in Hebrew means God-fearing, and sometimes described as ultra-Orthodox, though this term is not recognised by this community.
9. Kahn-Harris and Gidley, *Turbulent Times*, pp. 165–6.
10. Jonathan Boyarin, *Storm from Paradise: The Politics of Jewish Memory* (Minneapolis: University of Minnesota Press, 1992), p. 105 (Boyarin 1992).
11. Ibid., p. 83.
12. Ibid., pp. 82–3.
13. For a detailed discussion see Keith Kahn-Harris, *Uncivil War: The Israel Conflict in the Jewish Community* (London: David/Paul, 2014) and Ilan Zvi Baron, *Obligation in Exile: The Jewish Diaspora, Israel and Critique* (Edinburgh: Edinburgh University Press, 2015) (Kahn-Harris 2014; Baron 2015).
14. Moshe Lazar, 'The Lamb and the Scapegoat: The Dehumanization of the Jews in Medieval Propaganda Machinery', in *Anti-Semitism in Times of Crisis*, ed. Sander Gilman and Steven Katz (New York and London: New York University Press, 1991), p. 40 (Lazar 1991).
15. Bunzl, 'Anti-Semitism and Islamophobia'.
16. Fassin, 'In the Name of the Republic'.

17. Paul Silverstein, 'Comment on Bunzl', in *Anti-Semitism and Islamophobia*, ed. Matti Bunzl, pp. 64–5.
18. Silverstein, 'The Fantasy and Violence of Religious Imagination'; Mandel, *Muslims and Jews in France*; Katz, *The Burdens of Brotherhood*.
19. British Jews and British Muslims account for two sizeable, though numerically different, constituencies, with members identifying themselves as Jewish in the 2011 England and Wales Census accounting for 263,346 people and those identifying as Muslim for 2,706,066 people. In the UK, both Jews and Muslims present highly diverse communities. The first Jewish people arrived in England with the Normans in 1066 and then were expelled in 1290 under King Edward I. Jews started living openly in Britain again in the seventeenth century and now their congregations include members of both Sephardi and Ashkenazi descent, whose ancestors came to Britain mainly from different places in Western and Eastern Europe (with a small minority coming from Asia and Africa). In terms of religious observance, the community comprises people who describe themselves as secular, as practitioners of different forms of Judaism, and as practitioners of religions other than Judaism.

    The roots of the majority of British Muslims are in the post-war commonwealth migration, with the largest group (about 40%) identifying as Muslims of Pakistani descent, and other groups comprising Muslims from Afghanistan, Bangladesh, India, North Africa, the Middle East, Somalia and Eastern Europe. The religious denominations of British Muslims include Sunni, Shia and Ahmadiyya branches.
20. Meer and Noorani, 'A Sociological Comparison of anti-Semitism and anti-Muslim Sentiment in Britain', p. 210.
21. For instance, see the example of the Flemish Interest party discussed in Bunzl, 'Anti-Semitism and Islamophobia'.
22. Paul Silverstein, 'Immigrant Racialization and the New Savage Slot: Race, Migration and Immigration in the New Europe', *Annual Review of Anthropology*, 34 (2005), 367 (Silverstein 2005).
23. Silverstein, 'The Fantasy and Violence of Religious Imagination'; Mandel, *Muslims and Jews in France*; Katz, *The Burdens of Brotherhood*.

24. John Jackson, *Racial Paranoia, The Unintended Consequences of Political Correctness* (New York: Basic Civitas Books, 2008), p. 16 (Jackson 2008).

References

Baron, Ilan Zvi. 2015. *Obligation in Exile: The Jewish Diaspora, Israel and Critique.* Edinburgh: Edinburgh University Press.

Boum, Aomar. 2013. *Memories of Absence: How Muslims Remember Jews in Morocco.* Stanford, CA: Stanford University Press.

Boyarin, Jonathan. 1992. *Storm from Paradise: The Politics of Jewish Memory.* Minneapolis: University of Minnesota Press.

Bunzl, Matti. 2007. *Anti-Semitism and Islamophobia: Hatreds Old and New in Europe.* Chicago: Prickly Paradigm Press.

Fassin, Didier. 2015. In the Name of the Republic: Untimely Meditations on the Aftermath of the *Charlie Hebdo* Attack. *Anthropology Today* 31(2): 4–5.

Jackson, John. 2008. *Racial Paranoia, The Unintended Consequences of Political Correctness.* New York: Basic Civitas Books.

Jikeli, Gunther. 2009. Antisemitism Among Young Muslims in London. *International Study Group Education and Research on Antisemitism. Colloquium 1. Aspects of Antisemitism in the UK*, 5 December, London.

Kahn-Harris, Keith. 2014. *Uncivil War: The Israel Conflict in the Jewish Community.* London: David/Paul.

Kahn-Harris, Keith, and Benjamin Gidley. 2010. *Turbulent Times: The British Jewish Community Today.* London: Continuum.

Katz, Ethan. 2015. *The Burdens of Brotherhood: Jews and Muslims from North Africa to France.* Cambridge, MA: Harvard University Press.

Klug, Brian. 2013. Interrogating "New Anti-Semitism". *Ethnic and Racial Studies* 36(3): 468–482.

Lazar, Moshe. 1991. The Lamb and the Scapegoat: The Dehumanization of the Jews in Medieval Propaganda Machinery. In *Anti-Semitism in Times of Crisis*, ed. Sander Gilman and Steven Katz. New York and London: New York University Press.

Lewis, Bernard. 1984. *The Jews of Islam.* Princeton, NJ: Princeton University Press.

Ma'oz, Moshe (ed). 2010. *Muslim Attitudes to Jews and Israel: The Ambivalences of Rejection, Antagonism, Tolerance and Cooperation.* Brighton, Portland, Toronto: Sussex Academic Press.

Mandel, Maud. 2014. *Muslims and Jews in France: History of a Conflict.* Princeton, NJ: Princeton University Press.

Meddeb, Abdelwahhab, and Benjamin Stora. 2013. *A History of Jewish-Muslim Relations, From the Origins to the Present Day*. Princeton, NJ: Princeton University Press.

Nettler, Ron, and Suha Taji-Farouki. 1998. *Muslim-Jewish Encounters: Intellectual Traditions and Modern Politics*. Amsterdam: OPA.

Parfitt, Tudor (ed). 2000. *Israel and Ishmael. Studies in Muslim-Jewish Relations*. Richmond: Curzon Press.

Silverstein, Paul. 2005. Immigrant Racialization and the New Savage Slot: Race, Migration and Immigration in the New Europe. *Annual Review of Anthropology* 34: 367.

———. 2010. The Fantasy and Violence of Religious Imagination: Islamophobia and Anti-Semitism in France and North Africa. In *Islamophobia/Islamophilia: Beyond the Politics of Enemy and Friend*, ed. Andrew Shryock. Bloomington: Indiana University Press.

Stillman, Norman. 2003. *The Jews of Arab Lands in Modern Times*. Philadelphia, PA: Jewish Publication Society.

Index

Note: Page numbers followed by 'n' refer to notes.

J. Renton, B. Gidley (eds.), *Antisemitism and Islamophobia in Europe*, DOI 10.1057/978-1-137-41302-4

L

M

Printed by Printforce, the Netherlands